The Analysis of Burned Human Remains

THE ANALYSIS OF BURNED HUMAN REMAINS

EDITED BY

CHRISTOPHER W. SCHMIDT
STEVEN A. SYMES

ELSEVIER

AMSTERDAM • BOSTON • HEIDELBERG • LONDON • NEW YORK • OXFORD
PARIS • SAN DIEGO • SAN FRANCISCO • SINGAPORE • SYDNEY • TOKYO
Academic Press is an imprint of Elsevier

Academic Press is an imprint of Elsevier
84 Theobald's Road, London WC1X 8RR, UK
Radarweg 29, PO Box 211, 1000 AE Amsterdam, The Netherlands
30 Corporate Drive, Suite 400, Burlington, MA 01803, USA
525 B Street, Suite 1900, San Diego, CA 92101-4495, USA

First edition 2008

British Library Cataloguing in Publication Data
A catalogue record for this book is available from the British Library

Library of Congress Cataloging-in-Publication Data
A catalog record for this book is available from the Library of Congress

ISBN: 978-0-12-372510-3

For information on all Academic Press publications
visit our web site at books.elsevier.com

Printed and bound in Great Britain

08 09 10 11 12 10 9 8 7 6 5 4 3 2 1

Contents

LIST OF CONTRIBUTORS

Jeremy J. Beach, M.S. (137) Department of Anthropology, University of Indianapolis, Indianapolis, IN

Amanda Baker Bontrager, M.S. (211) University of Indianapolis, Archeology and Forensics Laboratory, Indianapolis, IN

Erin N. Chapman, B.A. (15, 137) Department of Applied Forensic Sciences, Mercyhurst College, Erie, PA

Della C. Cook, Ph.D. (95) Department of Anthropology, Indiana University, Bloomington, IN

A. Joanne Curtin, Ph.D. (201) Department of Anthropology, University of West Florida, Pensacola, FL

John D. DeHaan, Ph.D. (01) Fire-Ex Forensics, Inc. 3505 Sonoma Blvd., #20–314 Vallejo, CA 94590

Joanne B. Devlin, Ph.D. (109) Department of Anthropology, University of Tennessee, Knoxville, TN

Desina Rachael Gipson, B.A. (15) Department of Applied Forensic Sciences, Mercyhurst College, Erie, PA

Tammy R. Greene, Ph.D. (227) Department of Biology, Arkansas State University, Beebe, AR

Robert G. Hayes, Ph.D (95) Department of Chemistry and Biochemistry, University of Notre Dame, Notre Dame, IN

Nicholas P. Herrmann, Ph.D. (109) Department of Anthropology, University of Tennessee, Knoxville, TN

John S. Krigbaum, Ph.D. (75) University of Florida, Department of Anthropology, Gainesville, FL

Rachel A. Lockhart, B.S. (227) Department of Anthropology, Indiana Prehistory Laboratory, University of Indianapolis, Indianapolis, IN

Hugh G. McKenzie, Ph.D. (185) Department of Archaeology, University of Sheffield, Sheffield, UK

Jacqueline I. McKinley, B.Tech. (163) Wessex Archaeology, Portway House, Old Sarum Park, Salisbury, Wiltshire

Kevin W.P. Miller, Ph.D. (129) Department of Chemistry, California State University, Fresno, CA

Stephen P. Nawrocki, Ph.D., D.A.B.F.A. (211) University of Indianapolis, Archeology and Forensics Laboratory, Indianapolis, IN

Nicholas V. Passalacqua, M.S. (137) Department of Applied Forensic Sciences, Mercyhurst College, Erie, PA

Andrea L. Piper, B.A. (15) Department of Applied Forensic Sciences, Mercyhurst College, Erie, PA

Christopher W. Rainwater, M.S. (15) Office of Chief Medical Examiner, New York, NY; Department of Anthropology, Center for the Study of Human Origins, New York University and New York Consortium in Evolutionary Primatology, New York, NY

Gregory A. Reinhardt, Ph.D. (227) Department of Anthropology, University of Indianapolis, Indianapolis, IN

Rebecca Richman, M.S. (129) Department of Anthropology, University of California, Santa Barbara, CA

Christopher W. Schmidt, Ph.D. (55, 227) Department of Anthropology, Indiana Prehistory Laboratory, University of Indianapolis, Indianapolis, IN

John J. Schultz, Ph.D. (75) University of Central Florida, Department of Anthropology, Orlando, FL

Mark R. Schurr, Ph.D. (95) Department of Anthropology, University of Notre Dame, Notre Dame, IN

Steven A. Symes, Ph.D., D.A.B.F.A. (15) Department of Applied Forensic Sciences, Mercyhurst College, Erie, PA

Curtis Tomak, M.A. (227) Indiana Department of Transportation, Indianapolis, IN

Joachim Wahl, P.D.Dr. (145) Regierungspräsidium Stuttgart, Landesamt für Denkmalpflege, Arbeitsstelle Konstanz, Osteologie, Stromeyersdorfstraße 3, D-78467 Konstanz, Germany

Phillip L. Walker, Ph.D. (129) Department of Anthropology, University of California, Santa Barbara, CA

Michael W. Warren, Ph.D. (75) University of Florida, Department of Anthropology, Gainesville, FL

Misty A. Weitzel, Ph.D. (185) Department of Anthropology, Oregon State University, Corvallis, OR

Howard Williams, Ph.D. (239) Department of History and Archaeology, University of Chester, Chester, UK

PREFACE

SPECIAL CIRCUMSTANCES

In December 1998 a local news report announced that a large house had burned to its foundation the night before and that one of the residents – a woman thought to have been there at the time of the fire – had not been found. I decided to offer my assistance to the recovery effort and drove to the scene. When I arrived mid-afternoon, the ashes of the house were still steaming. Except for the fireplace and chimney standing erect, all that remained was a pile of charred rubble within the confines of a cement block foundation. The basement was nearly filled with burned wood and fragmented furniture floating in at least a meter of water.

As I approached the handful of firefighters still working the scene, I introduced myself as a biological anthropologist interested in helping recover the body. They looked at me with disbelief, seemingly surprised that someone would volunteer to help find a burned body – on a very cold day no less – but they were happy for the help and proceeded to explain the situation. On a makeshift shelf, they had placed at least eight badly burned pieces of flesh and bone. The firefighters had no idea whether they were human in origin because the house had a large walk-in freezer in the basement. Hundreds of bits of burned animal remains were strewn throughout the ash and debris. Among the recovered fragments was a femur, severed about 10 cm above the distal end. It was relatively large and the condyles were asymmetric in shape and size. Although thoroughly blackened, the fragment was unmistakably human.

I related the news to the firefighters who, until that point, had become frustrated with their search. Within a few moments, their backhoe exposed two femora protruding from a pile of debris near the east wall. I climbed into the mess and began exposing what I could. The body had fallen some distance and landed on a dresser. The forelimbs had detached at the elbow and the lower limbs were missing roughly below mid-thigh. The corpse was facedown and the head seemed oddly small. I continued to search for the forelimbs and feet. Frankly, it seemed like a cruel joke to have so many pig, chicken, and cow bones intermingled with the human fragments. The burning tested my anatomical knowledge because charred meat looks remarkably similar regardless the animal of origin.

After searching for several hours, I had exposed as much of the torso as I could and prepared it for removal. Deftly I lifted the victim around her

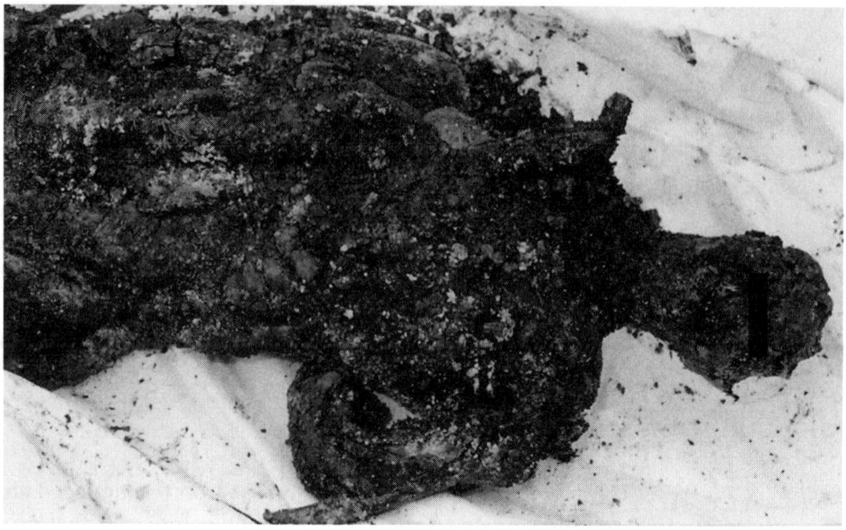

FIGURE 1 Body recovered from a house fire, ventral view. Notice the exposed organs, missing forelimbs, and reduced cranium; otherwise, the body is intact despite having been in a fire that destroyed the structure within which it was found. (Photo courtesy of Ken Lightle, Lawrence Fire Department, Lawrence, Indiana.)

ribcage and raised the remains to the firefighters standing at the foundation's edge. As I lifted, the torso turned toward me exposing a remarkably well-preserved abdomen and thorax. The abdominal wall was burned away as were the soft tissues of the chest, but the underlying organs were in pristine condition. Upon situating the remains on the body bag, it became evident that most of the cranial bones were gone. The brain was significantly reduced and a few facial elements, including some teeth, were imbedded in it. Perhaps in an attempt to bring morbid humor to a gruesome sight, I thought for a moment that it looked akin to a South American shrunken head stuck on a normal body. The remains were delivered to the morgue where they were later identified as those of the missing woman (Figure 1).

I share this particular experience to illustrate important aspects, five in particular, of studying burned human remains. The first three are addressed thoroughly in various chapters throughout this volume, while the latter two are important caveats:

1. The study of burned bone is a unique enterprise. Those of us who deal with the dead as a part of our profession should not underestimate the value of understanding combustion and the changes the body undergoes as it is heated. It is asking much to expect firefighters and law enforcement to scientifically recover seriously burned bodies, especially those that have fragmented. We in forensic anthropology (and related fields) need to make it clear to the medico-legal community that biological anthropology can make significant contributions to the recovery of human remains from many contexts. It has been my experience that most firefighters and police officers are unaware that among the ranks of anthropologists are those both trained and willing to help with the recovery of burned human remains.

2. Many variables can affect the appearance of a burned body. Despite the fact that the house had completely burned in the aforementioned case, the body recovered had large areas only modestly heat altered; only the head, limbs, and the outermost aspects of the torso had been heavily damaged in the fire. This suggests that analysts of burned human remains must be familiar with taphonomy, the study of the chemical and mechanical alterations an organism experiences after death. To do this, they must have an intimate understanding of human anatomy as well as the changes it endures when particular agents enhance or hinder fire-related damage.
3. Not only are there numerous variables to consider when it comes to burned human remains, there is also an analytical history to consider. For example, anthropologists and pathologists have assumed for decades that intact human skulls explode when exposed to heat and flames, and this notion continues to be believed despite growing evidence to the contrary. In fact, in the fire death example cited above, the skull did not appear to follow a pattern of destruction outlined in forensic anthropology textbooks. While anthropologists have studied burned remains for about a century, there are still numerous misconceptions and misinterpretations regarding field recoveries and laboratory analyses.
4. Burning can make many objects look like bone that are not even biological in nature. In the case here, the most challenging items to distinguish from burned human bone were nonhuman bones. However, a number of materials can thermally change in ways that make them look like burned bone or soft tissue. For example, certain plastics and ceramics can mimic bone once they have melted. Other petroleum-based products and fabrics can look like charred soft tissue as they darken and become pitted. Even iron slag can look like burned trabecular bone. Care must be taken to ensure that time is not wasted expertly excavating burned conduit instead of a victim's body.
5. Working a fire scene can be dangerous! This particular case is a good example of an apparently benign scene that was actually unsafe in many ways. Numerous hazardous materials were in the water and debris, and the air was contaminated. Some of the rubble was still smoldering; at times my hand would sink through the ashes and I could feel the heat. Thus, while I was excavating the body, some substances were still burning and emitting toxic gases. Emergency personnel are often highly trained to work in hazardous environments, but well-meaning anthropologists may be putting themselves – and their students – at risk if they do not adopt similarly rigorous procedures when recovering burned bodies. Many anthropologists have developed extensive biohazard protocols, especially in their labs, but they also need to establish field protocols that address safety in situations where the dangers go beyond pathogen exposure and include air and ground toxicity.

Christopher Schmidt, Indianapolis, Indiana, 2007

MISCONCEIVED NOTIONS

Like much of the rest of science, our knowledge of burned bone has not developed in an intellectual vacuum; we carry with us the baggage of earlier studies, many of which have proven to be quite valuable and others that have led us astray. Yet, some of the older studies that recently have been refuted continue to be cited simply because they have been a part of the history of the discipline for so long. This has allowed certain misinterpretations and misconceptions to persist to this day. The following is a case study where I had a first-hand opportunity to confront and challenge one of these out-of-date notions: that the human skull explodes when exposed to heat and flames.

It was 1977. The scene was a rural area near Memphis, Tennessee, where a drug deal had turned deadly for two victims, each shot in the chest with an AK-47 assault rifle. One victim was tossed in the backseat of a car; the other was situated in the front of the car with his upper body draped over the passenger seat. The vehicle was driven to a deserted farm, parked in an isolated shed, and set afire. Three months passed before the car and its contents were discovered.

The advantage of motor vehicle crime scenes is that they tend to be self-contained. In this case, the vehicle was transported to a crime laboratory where I assisted anthropologist Pat Peden systematically excavate the burned and decomposed victims from the car. Burning was extensive and only the basal aspects of the skulls remained perched upon the torsos – the top of each head was gone. Assuming that we were dealing with the textbook 'exploded skull victim,' my concern was that fragments of postcranial bone recovered from the middle of the car and the floor of the backseat would be littered with commingled cranial remains. But, careful excavation revealed that all cranial fragments of both victims were slowly compromised in a uniform fashion, eventually fracturing and falling straight down. There was no commingling; in fact, no fragments strayed any farther than gravity allowed. It was clear that the heads did not heat up and compromise all at once.

While these results were reported in a conference in 1999, anthropologists have not been swift to change old beliefs. Chapter 2 in this text details why these common misconceptions remain and emphasizes the point that scientific studies should confirm or reject previous concepts and that rejected ideas should be replaced by those that stand the tests of rigorous scientific scrutiny. It may be that anthropologists have studied burned remains for a century, but thermal destruction of human remains, particularly those with potential for forensic significance, remains misunderstood.

Steve Symes, Erie, Pennsylvania, 2007

PURPOSE

The editors of this volume are well aware that cases like the ones mentioned above are not uncommon. Unfortunately, there has not been a single volume that unifies our current knowledge of burned human remains. Our purpose, therefore, is to provide insight into the recovery and analysis of burned human remains from virtually any context, ancient or modern. This volume is meant to offer information relevant to students and professionals who are crime- and fire-scene investigators, coroners and medical examiners, forensic/biological anthropologists, and archaeologists.

Our objective is to provide a comprehensive reference volume that can be used any time professionals encounter burned human remains. However, this volume's overriding emphasis is on interpreting burned bones and teeth. The chapters range from how to determine from visual inspection if bone is burned to chemical studies of burned remains that can lead to positive identification. Some articles provide advice while others offer rich descriptions of cremation mortuary practices. Although this book covers a broad spectrum of topics, it is certainly not the last word on burned bone studies. In fact, we hope that later editions will develop as more investigators are inspired to share their experiences.

Christopher W. Schmidt, Ph.D.
Steven A. Symes, Ph.D.
2007

1

FIRE AND BODIES

JOHN D. DEHAAN, PH.D.

Fire-Ex Forensics, Inc. 3505 Sonoma Blvd., #20–314 Vallejo, CA 94590

Fire and death are frequent companions. Fire can be the weapon that actually causes death or the device used to conceal the cause of a homicidal death or to destroy other physical evidence. Fires are also a major cause of accidental deaths in the United States. No matter what is the cause and manner of death, bodies may be exposed to fire, causing a wide range of postmortem effects. Just as modern fire science has provided assistance to investigators of structure, vehicle and wildland fires and helped refute some of the myths and misconceptions upon which investigators once relied, the scientific process also assists those investigators concerned with human bodies affected by fire. This chapter begins with a discussion of fire itself, then describes the interaction of bodies and fire, and then offers some guidelines for processing fire-damaged remains. Case examples are used to illustrate primary concepts.

A FIRE PRIMER

For a phenomenon that has been a part of the human experience for so long, fire is still misunderstood by most who experience it. Fire is an exothermic, oxidation reaction between a fuel and an oxidizer (most often the oxygen in the surrounding air) that generates sufficient heat to be self-sustaining and yields readily detectable heat, and often light. Fire requires four basic ingredients – fuel in a suitable form, oxygen, heat, and a chemical oxidation that causes the reaction to be self-sustaining. Without all four, there cannot be a fire. Fire occurs in two basic forms – flaming and smoldering. A flame is the visible product of a fuel in a gaseous state burning in the presence of oxygen. It is a gas–gas reaction, made visible by the effects of the heat produced. A smoldering fire, by contrast, is the oxidation of a solid fuel in direct contact with oxygen. It is a solid–gas reaction that occurs on the surface of the fuel (and within the matrix of a porous solid fuel like charcoal). If this reaction proceeds quickly enough that the temperature of the fuel exceeds about 500°C (950°F), the incandescence of the surface makes it visible to the unaided human eye. We perceive this as a glowing combustion, whose visible

TABLE 1.1 Visual Color Temperatures of Incandescent Hot Objects

Color	Approximate temperature (°C)	Approximate temperature (°F)
Dark red (first visible glow)	500–600	930–1100
Dull red	600–800	1110–1470
Bright cherry red	800–1000	1470–1830
Orange	1000–1200	1830–2200
Bright yellow	1200–1400	2200–2550
White	1400–1600	2550–2910

DeHaan, J.D. (2002) *Kirk's Fire Investigation*, 5th ed. Brady Publishing, Upper Saddle River, NJ, p. 27.

Source: Data taken from Turner, C.F. and McCreery, J.W. (1981). *The Chemistry of Fire and Hazardous Materials*. Allyn and Bacon, Boston, MA, p. 90. See also: Drysdale, D.D. (1999). *An Introduction to Fire Dynamics*, 2nd ed. John Wiley & Sons, Chichester, United Kingdom, p. 53.

color is related to the temperature of the glowing surface, as in Table 1.1. A smoldering cigarette is a good example. If a cigarette is 'puffed,' the ash sustains a high-enough temperature to be visible as a yellow glow even in a lighted room. When it is allowed to rest, the surface temperature of the ash drops to the point where its red glow will be visible only in near-total darkness. In order for nearly all fuels to support a smoldering fire in normal oxygen levels, they must be porous and must maintain a rigid porous char as they burn. Thermoplastics melt as they burn, so they do not create a rigid porous char and will not smolder (except briefly at their surface or when exposed to a very intense radiant heat from another flame).

The flaming fire is the most common form of destructive fire and is capable of spreading (propagating) at very high speeds and extremely high rates of heat release. Since a flame requires that its fuel be in a gaseous form, the flame can be supported by a fuel gas such as methane, by vapors from a liquid fuel such as pentane or gasoline, or by vapors driven off by some form of degradation of a solid fuel. The fuel gas is the simplest case. Since the fuel is already in a gaseous state, heat serves only to trigger the separation of atoms (typically hydrogen and carbon) in a molecule. Although methane seemingly undergoes the simple reaction

$$CH_4 + 2\,O_2 \rightarrow CO_2 + 2\,H_2O \tag{1.1}$$

there are over 100 intermediate steps. A gas flame may be premixed as in a gas stove burner, or a diffusion flame. With a liquid fuel, some of the heat generated by the oxidation must heat the fuel to produce vapors and then break up these molecules to the point where they can combine with oxygen. The temperature to which a liquid fuel must be raised to be able to support even a brief flash of flame across its surface is called its flash point. A slightly higher temperature (called the flame point or the fire point) is needed for the flame to self-sustain in the vapor layer above the liquid (both such temperatures are far below that required for autoignition of the vapors, i.e., in the absence of a pilot flame or an igniter.). Fresh animal fat (lard), for instance, has an autoignition temperature of ~350°C (660°F), but its vapors can be ignited at much lower temperatures (~250°C) (Babrauskas, 2003). Gasoline vapors have a flash point of −40°C but an autoignition temperature of 280–456°C (536–853°F) (depending on the grade) (DeHaan, 2002).

With the exception of reactive metal fuels such as sodium or potassium, solid fuels must be changed into a vapor form that will undergo flaming combustion. This can be accomplished by sublimation (e.g., naphthalene), melting and evaporation (e.g., candle wax), or thermal decomposition (pyrolysis). Pyrolysis is the fundamental process, which is a required step for nearly all fires with solid fuels. The molecules of the solid fuel must be broken into small enough 'bits' to allow their combination with oxygen. Some of the heat produced in the fire (either flame or smolder) goes back into this pyrolysis process. If too much energy is needed (i.e., more than the fire can produce), the fire will self-extinguish.

Flames may be either laminar or turbulent. Small flames such as that of a candle are laminar, with overlying zones or layers (like an onion) of chemical reaction. Some zones in a laminar flame can reach very high temperatures. A wax candle flame has a zone that can reach 1400°C (2550°F), and a premixed flame in an acetylene welding torch will have a zone with temperatures greater than 3000°C (5400°F).

Once a flame becomes too large, the laminar structure breaks down due to turbulence. Most fires we see in everyday life (fireplaces, trash fires, structure fires, and wildland fires) are turbulent (sometimes to extremes). These are also diffusion flames where fuel vapors generated from a fuel surface (liquid or solid) diffuse outward, mixing with oxygen in air until they are mixed in the right concentrations to combust, as in Figure 1.1. The zones where that mixture occurs vary somewhat randomly, giving the movement and 'flicker' we normally associate with a flaming fire. Because of this turbulence, when we try to measure the temperature of the flame at any one point, we find that the temperature rapidly goes up and down. The best we can do is measure an average flame temperature (this average also reflects the effects that flame will have on a solid surface in contact with it). Even though fuels vary considerably in their measured 'heat of combustion' (see Table 1.2), the average flame temperatures when most fuels burn in air as a turbulent fire are nearly all in the range of 800–1000°C (1475–1832°F). The most common exceptions are methanol whose flame temperatures can be in the

TABLE 1.2 Heat of Combustion

Wood or paper	16 kJ/g
Methanol	20 kJ/g
Polyurethane foam	24 kJ/g
PMMA (acrylic plastic)	25 kJ/g
Nylon	30 kJ/g
Animal fat	32 kJ/g
Charcoal or coal	30–33 kJ/g
Polyethylene/polypropylene	44 kJ/g
Gasoline	46 kJ/g

DeHaan, J.D. (2002). *Kirk's Fire Investigation*, 5th ed. Brady Publishing, Upper Saddle River, NJ, p. 69.

range of 1200°C (2200°F), charcoal, and some plastics such as styrene and polyurethane whose measured flame temperatures can be on the order of 1400°C (2550°F). It was once thought that gasoline-fueled flames were much hotter than those of ordinary combustibles such as wood or plastics, and therefore any supposed 'high temperature' effect was due to such a petroleum product. This has now been demonstrated to be untrue (see Table 1.3 for examples). A flame from a burning pool of gasoline is indistinguishable in its average temperature from that of a pile of wood. The other exceptions are more rare – pyrotechnic mixtures or reactive metals (sodium, magnesium, and the like) produce flames that can reach 3000°C (5400°F) or higher. Fires in oxygen-enriched atmospheres can also produce very high flame temperatures.

Fires do vary in their heat release rate (HRR) (or heat output or power). Fires of all sizes and types can be characterized by their HRR – measured in kilojoules per second (kJ/s) or kilowatts (kW) (see Table 1.4). The HRR of a flame or a fire is the basic property of that heat source. The HRR allows us to predict the height of the flame, how quickly temperatures will rise in a room containing it, how quickly the room will be filled with smoke and, most importantly, the effect that fire will have on materials nearby or in contact with it.

Heat is transferred by conduction, convection, and radiation (or a combination of them). For a small fire (wastebasket fire), most of the heat will be

TABLE 1.3 Maximum Flame Temperatures (Measured in Air, Diffusion Flames)

Wood	1027°C
Gasoline	1026°C
Methanol	1200°C
Kerosene	990°C
Animal fat	800–900°C
Charcoal (forced draft)	1390°C

DeHaan, J.D. (2002). *Kirk's Fire Investigation*, 5th ed. Brady Publishing, Upper Saddle River, NJ, p. 69.

TABLE 1.4 Typical Heat Release Rates (HRRs)

Glowing cigarette	5 W
Kitchen match	50 W
Small wastebasket fire	50–150 kW
Small upholstered chair	150–250 kW
Cotton mattress	100–970 kW
Upholstered (modern foam) easy chair	350–750 kW
Recliner (PU foam, synthetic upholstery)	500–1000 kW
Sofa	1–3 MW
Gasoline pool on concrete (2 l)	0.9–1.0 MW
Dry 6 ft. Christmas tree	1–3 MW
Queen-size (PU foam) mattress with covers	2–4 MW

DeHaan, J.D. (2002). *Kirk's Fire Investigation*, 5th ed. Brady Publishing, Upper Saddle River, NJ, p. 33.

transferred via convection (moving hot gases) and only surfaces very close to the flames will be affected by radiant heat from the flames. The larger the fire is (the higher its HRR), the more impact radiant heat will have on nearby surfaces. Objects in direct contact with the flames will be heated by both radiant and convective processes, and their temperatures will rise very quickly (conduction plays a role mostly in ignition and flame spread processes, factors we are not concerned with in this chapter). The reader is referred to other texts for a full discussion of heat transfer (DeHaan, 2002).

When heat is transferred to an object, the temperature of this object rises, first at the heated surface, then as heat penetrates (by conduction) further into the mass. The intensity of the heat is the heat flux, the rate at which heat is striking a surface per unit area. It is measured in $kJ/s\,m^2$ or kW/m^2. The rate multiplied by time (duration of exposure) is a measure of how much heat is transferred to an object. If we know the mass of the object, its heat capacity and insulation factors that control the rate at which the heat can be lost from the object (by conductive, convective, or radiative processes), we can estimate the equilibrium temperature of the object. The physical and chemical effects we see after fire exposure are then the result of the intensity of heat transfer (heat flux), time (duration), and physical parameters of the object. As an example, if we place a wooden stick 4 m from a large campfire, the radiant heat falling on it will not raise its temperature to any measurable degree because it can lose heat to its surroundings as quickly as it is added. If the stick is moved to 2 m from the same fire, we do not see any change, but if we touch the side of the stick facing the fire (or measure its temperature), we find it is warm. If we move the stick to 1 m from the fire, the surface gets warmer, and we may see wisps of white 'smoke' coming from it (which is actually water being driven off). Moving it to 0.5 m, we see scorching as the surface temperature is now high enough to cause pyrolysis of the wood, releasing vapors and leaving char behind. At 0.25 m we see charring and the vapors being generated may ignite. Moving the stick to near contact with the flames, convective heat transfer becomes significant (although still only about 10% of that from radiative transfer) and the heated surface ignites. Taking the same stick from a distance and plunging it straight into the flames, we see it scorch, char, and ignite in a matter of seconds. Even though we have not changed the stick or the source fire, we have changed the conditions of exposure, so we have ignition in a much shorter time of exposure.

Once flaming combustion is established on a fuel surface, it reaches an equilibrium condition where its temperature stabilizes no matter what the temperatures of the flames above are. For solid fuels like wood or plastics, the surface temperature of the horizontal fuel is on the order of 350–400°C (650–750°F) (Drysdale, 1999). This temperature is achieved by radiant heating from the flames above and any smoldering combustion occurring in the surface are balanced against radiative and convective losses and 'evaporation' of volatile pyrolysis products. Figure 1.1 illustrates the physical processes involved.

The temperature in a turbulent flame is a function of its position. Flame temperatures are at their maximum along the centerline (axis) of the flame plume since there is some mixing and cooling near the edges from entrainment of cool atmospheric air. The height above the burning fuel surface also determines the flame temperature, as in Figure 1.2. The average temperatures

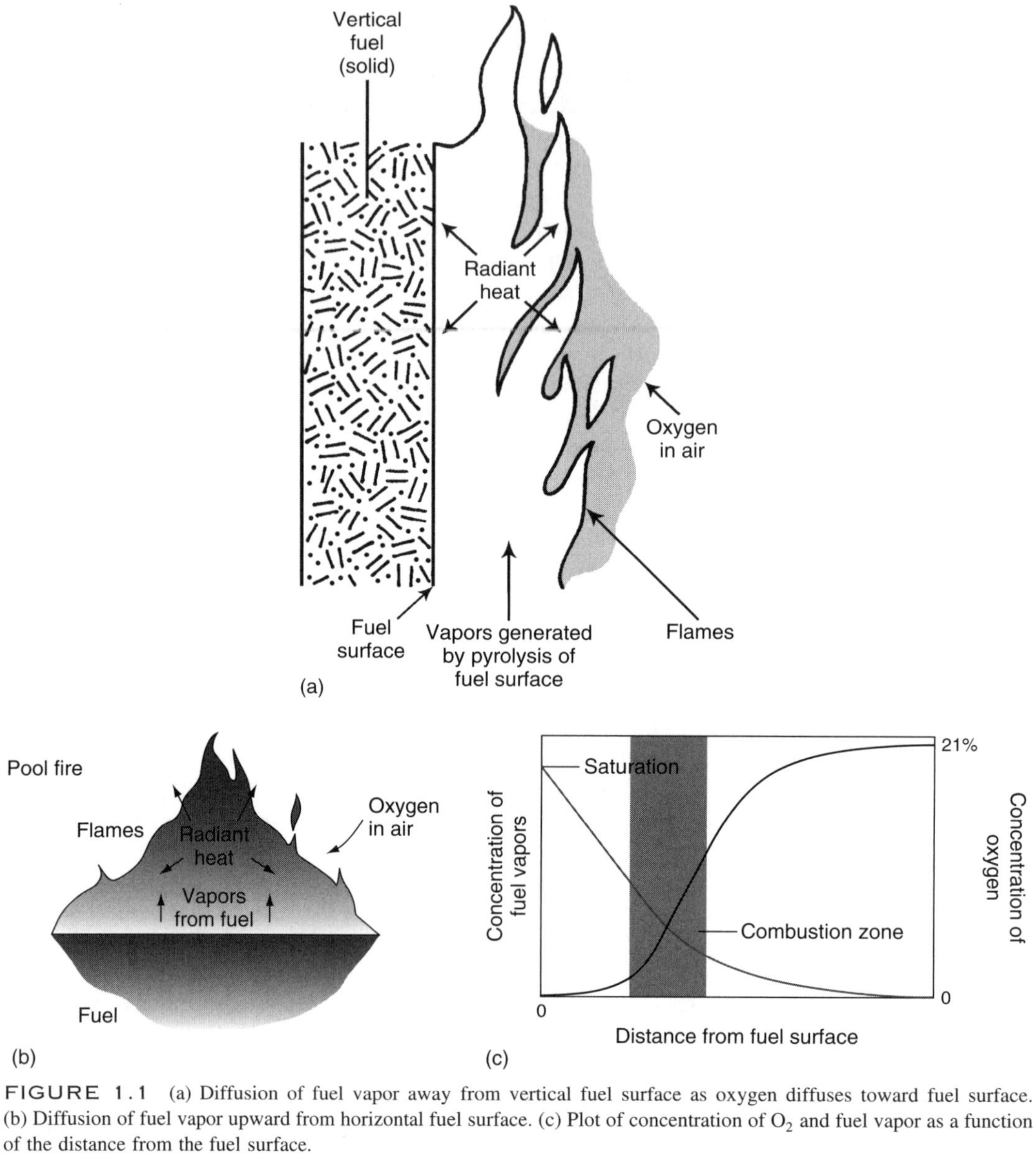

FIGURE 1.1 (a) Diffusion of fuel vapor away from vertical fuel surface as oxygen diffuses toward fuel surface. (b) Diffusion of fuel vapor upward from horizontal fuel surface. (c) Plot of concentration of O_2 and fuel vapor as a function of the distance from the fuel surface.

in the steady (continuous) flame are at a maximum just above the fuel surface (800–900°C, 1470–1650°F). The gases lose heat as they rise and cool to the point where the flames at their very tip become very intermittent and the average temperature drops to about 500°C (950°F). Below this temperature the pyrolysis products and the soot in the flame plume are no longer incandescent and are visible to the unaided human eye only as smoke. Above the tip of the flame, we see the column of buoyant hot smoke continuing to rise. Eventually the gases cool to the point where they are no longer buoyant and form a horizontal layer.

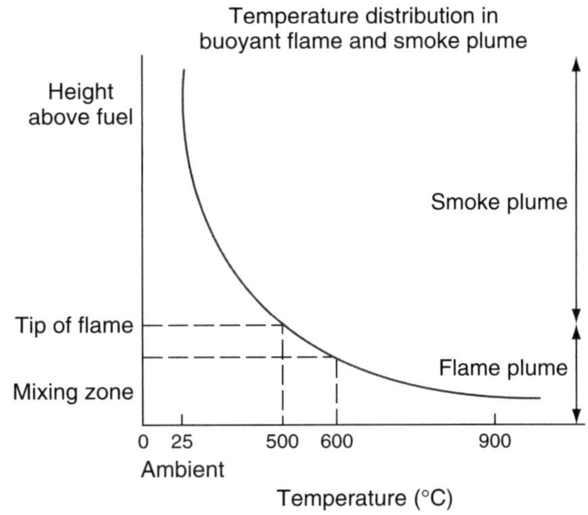

FIGURE 1.2 Temperature distribution in buoyant flame and smoke plume along vertical centerline.

ROOM FIRES

Since most fires of forensic interest involve fires in rooms (or enclosures) of some type, there must be some familiarity with fire growth in such conditions. The buoyant gases rise from the initial fire and form a discrete layer at the ceiling, which fills the room from the top-down. This layer can, of course, spill out the tops of window or door openings (or leak out through any ceiling vents). The larger the fire is (the higher the HRR), the faster the layer develops, fills the room, and the faster its temperature will increase. While this is happening, most effects of the fire on the room contents will be immediately above and around the initial fire. The size of the fire may be limited by the nature of the fuel source (as in Table 1.1) or by the size of ventilation openings. A normally sized room where all the doors and windows are fully closed will support only a very limited fire. A single open door can allow the entry of enough fresh air (and release of hot smoke) to support a fire of some 3 MW (megawatts), i.e., a large enough fire to cause full room involvement. A fire like that can grow by direct flame contact or direct radiant heat ignition of other fuels to the point that all fuel packages in the room are on fire. More commonly, the temperature of the hot gas layer increases to the point where radiant heat from it is enough to cause ignition of all other exposed fuel surfaces in the room. This transition is called flashover. In a normal-size room (2.4 m ceiling), when the average ceiling layer temperature reaches ~600°C (1150°F), the radiant heat flux (intensity) reaches 20 kW/m^2, which is enough to cause most furnishings, wall coverings, and floorings to ignite. When this condition is reached, a post-flashover room fire is created in which temperatures can range from 800 to 1000°C (1470–1830°F) from floor to ceiling and heat fluxes can reach 150 kW/m^2. These are, of course, far in excess of the conditions found in a simple flaming fire and are caused by the intense turbulence and mixing of incoming fresh air and hot combustion gases. When evaluating the exposure to fire conditions, the presence of post-flashover conditions (full-room involvement) must be established by scene investigators or reliable eyewitnesses such as firefighters.

EFFECTS

When we assess the effect of fire on a material like tissue or bone, it is not only temperature that matters, but also the atmosphere around the heated area (actually the atmosphere that can come into contact with the heated surface) and the duration of the contact. If a laminar flame (like a candle or a gas jet) plays against a surface, the impact of the heated gases drives away the combustion products and prevents atmospheric oxygen from reaching the heated surface. Any chemical reactions on the heated surface, then, take place in the absence of oxygen. At the other extreme, in a turbulent fire, if one blows on a burning wood stick, the air stream displaces the pyrolysis products that are keeping oxygen from the hot surface, and the enhanced oxygen content causes a sudden rise in surface temperature of a smoldering solid fuel. This is evidenced by the bright 'glow' seen on a burning wood surface when a bellows is directed against it. In a typical turbulent fire, the surface temperature of the burning fuel rises and falls and the atmosphere can range from reductive to occasionally oxidative. These conditions must be kept in mind when estimating or recreating fire conditions. Placing bone in a laboratory oven at 500°C does not duplicate fire exposure in temperature or atmosphere. Bones in a commercial crematorium may come closer to real fire conditions, but it should be remembered that this exposure is at relatively constant heat exposure, rather than the on-and-off exposure in most fires.

Variables of fire exposure, then, can be summarized as:

(1) Size of the fire
  - (a) Single item burning
  - (b) Multiple items burning
  - (c) Full-room involvement (flashover)
  - (d) Sustained post-flashover burning

(2) Exposure of the body
  - (a) On noncombustible floor for duration of fire
  - (b) On top of burning item(s)
  - (c) On combustible floor that collapses during fire
  - (d) In suspension on metal 'framework' (e.g., car seat)
  - (e) Exposed to fire on all sides (commercial cremation)

(3) Duration of exposure
  - (a) Antemortem
  - (b) Postmortem

(4) Condition of the bone
  - (a) Fresh or green
  - (b) Dried

Effects of fire exposure on bones are covered in great detail elsewhere in this text, but one point needs to be emphasized here and that is the complexity of the material as it responds to heat and flame. A living bone contains water, blood, fats, and other tissue in a complicated matrix. As the bone is heated, each of these components responds – sometimes by evaporating, or by contracting, liquefying, expanding, burning as a 'pool' of liquid fuel from a surface, or burning as a fuel from a porous rigid wick. Body fat is the best fuel in the body, but it will burn only when rendered and released to come into contact with air. The bone itself dehydrates, calcinates, shrinks, delaminates, and fractures. Dehydration occurs first, followed by charring of the organic constituents. With sustained exposure to direct flame contact temperatures (550°C or higher), the char oxidizes away. This produces what is called a 'clean burn' of the noncombustible substrate. The bone can also spall as internal moisture turns to steam, much like concrete fails under severe heating. Christensen (2002) has demonstrated that bone from osteoporosis victims degrades more quickly and more extensively than does healthy bone when exposed to fire. Bohnert *et al.* (1998) have published observations of the sequence of effects in legal cremations. Exposures to room or vehicle fires can vary from sustained and continuous to brief and intermittent, with cycles of sudden heating and cooling. There may be oxidative as well as reductive effects. The interpretation of fire damage must take these variables into consideration, and care exercised when using test data to ensure that test fire and exposure conditions are similar to those experienced by the remains at the scene.

It must also be remembered that fire suppression can affect burned bones and that not all observed damage may be due to heat and flame. Sudden cooling by hose streams may fracture or spall bones that are hot, particularly when they have been heated to the point of delamination or calcination. Mechanical damage can also occur from direct hose stream impact, falling debris, or post-fire overhaul or salvage operations.

CASE STUDIES

CASE 1

The author has been privileged to undertake several studies of the effects of fire and combustion on human bodies, usually driven by the investigation of criminal cases. Some of these focused on the combustion of soft tissue, incurring limited damage to bones. These studies revealed several important features of fires fueled by body fat and tissue alone. The best 'fuel' in the human body is the subcutaneous fat. It has an effective heat of combustion of 30–32 kJ/g, and an autoignition temperature of ~350°C (indistinguishable from that of pig fat) (DeHaan and Nurbakhsh, 2001). Such fuel will not smolder, will burn only as a flame, and has a high-enough flash point that it normally requires a rigid porous wick to maintain a flame. This 'wick' can be charred wood, clothing, carpet, or cellulosic upholstery (not synthetics that melt and do not form a porous char). Bone and even porous volcanic stone have been known to act as a wick for melted body fat. The flames thus produced have average temperatures on the order of 800°C (1475°F) and are turbulent. As these flames play directly on exposed skin or muscle tissue or bone, they can degrade them, even ignite tissue if it dries out sufficiently. The size of a fire that can be supported by body fat combustion is often controlled by the surface area of the available wick. With charred cotton cloth acting as a wick, the HRR (per area) of human body fat has been measured at ~250 kW/m^2 (compared to a pool fire of gasoline at 1500–2000 kW/m^2). The maximum size of the fire observed in a test of 50 kg of pork fat burning on a charred carpeted wood platform was less than 150 kW. A fire supported solely by the combustion of body fat is more typically in the range of 40–80 kW, the equivalent of a wastebasket fire. It can be sustained, however, as long as fat is rendering into a suitable wick in a manner to fuel the flames. The rates measured range from ~3 to 14 kg/h, so over a period of many hours a substantial portion of a normal human body can be destroyed by this mechanism (fluids drain out or evaporate and muscle burns reluctantly with a very low effective heat of combustion). If the fire does not spread to adjacent furnishings to create a large room fire, it can go undetected for many hours. It should be noted, however, that human fat is not susceptible to self-heating, and an external flaming fire of at least 5–10 min duration is required to split the skin and to induce sufficient rendering to support self-sustaining combustion. If the body represents the only significant fuel package involved in the fire, it may be possible to estimate (very roughly) the duration of the fire.

With regard to destruction of bones, one case study involved the recreation of a fire in a camping van that had caused considerable destruction to the body of an adult female trapped in the van (DeHaan, 2003). This fire was reported by a farmer who saw a vehicle fully involved in flames at a remote camping area. Due to its location, fire crews could not respond quickly and the fire was known to have burned for at least 25 min before it was suppressed. An adult male was found outside the vehicle who reported that he and his girlfriend were sleeping in the van when he awoke to find the interior heavily charged with hot smoke. He reportedly crawled to a rear window and broke it (by butting it with his head) to escape. He had smoke deposits, singed scalp hair, scalp and leg abrasions, and glass fragments in his hair to support his statement. His girlfriend was found seated on the rear (cargo) floor of the van with her back against the rear of the passenger seat and leaning against the (locked) right cargo door of the van. The right half of her body (torso and limbs) was rendered to calcined bones, exposing the spinal column and burned internal organs. Her ribs, shoulder, arm, and leg bones were denuded of all connective tissue and calcined. The right side of the skull was calcined, delaminated and heat-fractured. Postmortem examination revealed a COHb saturation of 41% with no significant drugs or alcohol or evidence of physical trauma. Scene investigators opined that the fire was so intense to destroy so much of the body and do so much damage to the van that it had to have been deliberately ignited with flammable liquid. Tests for ignitable liquids in the carpet found under the victim were negative. The van had been parked since the previous afternoon, was not running at the time, the fuel system was intact, and there were no signs of a grass fire extending under the van from outside. A visitor from the previous evening had said that they had no campfire, were not cooking and not using the propane Coleman lamp stored in the van.

Fortunately, the interior of the van had been photographed by the responding fire chief immediately after the fire in the van was knocked down, before overhaul disturbed the extensive fuel load in the van or the body. The victims planned to 'run away' from their domestic problems, and the van was heavily packed with bags, crates, and boxes of clothing, blankets, food, tools, and camping supplies. It had fiberboard paneling on the sides and the roof, carpet on the floor, and normal seats at the front. The van was a Tradesman Camper-style van with large windows all the way around. It was calculated that if these windows failed, a fire of some 3 MW could be supported inside the van. There was more than enough fuel with all the clothing and bedding present to sustain a very large fire within.

A similar test vehicle was obtained and loaded with an approximation of the fuel load identified in the original van (same type and arrangement of fuels but less total mass – meaning a similar size fire but for a shorter duration). Temperatures (measured by thermocouples located in the center of the cargo compartment) and smoke conditions were monitored inside the van as a fire was started by direct flame application to cardboard inside, and doors and windows were closed. The rear window was broken at 3 min and 45 s after the conditions inside had deteriorated to the point that the hot smoke layer temperature was over 100°C and there was heavy smoke making it untenable for the survivor to escape. The broken window simulated his reported escape and provided ventilation to support a growing fire. Temperatures climbed rapidly after the window was broken. The ceiling layer temperature exceeded 600°C (1150°F) at about 6 min, causing flashover inside the van and rapid shattering of all the remaining windows. All windows failed between 6 and 8 min, with each failure allowing a bigger fire to exist inside the van. Temperatures exceeded 1000°C (1830°F) inside the van at 8 min and peaked at nearly 1200°C (2200°F) during post-flashover burning. These temperatures and conditions exceeded those of a commercial crematorium and would have created similar damage to a body within. The reduced fuel load in the test van limited the test to less than 30 min (from ignition) at which point all combustibles within the van had been consumed (including the dashboard upholstery). The pattern of fire damage to the test van itself closely matched that observed on the van from the scene. The pattern of most intense damage to the victim's body matched the area of most intense ventilation-driven fire in the center of the van. Based upon the creation of a 'crematorium on wheels' without the presence of any accelerant, the district attorney dropped the murder charges pending against the survivor.

CASE 2

House fires can readily reach flashover conditions today due to the common presence of large furnitures filled with polyurethane foam. Although polyurethane would not appear to be a major problem based on its heat of combustion (see Table 1.2) alone, its physical form (a low-density foam) and its chemistry allow it to produce a very fast-growing, hot and smoky fire. A single polyurethane sofa or bed can become fully engulfed in flames in less than 3 min after ignition with a high-enough HRR to trigger flashover in a typical living room or bedroom. Fortunately, these items burn up so quickly that if they are the only major fuel package in the room, their fires seldom last long enough to burn a body much beyond partial thickness or soft tissue damage. In addition, their combustion products are themselves lethal. Their major advantage is that they require an open flame ignition source and cannot be ignited by a discarded cigarette except under the most unusual circumstances. In a normally furnished room, however, once alight, such furnishings will usually cause ignition of other fuel packages in the room, ultimately transitioning to flashover conditions. At flashover, all furnishings, carpet, and wall coverings will ignite creating a nonsurvivable inferno (even a firefighter in full turnout gear has only an estimated 8 s to escape from a post-flashover room fire). A body lying on the floor some distance from the original fire will be exposed to extremely high temperatures and heat fluxes far beyond what are encountered in a single-item fire.

Human victims in structure fires can be exposed to a wide variety of thermal and mechanical traumas. A body lying on a noncombustible floor will be exposed to the most intense radiant heat from above in addition to the flames around its perimeter. The side against the floor will be exposed to considerably less thermal insult and suffer proportionally less damage. A victim on a combustible (wood) floor will be offered the same pattern of protection until the floor is penetrated (a 3/4-in. plywood floor can be penetrated from above in 10–15 min of post-flashover fire). If the victim is on a wood floor above the fire room, the penetration time may be faster depending on the size of the fire and the height of the ceiling. An intact fire-resistive covering such as plaster or drywall (1/2 in.) will typically last 15–20 min before collapsing and allowing flame contact. Once the floor deck is penetrated, the portions of the body suspended in the open to direct flames will be destroyed very quickly. It is not uncommon to find very badly destroyed remains suspended by floor joists, plumbing, or even electrical wires. The same will occur in a vehicle fire if the body is supported by the steel seat structure, exposing it to fire impact on all surfaces.

The position of the body or skeletal remains must be carefully documented in relation to fuel sources and structural ventilation openings. In a post-flashover fire, the highest temperatures are usually found adjacent to door or window openings where fresh air can enter and support the most energetic combustion. Portions of the body closest to such openings are more likely to be more completely destroyed. Tests by this author in a wide variety of structures have demonstrated that floor-level temperatures of 900–1000°C are frequently achieved in post-flashover rooms, particularly near vent openings. Patterns of 'clean-burn' and intense burning near such openings can be found on portions of bodies nearest to these openings.

People often assume that gasoline alone will accomplish great damage to a human body. Gasoline burns off very quickly, however, and damage from the flames will only rarely exceed splitting and charring of the dermis. In one case in the author's experience, a woman was doused with a gallon of gasoline and set alight while dressed in a shirt, denim pants and cotton socks, lying on a sandy soil surface. She died from shock but her skin was nearly intact, penetrated to any degree only at her ankles where the cotton socks, secured by a leather belt, absorbed enough gasoline to continue to burn for some minutes (Icove and DeHaan, 2004).

Radiant heat alone can induce severe thermal damage. A well-ventilated wood fire produces a high heat flux (as evidenced by how quickly it becomes 'too hot' to sit close to a burning fireplace). An example comes from an apartment fire where the victim apparently awoke to find his bedroom well involved in a fire that had already spread through an open door to ignite the underside of the overhanging deck of the apartment above. He fled through the nearest door but the radiant heat from the burning wood deck above was enough to incapacitate him as he reached the rear fence. Although the fire was suppressed in about 10 min after being reported (by the neighbor above), the radiant heat from the overhanging deck and wood fence was enough to consume the exposed cutaneous and subcutaneous layers of his exposed back down to the musculature.

CONCLUSIONS

When assessing fire damage to human remains, the relationship between the remains and the fire is critical. The physical and thermal relationships and conditions are not static through a fire. No two fires are exactly alike. Temperatures can vary enormously, heat fluxes change often and rapidly, ventilation conditions change continuously and sometimes dramatically. Bone is a complex composite material that can vary even with its basic starting conditions. Careful documentation of the fire scene and remains prior to and during recovery is critical. If possible, the remains should be first examined in the fire debris as found. It is important to avoid some misconceptions (like 'flammable liquid' fires are always hotter than 'ordinary combustible' fires) and over-simplifications (a fire is a single set of condition that stays the same throughout). The reader is encouraged to review other texts dedicated to fire and its investigation (e.g., *Kirk's Fire Investigation*, 2002). Of course first-hand experimentation, when feasible, is the perfect answer to the question 'How did this happen?'

LITERATURE CITED

Babrauskas, V. (2003). *Ignition Handbook*. Fire Science Publishing, Issaquah, WA.

Bohnert, M., *et al.* (1998). The degree of destruction of human bodies in relation to the duration of the fire. *Forensic Science International* **95**, 11–21.

Christensen, A.M. (2002). Experiments in the combustibility of the human body. *Journal of Forensic Sciences* **47**, 466–470.

DeHaan, J.D. (2002). *Kirk's Fire Investigation*, 5th ed. Brady Publishing, Upper Saddle River, NJ.

DeHaan, J.D. (2003). Reconstruction of a fatal fire in a parked motor vehicle. *Fire & Arson Investigator* **January**, 42.

DeHaan, J.D. and Nurbakhsh, S. (2001). Sustained combustion of an animal carcass and its implications for the consumption of human bodies in fires. *Journal of Forensic Sciences* **46**, 1076–1081.

Drysdale, D.D. (1999). *An Introduction to Fire Dynamics*, 2nd ed. John Wiley & Sons, Chichester.

Icove, D.J. and DeHaan, J.D. (2004). *Forensic Fire Scene Reconstruction*. Brady Publishing, Upper Saddle River, NJ.

2

Patterned Thermal Destruction of Human Remains in a Forensic Setting

Steven A. Symes, Ph.D., D.A.B.F.A.[1], Christopher W. Rainwater, M.S.[2,3], Erin N. Chapman, B.A.[1], Desina Rachael Gipson, B.A.[1], and Andrea L. Piper, B.A.[1]

[1]*Department of Applied Forensic Sciences, Mercyhurst College, Erie, PA*
[2]*Office of Chief Medical Examiner, New York, NY*
[3]*Department of Anthropology, Center for the Study of Human Origins, New York University and New York Consortium in Evolutionary Primatology, New York, NY*

Introduction to Burned Bone Research

Out of necessity, anthropologists have concerned themselves with fire modification studies because of the significant impact that fire and heat alteration of physical evidence and human remains have on their analyses and interpretations. Fire is a destructive force that can damage, alter, or destroy valuable evidence essential to biological and behavioral reconstructions. The ramifications of heat modification for anthropological and forensic analysis are numerous and diverse; fire can hinder personal identification, make demographic assessments difficult, and conceal evidence of a crime. Interestingly, the same modifications that obstruct standard analyses may have evidentiary value by themselves. The destructive forces altering remains and physical evidence also create recognizable patterns and *process signatures* Piper 2007 that, when viewed as evidence, can lead investigators to more accurate and comprehensive interpretations of fatal fire scenes. The correct interpretation of these artifacts is essential to modern forensic investigations, conferring a new impetus to forensic anthropological research in the area of burned bone. Yet, a review of the anthropological literature reveals that the current methodologies for the analysis of burned human bone remains are, at best, confusing.

The main goal of forensic anthropologists in this area is to ascertain enhanced methodologies for the collection, preservation, and interpretation of bones altered by thermal processes, in order to establish identity and recognize potential evidence of criminal activity. Research studies focus on whether perimortem and antemortem trauma can be detected postcremation, whether it is possible to determine the body's position and state of decomposition at the time of cremation, how and to what extent fire consumes bone in a fatal fire setting, and whether biological parameters are obtainable after bone has been thermally altered (Stewart, 1979). Moreover, in cases where normally diagnostic bone characteristics have been fire-altered, forensic anthropologists strive to identify these methods providing the most accurate diagnoses (Buikstra and Swegle, 1989). The importance of effective collection and preservation methods underlies all of these studies (Krogman and Işcan, 1986; Dirkmaat and Adovasio, 1997; Mayne Correia, 1997; Dirkmaat, 2002; Mayne Correia and Beattie, 2002).

EARLY STUDIES AND CLASSIFICATION OF RESEARCH

Most of the initial research on thermal modification of human skeletons was developed within an archaeological framework, aimed at the interpretation of cultural cremation patterns. Early archaeological research in the area still provides forensic anthropologists a procedural baseline for the analysis of burnt bone today. Archaeologists studied burned human remains primarily to elucidate the conditions in which bone was exposed to fire (i.e., burning methods, temperatures achieved, and whether the remains were dry, fresh, or fleshed at the time of burning). They based their assessments in the examination of factors such as color changes, fracture presence, and fracture patterns in relation to the duration of heat exposure. As research advanced, quantitative traits, such as bone dimension changes from thermal processes, also began to be investigated (Binford, 1963; Van Vark, 1970; Stewart, 1979).

Ironically, one of the first anthropological studies on cremated bone, largely responsible for generating the classic archaeological studies on the subject, arose from the forensic field. In 1943, Wilton M. Krogman published an article in the *FBI Law Enforcement Bulletin*, discussing variations in heat-induced alteration to wet and dry bone. In this paper, Krogman examined the differences confirmed by bones in physical characteristics such as bone coloration, fracture morphology, and surface alteration, depending on the thickness of the overlying soft tissues.

Other researchers soon realized the potential for cultural inference of Krogman's observations. Shortly after Krogman's publication, Webb and Snow (1945) used their approach to study the cremation practices of prehistoric peoples in the Ohio River area. Webb and Snow aspired to differentiate Adena cremation practices from those of the earlier Hopewell. They sought the opinion of Krogman (Stewart, 1979:60), who concluded that, although some overlap was present, the Hopewell cremations were preferentially performed on defleshed, dry bones, while the Adena people practiced cremation mainly on fleshed bodies (Webb and Snow, 1945:189).

Webb and Snow's results were tested by Raymond Baby (1954), who examined cremated remains from different Hopewell sites in Ohio. Based on the heat alteration patterns shown by the skeletal materials, Baby also sought to determine whether dismemberment was part of the Hopewell ritual. In addition to the observational study, Baby pioneered experimental research in the area by conducting a study, using a crematory furnace, to compare heat alteration patterns in two unembalmed cadavers (fleshed and defleshed) and a sample of dry bone. Based on his experimental results, and the patterns present in the archaeological samples, Baby concluded that the Hopewell practiced cremations mainly on fleshed bones, contradicting Krogman's earlier determination for Webb and Snow.

Baby's (1954) and Webb and Snow's (1945) conflicting interpretations serve to illustrate the necessity of examining multiple variables in the analysis of thermally altered bone as well as the importance and need for experimental work in the area. This was also the first of many debates to arise regarding the interpretation of burned bones. Other early cremation studies considered variables such as body position, firing temperatures, and the characteristics of perimortem versus postmortem fractures (Krogman, 1949; Wells, 1960; Van Vark, 1970; Binford, 1972; Stewart, 1979; Trotter and Peterson, 1995). Surprisingly, varying anthropological interpretations in these and other related areas still lead the debate among current researchers in thermal skeletal modification, sometimes with little departure from the points of contention bandied about more than 50 years ago.

In fact, a thorough review reveals that the dominant theme in the related literature is precisely the lack of consensus on the methodology required for proper forensic interpretation of thermal modifications to bone. As posed by Mayne Correia (1997:275), 'the search for an understanding of how bone cremates has resulted in the application of a wide range of research techniques in various experiments, usually producing incompatible results. Inconsistencies in terminology and experimentation with a variety of skeletal materials have also produced disparate conclusions. Demographic methodology developed for non-cremated bone was applied to the analysis of cremated bone, although there is little information on the reliability of the results.'

The absence of standards to recognize *normal burn patterns* in fire-altered skeletal remains, and the misinterpretation of thermal process signatures, is at the root of this lack of consensus. A refinement of research practices in these areas is therefore indispensable to improve the interpretation of skeletal burn trauma. Given the complexity of the processes involved, and the large number of variables influencing them, research in this area must follow a holistic approach, being particularly cautious with interpretations based on episodic observations. Additionally, in view of the forensic implications of the subject, the need for forensic standards adds to the fundamental necessity of producing and employing a consistent descriptive terminology. Traditionally this has been impeded by a lack of consensus on the goals and problems addressed by bone heat alteration research, and the diversity of research approaches derived from the complexity and variety of the variables and processes involved. This has resulted in a general difficulty to perceive the subject as a cohesive field of study, which has developed instead as a set of poorly interrelated, almost independent, and sometimes even contradictory lines of research.

The present study proposes a set of general goals and guidelines to orient future research on thermal destruction of human remains, in order to promote the creation of consistent, Daubert-compliant standards and terminology for the analysis of this type of evidence in forensic settings. In order to attain this goal, it is necessary to review and organize different approaches taken by past and current studies and to define the goals of the field as they relate to the problems and questions confronted by the anthropologist practicing forensic science.

RESEARCH APPROACHES: SAME PROBLEM, DIFFERENT PERSPECTIVES

There have been different attempts to classify burned bone studies from different points of view, usually not to promote commonalty of research, but rather to exemplify the trends and lines of research dominating the field. Warren and Maples (1997) provide a classification system that organizes the literature into four categories based, in general, on the type of data sources: (1) analysis and experimentation, (2) studies on early cremation practices, (3) histological studies, and (4) contemporary commercial cremation analyses. Mayne (1990) and Mayne Correia (1997) propose a methodological classification, focused on data-analysis techniques: (1) analysis of visual characteristics, (2) histological analysis, (3) demographic analysis, and (4) methodology as reported by medicolegal personnel. A comprehensive classification should therefore take into account and combine each element, data source, and means of data analysis, and should include data collection during forensic recovery. The proposed new classification takes into account, for example, the necessary distinction between trauma interpretation and remains recovery and handling, derived from the legal implications and evidentiary value of each of these key components (Berryman and Berryman, 2007).

The importance of developing, testing, and applying efficient forensic recovery protocols cannot be underestimated. Methodical collection, handling, and preservation techniques must be implemented and well documented not only for reconstruction efforts but also to withstand courtroom scrutiny. The same is true for trauma analysis techniques. Osteological trauma analysis involves sophisticated methods that extend beyond gross visual examination. A detailed and meticulous examination under magnification is required, with methods often involving casting of skeletal material, particularly to aid in the interpretation of sharp force trauma (Symes *et al.*, 1999a, b). These methods afford expert examination of minute evidentiary details. The medicolegal community commonly relies upon the results of trauma analyses for assistance in determining cause and manner of death, sequence of trauma (ante-, peri- or postmortem), and the potential tool or tools used in a crime. Due to potential legal ramifications, trauma assessments are one of the most heavily scrutinized aspects of a forensic anthropologist's report. The methods employed and the examiner's skills are subject to critique. For this reason, anthropologists must be able to clearly defend their findings and clarify evidence. Streamlined, consistent terminology and well-defined, well-supported cremation research would aid toward this end. Therefore, we contend that burned bone research can and should be categorized and approached systematically in order to support professional concordance and to clarify its evidentiary validity.

The following six (6) interrelated categories are a proposed revision for thermal modification research classifications:

1. *Historical research.* Early research (up to the mid-1970s) examining cremated bone from an archaeological perspective, or for the purposes

of archaeological inquiry. Works such as Krogman (1943a, b), Webb and Snow (1945), Baby (1954), and Binford (1963, 1972) exemplify this category.
2. *Histological research.* Studies examining heat-altered bone at the histological level such as those of Forbes (1941), Herrmann (1977), Bradtmiller and Buikstra (1984), Nelson (1992), Stiner *et al.* (1995), and Hiller *et al.* (2003).
3. *Identification and visual classification.* Studies oriented toward the identification of thermally altered skeletal materials and analyzing visual changes of the bone. This area has been the focus of many studies, among them Wells (1960), Gejvall (1969), Van Vark (1975), Richards (1977), Dunlop (1978), Stewart (1979), Thurman and Willmore (1980), Eckert (1981), Bass (1984), Heglar (1984), Shipman *et al.* (1984), Eckert *et al.* (1988), Buikstra and Swegle (1989), Holland (1989), Mayne (1990), Fairgrieve and Molto (1994), DeHaan (1997), Mayne Correia (1997), Grevin *et al.* (1998), DeHaan *et al.* (1999), DeHaan and Nurbakhsh (2001), Delattre (2000), McKinley and Bond (2001), and Thompson (2004).
4. *Cremation studies.* Early and modern cremation studies such as those of Murray *et al.* (1991), Murray and Rose (1993), Owsley (1993), Warren and Maples (1997), Bennett and Benedix (1999), Warren and Schultz (2002), and Bass and Jantz (2004).
5. *Recovery and handling.* Reports addressing modern archaeological protocols and procedures for the recovery and preservation of skeletal remains. This includes literature by Sigler-Eisenburg (1985), King and King (1989), Ubelaker *et al.* (1995), Dirkmaat and Adovasio (1997), Mayne Correia (1997), and Dirkmaat (2002).
6. *Trauma interpretation.* Research geared toward the detection and distinction of antemortem, perimortem, and postmortem trauma associated with blunt, sharp, or ballistic force in heat-altered bone. This includes articles such as those by Herrmann and Bennett (1999), de Gruchy and Rodgers (2002), Symes *et al.* (2002), and Pope and Smith (2004).

FORENSIC PRACTICE AND PERSPECTIVE

CLASSIC APPROACHES TO FORENSIC ANTHROPOLOGY

The majority of forensic anthropologists are based in universities and museums, acting as consultants on an as-needed basis, while only a small percentage work directly within the medicolegal community (Kerley, 1980; Mayne, 1990). No matter the nature of their day-to-day employment, forensic anthropologists traditionally have been charged with the responsibility of assessing the human condition of recovered skeletal remains, as well as creating biological profiles to aid law enforcement in victim identification (Stewart, 1979; Işcan and Kennedy, 1989). These traditional assignments no longer suffice to describe the role of forensic anthropologists in modern criminal investigations. The days when a physical anthropologist could occasionally don a forensic hat without extensive training or experience are over. While the traditional practice has allowed for academically trained physical anthropologists to assist in forensic cases, modern forensic anthropology has evolved into a highly specialized field laden with pitfalls for the unwary 'expert.'

The potential legal repercussions for an inexpertly recovered fire scene or, worse, inappropriate or inaccurate comments in an official report distinguish today from 25 years ago. The days of the 1-hour shovel recovery, 5-minute biological profile made at the scene without statistical analysis, and offerings of trivial imaginative scenarios based on obscure or misinterpreted evidence are history. We might even suggest that the days of law enforcement agencies sending FedEx packages full of unexplained, 'picked up' bones to the desks of anthropologists are numbered. If modern forensic cases are not handled appropriately from beginning to end (i.e., from in situ scene documentation to case report), not only does their legal value in court become severely handicapped, but also the potential alteration of evidence derived from the recovery techniques applied and the appropriateness of the 'blindfolded analysis' become susceptible to legal scrutiny.

This new era of forensic anthropology – some have labeled it the Fourth Era of Forensic Anthropology (Sledzik *et al.*, 2007) – has a direct influence on the analysis of thermally altered skeletal materials. Past criticisms reveal that crime scene investigators, first responders, forensic pathologists, prosecutors, and traditional academic physical anthropologists may often overlook observable data produced by burned bone. As the first editor (C.W.S.) comments in the preface, it should take only one case where recovery attempts are hampered by lack of experience and cooperation, before anthropologists can identify with the complexity and difficulty involved with a complete recovery of remains presumed 'obliterated' by fire. Anthropological analysis does not end as the fire gets hot. On the contrary, everything, from the fire scene itself to the pile of recovered ashes on the laboratory table, offers an abundance of evidentiary data for the trained forensic scientist.

NEW INNOVATONS IN FORENSIC ANTHROPOLOGY

While there have been many recent innovations in the field of forensic anthropology, three emerging subfields of methodological study have proven essential to burned bone analysis: (1) forensic archaeological recovery, (2) forensic taphonomy, and (3) bone trauma interpretation. While the importance of these three areas needs little justification, they are described briefly below.

Although often disregarded by forensic investigators, archaeological search, recovery, and collection are integral to forensic case scenarios. In such events, anthropologists apply archaeological principles to preserve the contextual setting of the scene and evidence. The implementation of an archaeology-based recovery affords maximization of evidence recovery and precise and accurate scene reconstruction, supporting subsequent identification efforts, and the reconstruction of the events surrounding death and deposition (Sigler-Eisenberg, 1985; King and King, 1989; Dirkmaat and Adovasio, 1997; Dirkmaat, 2002). A common practice among fire investigators is raking through the debris of a fatal fire crime scene in order to locate the human remains as quickly as possible. This leads to a heightened risk of displacing, concealing, and even destroying fragmented human remains and other evidence, as well as inflicting further traumatic modifications to the remains. The methodical excavation of a scene and incorporation of an *in situ* map of the evidence in its context greatly enhances the interpretation and reconstruction of the case. A map of the elements provides the utmost opportunity to define the original deposition of the body. Subsequently, patterns may emerge revealing suspicious or suspected activity at the scene.

Anthropological taphonomy research was formerly of prime interest in the paleontological and archaeological realms. As forensic anthropology embraces taphonomy, scientific inquiries into and interpretations of postmortem processes take on new implications. Though taphonomic interpretations have become an essential part of forensic anthropological investigations, most postmortem influences are of little interest in a court of law. The fine line that separates perimortem bone trauma from postmortem taphonomic influences is easily confused, and postmortem influences on the body are likely to be misinterpreted as trauma criminally inflicted on a victim. Inappropriate interpretations in these gray areas have monumental consequences for criminal investigations and the legal proceedings that depend upon them. Thus, the forensic taphonomist must be familiar with perimortem bone trauma, and those who interpret trauma to bone in an anthropological setting must be experienced in taphonomic influences. The two areas of forensic investigation cannot be reliably practiced independently of one another.

'A forensic anthropologist should simply describe any evidence of bone damage, point out its location in relation to vital centers, explain the possibility of it having been sustained at the time of death or otherwise, and discuss the likely types of objects that produce damage' (Stewart, 1979:76). In Stewart's classical view, a forensic anthropologist should simply 'describe' any bone damage. As a matter of fact, anthropologists have proven to be very good at describing skeletal defects *ad nauseam*. Since the time of the field's earliest cultural studies, this descriptive approach became so natural to anthropologists that even those working in forensics have essentially relied upon this admittedly comfortable approach for decades. Such Boasian descriptive procedures are perpetuated by the maxim (almost a mantra in the profession) that anthropologists are not allowed to contribute official opinions as to the cause and manner of death. Because we are not medical doctors, we are legally unable to tackle problems that involve the cessation of life. Taken at face value (as it appears most have), this means that anthropologists must simply describe trauma to bone without offering interpretations as to the implications of that trauma. The authors disagree with this approach and advocate a more holistic line of attack. With a better understanding of bone fracture biomechanics and trauma indicators and an increased exposure to medical examiner/coroner procedures, anthropologists can and do contribute to the final medical interpretation of the cause and manner of death. Determining the extent of this increased contribution requires the examination of the questions, goals and objectives most commonly faced by forensic anthropologists when confronted with a fatal fire scene in their everyday practice.

FORENSIC IMPLICATIONS: PROBLEMS AND GOALS

THE CONCEPTS OF 'PERIMORTEM' AND 'POSTMORTEM' IN BONE TRAUMA

While anthropologists attempt to ascribe a perimortem or postmortem temporal classification in skeletal traumatic interpretations, the timing of the defect in relation to the moment of death is the most perplexing objective of trauma and taphonomy analysis. Accordingly, this classification relies almost entirely on the determination of whether a defect occurred when the bone was fresh, or dry and degraded. After death, the biochemical composition of bone changes with time, especially in terms of the amount and preservation of its organic matrix. From a biomechanical point of view, the main consequence of these changes is a reduction in the elasticity of bone materials, in transition from fresh (perimortem) to dry (postmortem/taphonomic), classic anthropological bone categories.

This results in a temporal discrepancy between the concepts of *antemortem, perimortem*, and *postmortem* (taphonomic), when anthropologists consult with pathologists in medicolegal settings. *Antemortem* occurs before death, or more specifically long enough before death to allow for an identifiable vital reaction from the living tissue. The concepts of *perimortem* and *postmortem*, however, are not so easily delineated by the medical and anthropological communities. In the case of sharp force trauma examination, an anthropologist's assessment of a dismemberment case is performed essentially within the theoretical perimortem context, as the skeletal material will retain nearly all the same properties as it showed in life (Symes *et al.*, 2002). On the other hand, the dismemberment of a body before death clearly would be an unusual circumstance and medical personnel would likely assume the cut marks were created postmortem.

Hence, anthropologists must consider skeletal trauma primarily in a taphonomic context. Defects occurring in bone must be excluded as taphonomic in nature before they can be considered to have occurred in the perimortem interval. While careful taphonomic interpretations can reveal information concerning circumstances surrounding death and postmortem interval (Dirkmaat and Adovasio, 1997), perimortem trauma interpretation can lead directly to a forensic anthropologist aiding the forensic pathologist in determinations of cause and manner of death. Thus, 'accurate and conservative interpretation of contextual taphonomic data ultimately reduces confusion by simplifying key variables... regarding cause and manner of death' (Symes *et al.*, 2002:405).

Burned bone trauma further confuses this temporally sensitive issue. There are a number of circumstances in which fire trauma (though not to bone) may be directly related to the cause (e.g., smoke inhalation/asphyxia) and manner (e.g., homicide, accidental) of death and should be considered in the perimortem interval. If the thermal damage were to compromise bone (certainly, after death), medical personnel would consider this postmortem trauma. Although thermal damage and sharp force trauma derived from dismemberment or mutilation would conceivably be inflicted to bone in similar timeframes after death (anthropologically perimortem), the intense heat and subsequent drying of bone would result in bone fractures that would appear postmortem from a biomechanical point of view. This anthropological 'impression' of perimortem versus postmortem and wet versus dry lies within the rudimentary properties of bone itself. As bone degrades, it responds to forces in identifiably different manners. Herein lies the importance of applying burn pattern and process signature recognition to burned bone analysis.

THERMAL MODIFICATION OF BONE, CASE BY CASE: THE DYMANICS OF BURNING BODIES

The question: can anthropologists accurately differentiate perimortem bone trauma from bone modification due to fire? *The problem* is that anthropologists typically fail to observe and chart thermal destruction patterns in skeletal tissue. They can look at 'typically' burned bone and recognize the fire damage (Figure 2.1), but they often falter when burn patterns become complex and multiple fractures occur on the same element. They lack the background information to provide an accurate assessment of which fractures contribute to a homicide investigation. *The solution* involves the careful usage of process signatures and charted burn patterns as tools for thermal trauma interpretation. With the proper approach, perimortem trauma is distinguishable from postmortem fire damage.

Bone response to thermal alteration

In order to understand the effects of heat on bone, one must be familiar with its basic structure and function. Bone is a living connective tissue that continually repairs and remodels itself throughout life in response to stressors or injury. Its composition supports this function. Bone is a composite material consisting of organic (collagen and protein) components in an inorganic (mineral) matrix. This combination provides a strong, supportive, yet semi-flexible skeletal

FIGURE 2.1 This figure illustrates a burned distal femur. The epiphyseal end has fractured off and has been pulled up the shaft by shrinking muscles, while the distal shaft is left exposed. This photograph is an example of a bone shaft that exhibits obvious fracturing as a result of fire. (see Plate 1)

frame that is capable of withstanding tensile and compressive forces while allowing locomotion. The bones comprising the skeleton support the body, give it shape, protect organs, provide attachment for muscles, and act as levers for movement. In addition to these mechanical and structural functions, bone functions as a storage site for minerals and contains marrow that supports the development and storage of blood cells that supply the body.

When exposed to high temperatures, the chemical properties of bone alter and structural integrity is impaired or lost. This results from evaporation, organic degradation, and transformation of the inorganic matrix. The loss of organic and inorganic properties is a complex process not yet completely understood (Thompson, 2005). Researchers remain uncertain of the precise chemical reactions, timing and number of transitions that bone undergoes while burning. What has been ascertained from descriptive, experimental, and actualistic studies is that heat exposure causes significant chemical and mechanical changes to bone that result in discoloration, shrinkage, warping, fracture, and fragmentation. Although studies have thus far been unable to specifically pinpoint required exposure times or temperatures for causing these changes, some researchers have been successful in identifying temperature ranges with which they can be associated (Holager, 1970; Civjan *et al.*, 1971; Bonucci and Graziani, 1975; Rootare and Craig, 1977; Shipman *et al.*, 1984; Schultz, 1986; Thompson, 2004).

Dehydration, decomposition, inversion, and fusion are the terms currently describing stages of the heat-modification process. Shipman *et al.* (1984) introduce these terms, and they are strongly advocated by Mayne Correia (1997), who, based on her collation of previous studies, presented time intervals for stages of thermal modification. Thompson (2004) eventually modifies these time intervals. His revisions center on his own experimental studies that support a reduction of the temperature ranges. Thompson also points out that Mayne Correia's description of inversion inaccurately characterizes the conversion of hydroxyapatite as normal for this stage of thermal modification. It is important to note that the four stages 'in themselves do not explain all of the fundamental causal changes occurring within hard tissues, and to date are highly theoretical' (Thompson, 2004:203). Thompson suggests that all heat-induced changes can be classified within the four categories, but cautions that temperature is not a reliable variable for predicting these changes.

Deducing a pattern

There are two applicable systems for categorizing degrees of fire modification. Eckert *et al.* (1988) was first to propose a system that was later popularized by Mayne Correia (1997). This classification system defines degrees of thermal alteration by amounts of surviving tissue. The classifications include: (1) charring – where the internal organs remain, (2) partial cremation – where soft tissues remain, (3) incomplete cremation – where bone fragments remain and, (4) complete cremation – where only ash remains. The Crow–Glassman Scale (1996) classifies thermal alteration into five (5) levels: (1) recognizable for identification – typical of smoke death, with possible epidermal blistering and singeing of the hair, (2) possibly recognizable – with varying degrees of charring on elements such as the hands/feet, genitalia, and ears, (3) nonrecognizable – with major destruction/disarticulation of the head and extremities,

(4) extensive burn destruction – where the skull and extremities are severely fragmented or missing, and (5) cremation – where little or no tissue remains and osteological fragments are scattered and incomplete.

These classifications are useful for describing the state of bone after burning. It is important to acknowledge that a body or a single element may exhibit instances of all of these classifications. Such differential burning is common and illustrates the effects of exposure variability. When the totality of exposure variability in a body is considered, a burn pattern emerges.

Taphonomic filters

A significant percentage of forensic investigations focus on malicious fires set with the intention of destroying traces of evidence related to criminal death. In this day of modern technology, sophisticated crime laboratories and DNA analysis, it would appear that arson is one of the easiest ways to destroy evidence of a murder. Long ago, however, Dr. William Bass demonstrated that human bodies are not easily obliterated by fire and that, amid the ashes, identifiable material should remain even after prolonged exposure to extremely high temperatures (Bass, 1984). The modern technology and research utilized in a fire marshal's office suggest the same (Icove and DeHaan, 2004). Nevertheless, the fact that bodies cannot be destroyed completely under typical burning conditions does not negate the fact that there are numerous taphonomic complications that arise when dealing with the remaining material. The degree of cremation and the subsequent handling of the remains are two such taphonomic filters that influence the subsequent trauma interpretation of all fatal fire recoveries.

Process signatures

It is important that forensic anthropologists be able to recognize perimortem bone trauma after exposure to fire. To accomplish this, insight into patterned thermal destruction of bone and its associated soft tissue is required. Unfortunately, many investigators commonly fail to account for the fact that bones burn as a part of the fleshed body, not simply as a disassociated skeletal material (Thompson, 2005).

In fact, traditional taphonomic studies can be said to suffer from, for want of a better phrase, *the myth of flesh*. This bias manifests itself in experimental research and analyses that treat skeletal elements as though they had always existed without the encumbrances of skin, muscle, ligament, and other soft tissues (Haglund and Sorg, 1997:3, emphasis added).

With this in mind, we direct our focus to the examination of three major process signatures recognizable in human remains that have undergone normal burn patterning: tissue shielding, color change, and thermal fractures. By outlining normal process signatures, abnormal or unanticipated perimortem trauma becomes obvious, and aberrant criminal acts and unusual contextual situations become more easily discerned. Equally as important, typical postmortem taphonomic influences become recognizable as such, rather than being dismissed as a confusing mass of data that cannot be interpreted or mistaken as factors in the victim's demise. The following three case studies demonstrate normal process signatures and serve as examples of how fire typically destroys the human body.

CASE 1 (CAR WRECK)

Pulled from the wreckage of a Jeep Cherokee sheared in two and engulfed in flames, the charred body of a young man rests supinely before the forensic pathologist at the morgue. As she examines the badly burned remains, she recognizes numerous perimortem and postmortem indicators. The traumata she sees stem from a high-speed chase that ended in a collision with a tree, ejecting and pinning the driver under the flaming wreckage. As she scrutinizes the multiple trauma indicators, the pathologist notes a number of confusing signs. The right arm has been sheltered under the body and remains largely intact (Figure 2.2). The left arm, exposed to the air and severely burned, shows evidence of fracture at the wrist (see Figure 2.2, inset). The pathologist would like to know: was the fracture an impact injury resulting from the collision, or was it due to fire destruction after death? She solicits the expertise of an anthropologist working in the morgue. Neither can be sure of the answer. Is this the best we can expect from pathology and anthropology in this particular case .... indecision from representatives of two professions that have studied burned bone for decades?

We know the vehicle was traveling at a high rate of speed, so blunt trauma to the victim upon impact is anticipated. Unfortunately, the fire has essentially masked, or filtered, the data, making accurate interpretations difficult. So, how would one determine the cause of these fractures? Are they from the impact with the tree or are they fire alterations that occurred while the victim became trapped under the burning vehicle? If we classify burn trauma as postmortem, one of the questions we face is 'are the fractures peri- or postmortem trauma?'

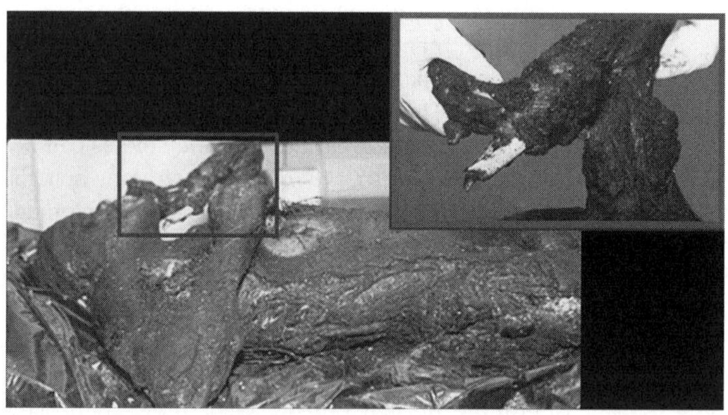

FIGURE 2.2 Seen in this figure is the victim of a violent motor vehicle crash. The victim was found facedown under burning wreckage. The inset represents the left forearm of the victim. The cause of the wrist fracture is questionable. The fracture was created either from the high-speed accident or from the postcrash burning. This is a question of perimortem versus postmortem injury.

CASE 2 (BURNED HAND)

While not easily recognizable, Figure 2.3, A through D, represents the remnants of hands that have essentially burned separate from the wrist. Those who work with burned bodies recognize this 'cartoon hand' as a common byproduct of severe heat damage to the distal upper limb. Dorsal wrists in a flexed 'pugilistic' position are highly exposed to burn trauma (see Figure 2.3B). The pugilistic posture initially protects the palmar side of the hand, as does the thick dermal ridged skin (see Figure 2.3C). Fire destroys the distal radius and the ulna; the metacarpals are obviously damaged and 'canoed' with the dorsal cortical bone of the shafts compromised early in the fire (see Figure 2.3B and D).

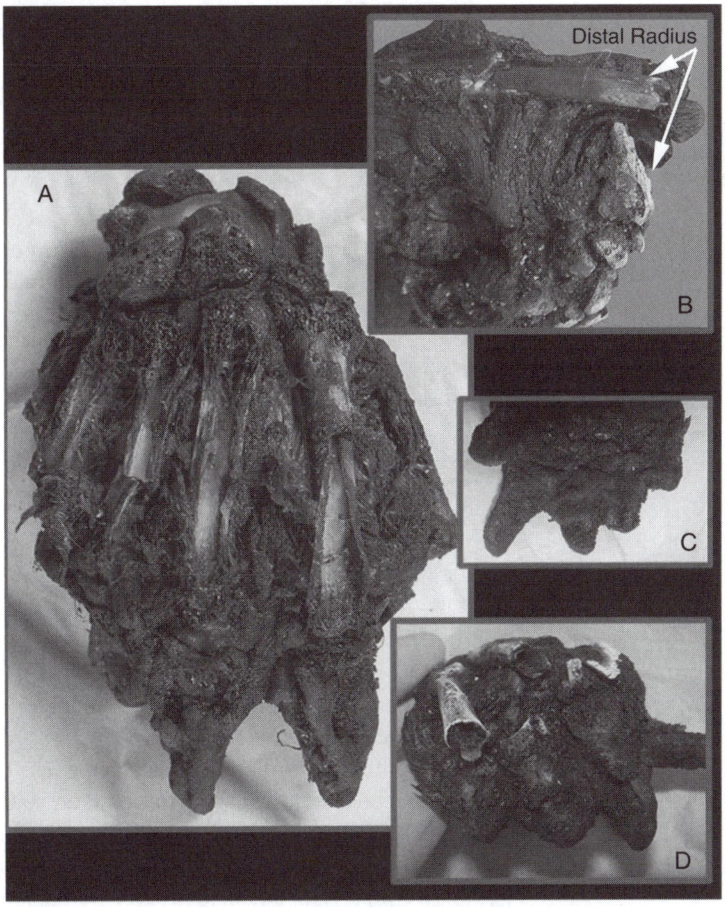

FIGURE 2.3 This illustrates four different views of extensively burned hands. Illustration A shows a typical burn pattern ('cartoon hand'); the fingers have extended out of the pugilistic posture. B illustrates how the wrist is flexed, allowing heat to compromise the distal forearm bones. C shows the palmar aspect of a hand with extended fingers. D demonstrates how the fingers are allowed to extend after the distal metacarpals and proximal phalanges are compromised.

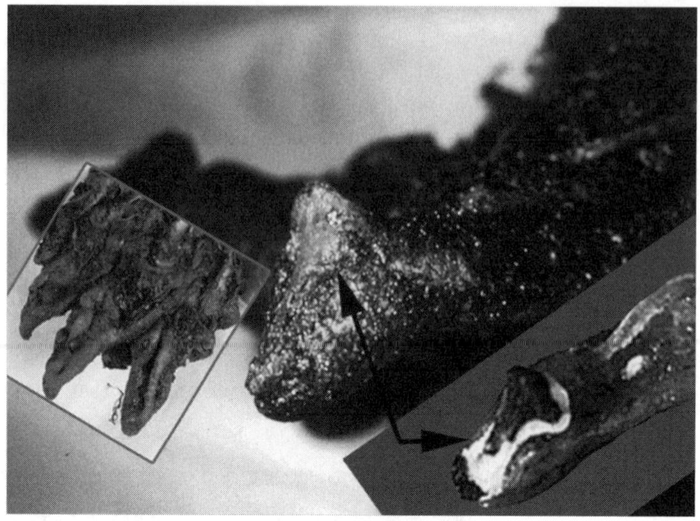

FIGURE 2.4 This figure is a close-up of a fingertip that is badly burned. The photograph illustrates the presence of distal phalanges when most medial and proximal phalanges and distal metacarpals have been destroyed. Arrows point to the phalanx before and after it has been processed. The inset illustrates the typical 'cartoon hand' posture.

How does one explain the fact that four of the five digits are present, when logically small appendages exposed to the air should be destroyed early in the fire? The distal forearm is already compromised, in this case fractured, but the distal phalanx on each digit is still present. This is evident in the laboratory once the soft tissues are removed (Figure 2.4). The destruction of the wrist with the preservation of the distal fingers needs further explanation, and obviously our understanding of burn patterns needs refinement.

CASE 3 (BODY IN CAR)

Firefighters arrive at the scene of an abandoned car burning out of control on a desolate country road. A passerby has called 911 to report that the car had literally exploded before his eyes. Upon reaching the scene, the fire marshal is quick to suspect arson, based on the intensity and enormity of the fire. Inside of what will later be determined to be a rental vehicle are the charred remains of a victim (Figure 2.5). The question begs answering: Was this a suicide or perhaps an attempt to camouflage a murder? Using our understanding of burning process signatures, can we assess the cause and manner of death? The skull shows evidence of extensive fracture patterns. Which, if any, of these fractures occurred before the fire, and how does a typical postmortem fire-fracture pattern look like? Is there any obvious perimortem lethal trauma? This is a test of sorting peri- from postmortem fractures.

FIGURE 2.5 This figure is a view of the interior of a burned vehicle revealing a facedown body (cranium) after flames were extinguished.

In order to comprehend these exemplars, and thermal destruction to bone in general, it is important to appreciate the normal patterns of burning tissue and the characteristics that are indicative of this patterning. As John DeHaan (Chapter 1, this volume) points out, the assessment of fire's effect on human remains is dependent on three variables: heat, atmosphere, and duration. Each variable is highly complex and often involves additional considerations that are out of the realm of the pathologists or anthropologist's training and expertise. Rarely do anthropologists consider all three variables in their analyses, and further much of the past research deals with temperature only, usually in an artificial environment.

NORMAL BURN PATTERNS

If expectable burning patterns under normal circumstances (totally engulfed) can be defined and recognized, abnormal burning due to perimortem trauma, criminal behavior, or other factors should become evident as departures from this normal patterning. On the other hand, if fire circumstances vary from one scene to another and factors such as heat, atmosphere, and duration are difficult to incorporate into anthropological analysis, where can we look for uniformity in a fatal fire scene? The proposed answer is human anatomy and physiology.

The remainder of this chapter emphasizes burn patterns and characteristics that are dependent upon the uniformity of human anatomy and physiology, and commonly occur in cases of fire consumption, independent of scenario-specific variations in temperature, atmosphere, and duration of exposure. In other words, bodies burn in a uniform, recognizable pattern if all external variables are similar. The anatomical features of the victim and the response of these features when engulfed by flames and heat are the key elements for this analysis. In order to understand the patterned thermal destruction of human remains, it is important to examine and recognize the characteristics generated by this destruction. When we accept the premise that all human bodies exposed to fire will share similar physical reactions, the question becomes: what are the visible patterns that are generic to this situation?

The authors' research has resulted in the recognition of three process signatures that are fundamental to the recognition of normal burn patterns in human bone. Remember, if normal burning is recognizable, then abnormal burning due to perimortem trauma, criminal behavior, or other factors should become evident. The three diagnostic process signatures of importance in this research are:

1. Body position and tissue shielding in bone
2. Color change in thermally altered bone
3. Burned bone fracture biomechanics

Body position and tissue shielding

Body position and tissue shielding refer to the pugilistic posture or pose induced by fire and heat and to the protection of bone from thermal destruction by other tissues. The heating and shrinking of muscle fibers create pugilistic posture. The encroaching flames eventually affect all the muscles in an exposed body, promoting their contraction. The final posture is a result of the overriding contractions of the most powerful muscles and ligaments. In this way, muscle contraction will result in increased exposure of some anatomical areas and shielding of others, depending not only on the depth of the immediately attached tissue, but also on how the area postures. As will be discussed below, the pugilistic posture of the distal hand phalanges (fingers) will become protected not only by the tissues directly attached to them but, as result of finger flexion, also by palmar soft tissues. Because of this, the degree of fire modification to bone is highly variable but predictable across the body. Since all human bodies have essentially the same anatomy and bone makeup, they tend to assume the same posture, shield the same tissues, and bone burns (dries) similarly, creating recognizable burn patterns. As heat exposure continues, soft tissues eventually disintegrate and body postures change.

Examine the forearm exposed to thermal destruction (Figure 2.6 inset) seen in this photograph of a badly burned individual. High temperatures surrounding the muscles produce adduction of the shoulder, and flexion of the elbow and wrist, bringing about the pugilistic pose. We can assume that bone will first burn in areas of minimal protection and increased exposure,

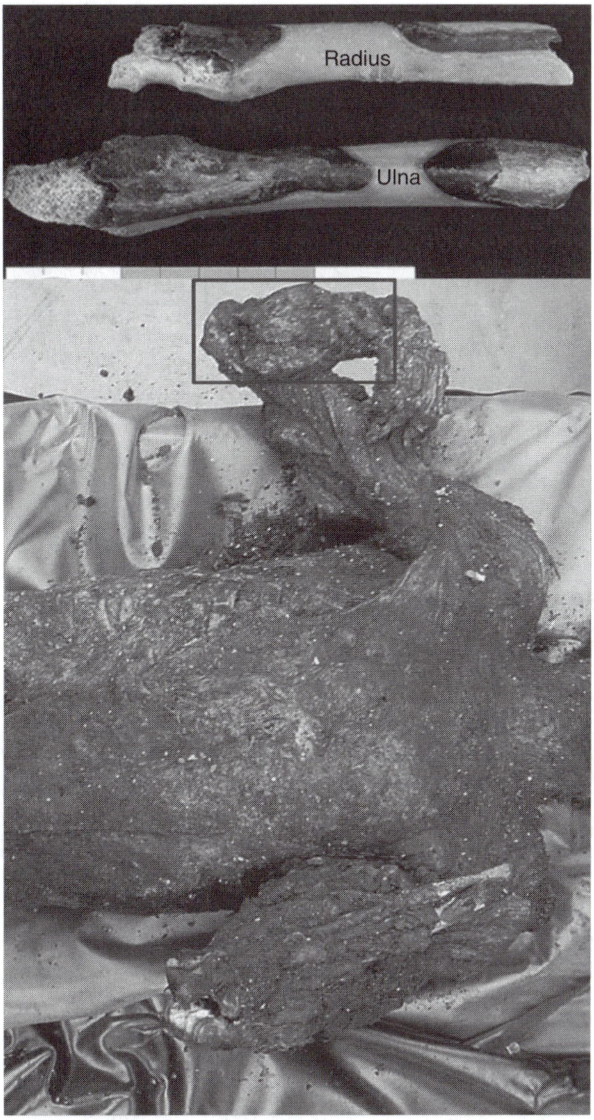

FIGURE 2.6 This figure illustrates a badly burned victim with the arms in typical pugilistic posture. The inset shows the right radius and the ulna after processing. Notice that the distal ends burn early while the proximal third of the shaft is the last place to burn.

as opposed to areas protected by greater tissue depth and pugilistic posture. As limbs draw farther away from the torso, their exposure to fire and oxygen increases and they suffer burning destruction. As the wrist flexes, the dorsal carpals and the posterior distal radius and the ulna are afforded little protection from the flames. The dorsal wrist is one of the first loci of thermal destruction in the forearm. The other locus is the posterior elbow, as the pugilistic posture draws that area away from the torso. The anterior elbow, however, experiences increased protection from the fire due to elbow flexure and the resultant increase in tissue depth.

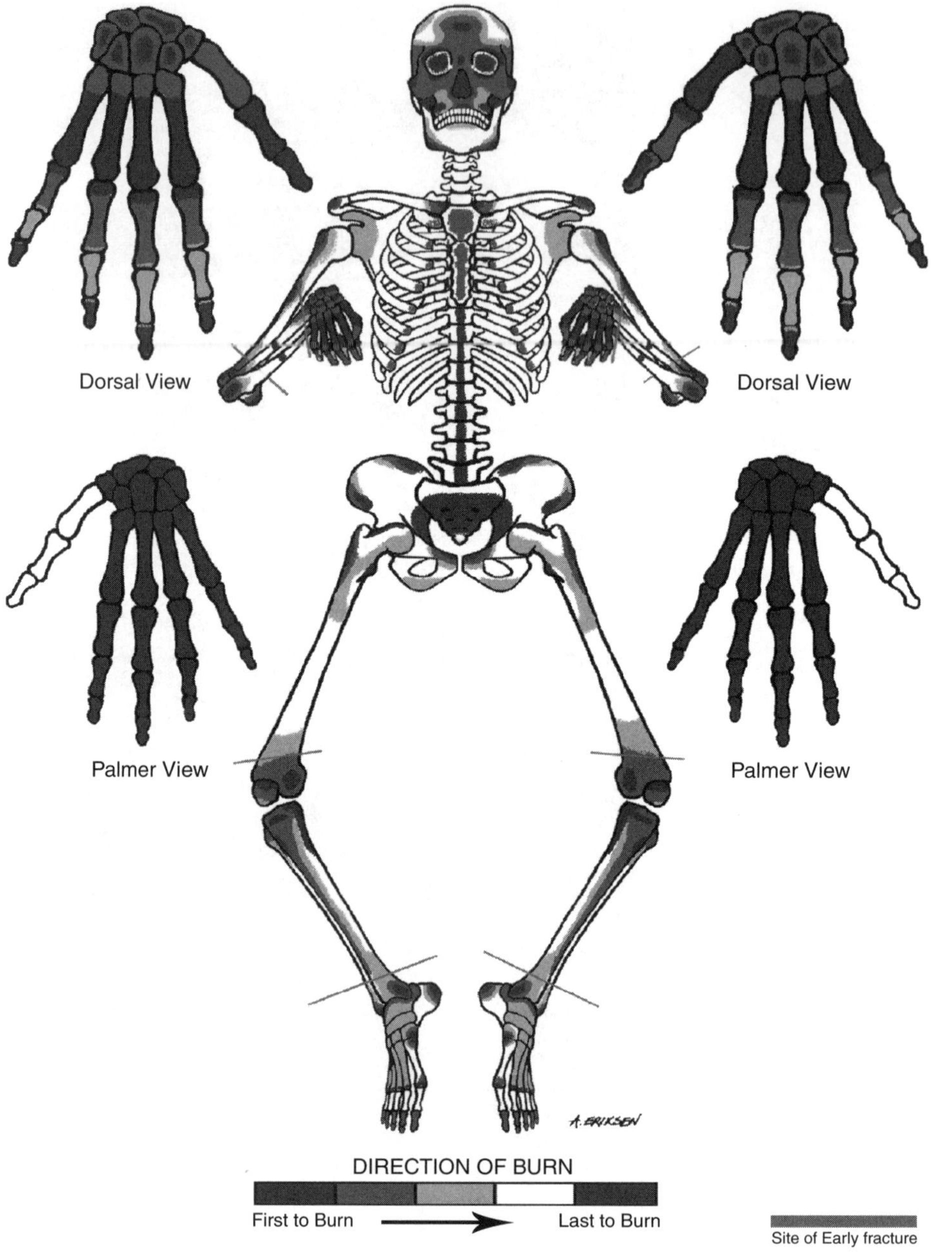

FIGURE 2.7 This is a diagram of an anterior skeleton in pugilistic posture highlighting the initial, secondary, and final areas to express burning on bone. The figure also includes dorsal and palmar views of the pattern of burning on the hand. The green lines indicate common areas of fracture. (see Plate 2)

Every area of the skeleton has its own signature of initial, secondary, and final destruction due to the dynamics of the tissue shielding the bone (Figures 2.7 and 2.8 chart these patterns). We can anticipate that the flames will destroy the distal and proximal ends of the radius and the ulna before

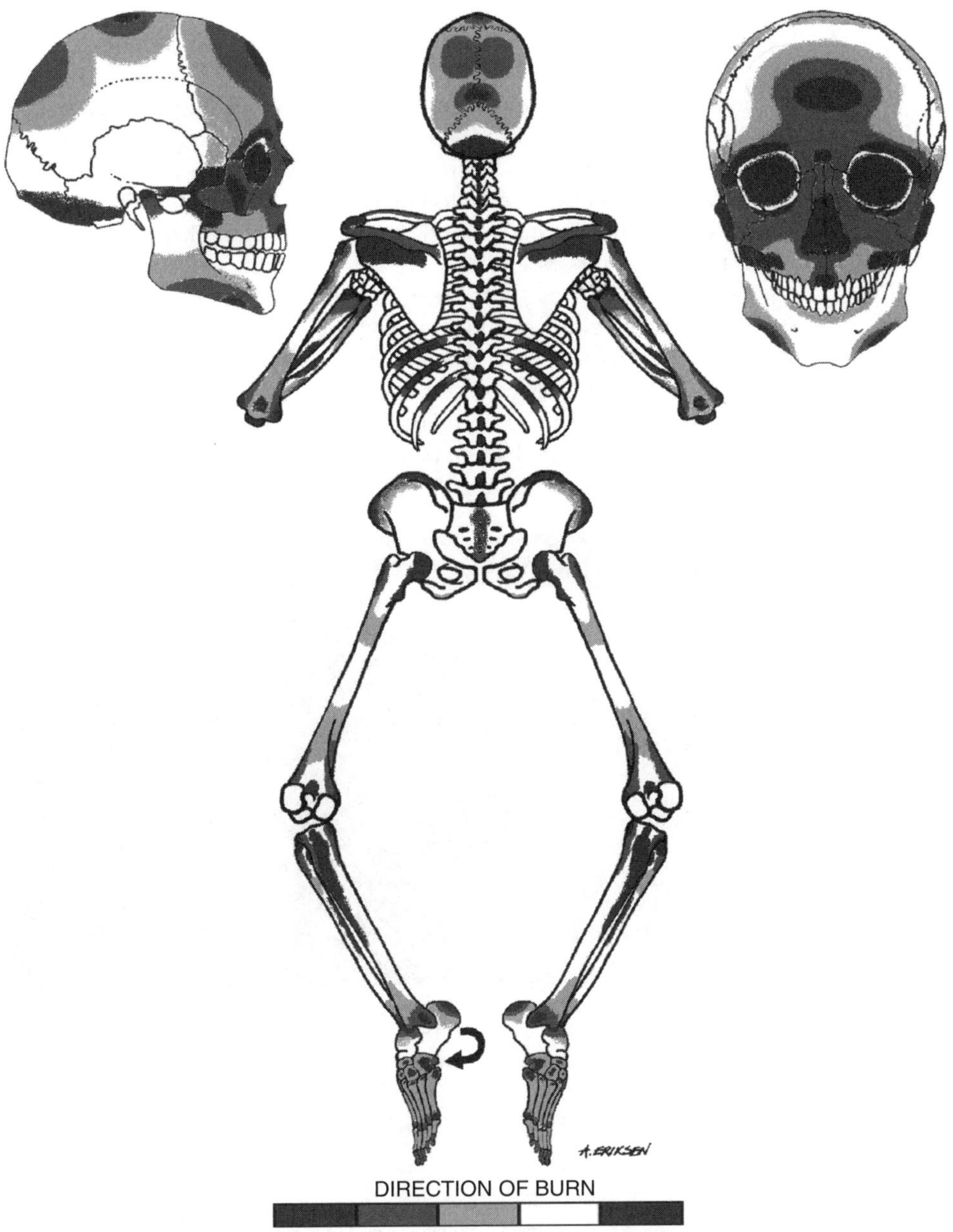

FIGURE 2.8 This is a posterior view of the skeleton diagram in pugilistic posture highlighting the initial, secondary, and final areas to burn on bone. This figure also includes a magnified view of the burn patterns on the frontal and lateral skull. (see Plate 3)

proceeding along the diaphyses. However, does that mean that the mid-forearm should be the last to burn? The proximal third of the forearm has the most muscle mass, so soft tissues – including the interosseous membrane of the radius and the ulna – produce the last resistance to total fire destruction of the forearm bones (see Figure 2.6, inset).

Case 1 (Car Wreck) Revisited

Let us go back to examine the victim of the fiery car crash. As we inspect these remains for process signatures, we see evidence of burn in the area of the wrist, where we would expect it (see Figure 2.2). The distal forearm has flexed into pugilistic posture with the dorsal wrist – an area susceptible to early fire destruction of bone – highly exposed. If we examine the processed bone, we can see burns on the dorsal aspect. At first glance, it appears that there is no fire damage on the anterior aspect of the wrist, but close examination and back lighting reveal that all surfaces are in fact burned. A conservative interpretation of thermal destruction of the arm should take precedence and therefore these injuries should be considered fire-related until more information is gained. Corroborating this pattern, the examination of another fire victim's distal radius revealed identical patterns (Figure 2.9).

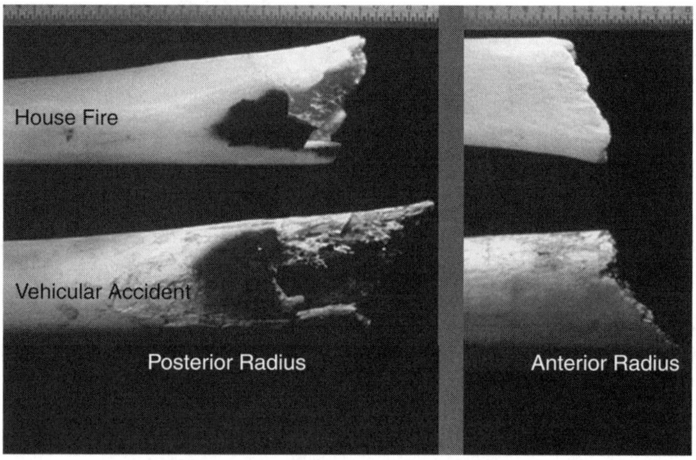

FIGURE 2.9 This figure represents two views (bottom) of the distal left radius from the vehicular accident in Figure 2. The radius on the top is from a victim of a house fire. The pattern displayed suggests that the bone damage is due to fire and not perimortem trauma.

Case 2 (Burned Hand) Revisited

It is also possible to explain the process signatures in Case 2 using body positioning and tissue thickness as a basis for interpretation. Severely burned victims commonly display the 'cartoon hand.' Logic would suggest that small appendages like fingertips would be the first tissue consumed in a fire, as is the case with toes. To explain this discrepancy, one must apply the progression of thermal destruction to a hand. Pugilistic posture causes the hand to make a fist as muscles shrink. Flexor muscles of the hand primarily originate on the proximal, anterior forearm. Once the muscle fibers are exposed to fire, there is a powerful contraction of the muscles, flexing the fingers. As one might guess, the flexed fingertips provide protection from heat and flames, even after the dorsal metacarpals have been destroyed. Along with dorsal metacarpal and shaft destruction, the articular surfaces of the distal metacarpals and proximal phalanges are exposed early in the fire. After proximal phalanx destruction, the flexors of the fingers no longer have fulcra for movement, and the shrinking muscles of the forearm no longer can flex the digits (see Figure 2.3). Contrary to what has been suggested by Pope (2007), simple tissue shrinkage cannot be assumed to cause joints to take on typical pugilistic posture after the bones supporting muscle attachments are compromised. Certainly, limbs move due to tissue shrinkage, but limb flexion is a product of the lever forces exerted by muscle fibers on joints and cannot occur without a fulcrum. Heated forearms force the hand into a closed fist posture. Without burning to the forearm, or without the stability of the radius and the ulna, there is little hand posturing. The 'cartoon hand' is present in burnt victims only after severe burning of the hand destroys the bone fulcrum or the knuckle area (distal metacarpals and proximal phalanges) and associated tendons. With the fulcrum destroyed, a fist can no longer be retained. Fingers two to five relax away from the protecting palm, while the thumb burns early due to its opposable position. Despite what logic would seem to suggest, one of the last bones to survive in a severely burned hand is the smallest bone of the arm, the distal phalanx. However, when the digits are directly exposed to heat, they quickly succumb to the same destructive processes as the rest of the dorsal hand, just later.

Color change in thermally altered bone

Color change occurring in fresh bone during burning is the second diagnostic process signature considered for this research. Heat produces a gradient of colors as bone dehydrates and becomes exposed to the gradual loss and shrinkage of the muscle tissue. Anthropologists have discussed and debated color changes in burned bone for decades, but have yet to chart effectively the patterns or recognize the full value of this variable and its implication for the interpretation of burned remains. We disagree with the previous literature which gives the impression that a single bone exhibiting numerous colors is of little use for diagnosing fire dynamics (see Mayne Correia. 1997: 276–277). For example, Figure 2.10 illustrates color changes in three bones. Figure 2.10A is a single radius shaft partially burned and fractured due to fire. Rather than suggesting that color change in burned bone is arbitrary and indecipherable, this varied coloration demonstrates how bones generally burn from the outside to the inside (before the shaft is compromised), with the lighter calcined bone being external and the black charred (protected) bone internal, (Figure 2.10B). Figure 2.10C is yet another proximal radius demonstrating color change.

The forearm example above (see Figure 2.6, inset) illustrates how the wrist is the first and the surfaces surrounding the interosseous crest, in the proximal third of the radial and ulnar shafts, are the last to burn. This dynamic pattern reflects the variation of heat exposure both with time and by anatomical area.

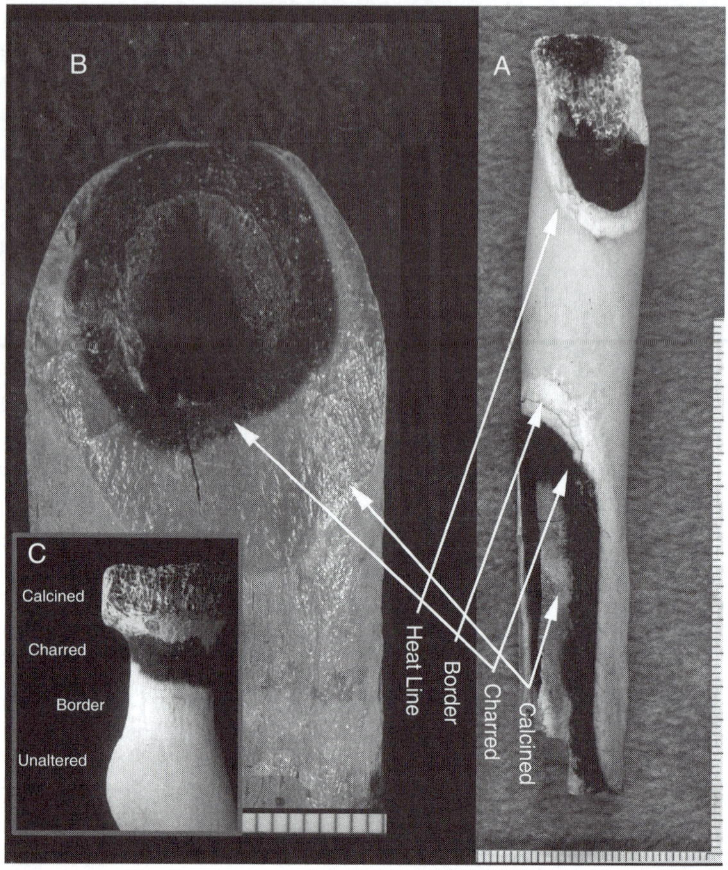

FIGURE 2.10 This figure illustrates color changes in a three partially burned bone shafts with fractures due to fire. Rather than suggesting that color change in burned bone is arbitrary and indecipherable, this varied coloration demonstrates how bones burn predictably. Inset A demonstrates color changes in a proximal radius. Inset B demonstrates color differences from the outside to the inside of this long bone (before the shaft is compromised); the lighter calcined bone is external and the black charred (protected) bone internal. Inset C is another example of color change in a proximal radius. (see Plate 4)

Bones burn within soft tissues, and this burning never uniformly encompasses the entire surface of any bone at any one time (see Figure 2.10 A–C). If we assume that color changes evolve with heat exposure, the dynamic nature of these exposure patterns will also translate in a series of color changes across the different bones. Therefore, if color exhibits a gradient across the bone, it has obvious value as an indicator of the progression of thermal damage.

Color changes exhibited by burned bone have numerous designations. Symes *et al.* (1996) have attempted to correlate chemical changes of burning bone with recognizable color bands. This analysis offers separate color classifications for unaltered and heat-altered bone. The categories proposed and defined for heat-altered bone are (1) calcined, (2) charred, (3) border, and (4) heat line (Figure 2.10A). The first two terms are common in the literature, while the latter two represent relatively new concepts (Symes *et al.*, 1999a, b).

Calcined bone is a thermally altered bone that has lost all of its organic material and moisture and exists simply in a framework of ashen, fused bone salts (Mayne Correia, 1997). This bone reduces to minute fragments that rarely survive evidence storage, much less transportation to the laboratory. Shrinkage and deformation in calcined bone can be extreme to the point of nonrecognition, depending on the skeletal element and the location of thermal damage to that element. In most cases, calcined bone is distorted, warped, and fractured beyond any potential for classification or identification: commonly calcined aspects of burned skeletons are neglected when emergency crews are tasked with recovery. Thin cortical bone supported by bone trabeculae resists deformation and remains recognizable and likely measurable.

Charred bone is black in color and represents a carbonized skeletal material in direct contact with heat and flames (Herrmann, 1970). Charred bone usually has microscopic residual burned soft tissues adhering. The black appearance remains even after laboratory processing. Charred bone is still somewhat durable compared to calcined bone, even though it is severely altered by reduced moisture content. Although charred bone is still recognizable and retains diagnostic features, it is difficult to differentiate from other tissues and debris.

An often ignored yet obvious signature of thermal modification is the heat-altered *border* – this feature varies in width and is distinct from charred areas of burned bone. The border is an off-whitish area protected from direct contact with smoke and flames by receding soft tissues. Yet the bone has still undergone some amount of dehydration and molecular alteration due to heat. This process can create heat shrinkage fractures that commonly occur between the charred and border margins. Another feature commonly associated with the border is flaking or distortion of outer cortical layers of bone.

The border is sometimes difficult to recognize with the naked eye, and it becomes more difficult to detect the longer the bone is subjected to taphonomic elements. Archaeological remains rarely exhibit this color gradient. One efficient way of recognizing the border in freshly burned bone is to backlight the bone with an intense, direct light source. Unaltered, greasy (wet) bone is translucent while heat-altered bone is usually opaque. The distinction between translucent bone and discolored, opaque bone makes areas of unaltered versus heat-altered bone readily apparent (Symes *et al.*, 2005a, b).

Adjacent to the border is an occasionally occurring feature called *heat line*. The heat line is peripheral to the border, occurring at the junction between unburned and burned bone. This line is generally narrower than the width of the border and appears to be an area of initial transition from unaltered to heat-altered bone.

The changes in the visual appearance of thermally altered bone result in a scale that gradually evolves from a translucent yellowish (unaltered bone), to an opaque white (heat line and border), to a blackened appearance (char), and eventually to a totally ash-colored, calcined condition. This color transition represents a recognizable process signature, delineating the progression of fire as it alters the burning bone. Color not only distinguishes burned from unburned bone, but also delineates stages of thermal alteration.

CASE 3 (BODY IN CAR) REVISITED

Let us return to our victim found in the burned rental car. After processing the skull free of soft tissues, interesting patterns emerge (Figures 2.11 and 2.12). Cranial material typically undergoes fracture and fragmentation in intense heat but, in this case, there appear to be fractures in unburned areas of the cranium as well. The basal skull exhibits numerous bilateral fractures of the temporal and occipital bones, and there are fractures that some would term 'Le Fort fractures' of the face (Le Fort, 1901) (Figures 2.13 and 2.14).

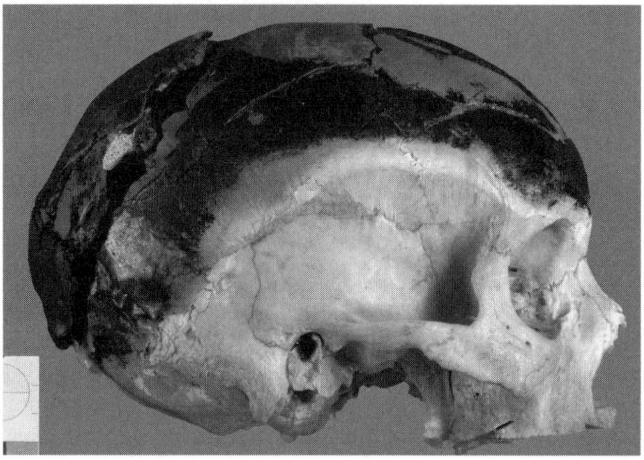

FIGURE 2.11 This figure is the right side view of the victim's skull discovered in a burned car. Notice the color changes from the superior to inferior sides of the skull.

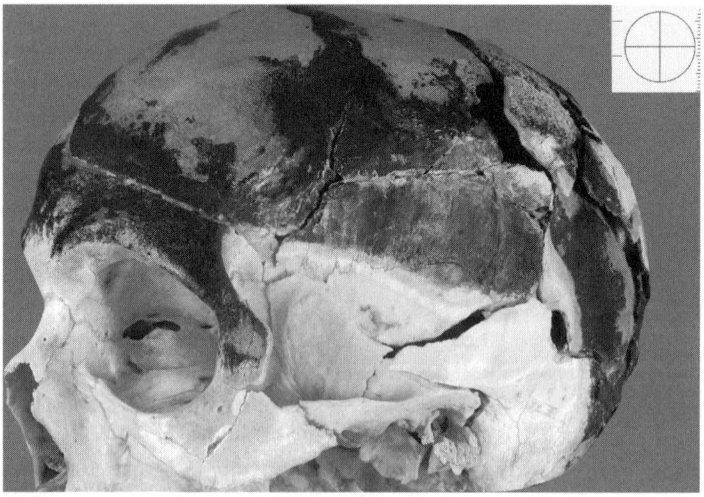

FIGURE 2.12 This figure shows the left oblique view of the victim's skull from the burned car.

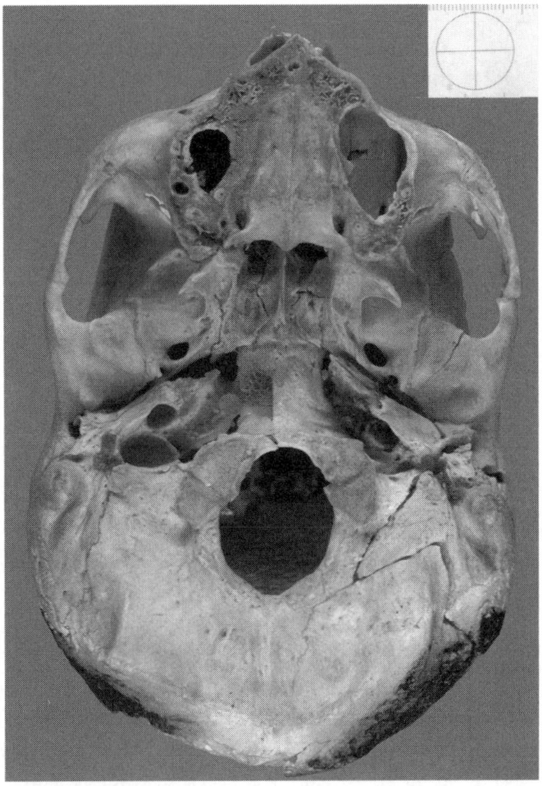

FIGURE 2.13 This figure shows the basal skull view of the fire victim found in a car. Notice the numerous fractures in areas of the basal skull where there is no fire damage.

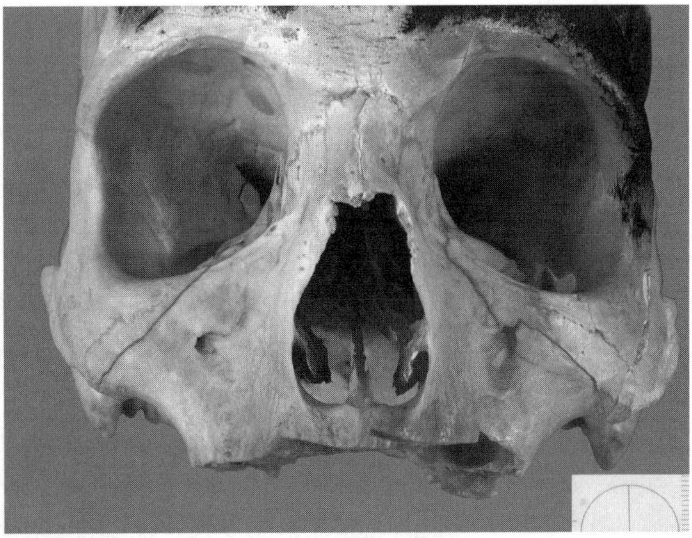

FIGURE 2.14 This figure shows the front of the skull (with dentition removed) of the victim who was found burned in a car. Notice LeFort-like fractures in the face, which are usually associated with blunt trauma.

How can we distinguish perimortem fractures from postmortem fractures resulting from the fire in this skull? Fractures occurring in unburned bone should be considered perimortem. Developing burn fractures do not have the energy to radiate out of burned areas into unburned bone. In Case 3, the presence of fractures outside of the burned superior skull appears to be a clear indication of perimortem trauma. Reconstruction of the cranium reveals a gunshot wound superior to the left external auditory meatus (Figure 2.15).

Figure 2.16 illustrates the diverse color changes that occur in the superior cranial vault. Here, unfractured areas appear calcined. The curvilinear fracture traversing the frontal bone is tracked back to the gunshot wound entrance (Figures 2.15 and 2.17, black arrows). The edges of this fracture are straight; its uninterrupted progression through both unburned and burned areas of bone and fracture location indicate perimortem trauma.

Color augments the gunshot fractures and cranial sutures. While the flat bones of the cranium exposed early to the heat and flames are calcined to an ashen gray, the gunshot fractures are strikingly black. Pope and Smith (2004). They describe these types of fractures as openings for fluids to vent from the brain case. This 'venting' is said to trap fluids and tissues on the surface of the bone, causing it to be imbued black rather than the typical ash/gray seen in calcined bone.

It is unfortunate that Pope *et al.* neglected their past research in color changes of burned bone (2001), otherwise they would have noticed that the black-lined sutures and radiating fractures transecting the islands of calcined bone simply correspond to areas subjected to different degrees and duration of heat as told by color changes. The perimortem fracture lines from the gunshot wound, unfused and diastatic sutures, allow escaping fluids to protect the bone in these areas. When subjecting the skull to intense heat, the cranial bones eventually become calcined, with the anterior skull vault commonly being the area first susceptible to heat damage. The darker color exhibited by the perimortem fractures and sutures is explained by the fact that they are

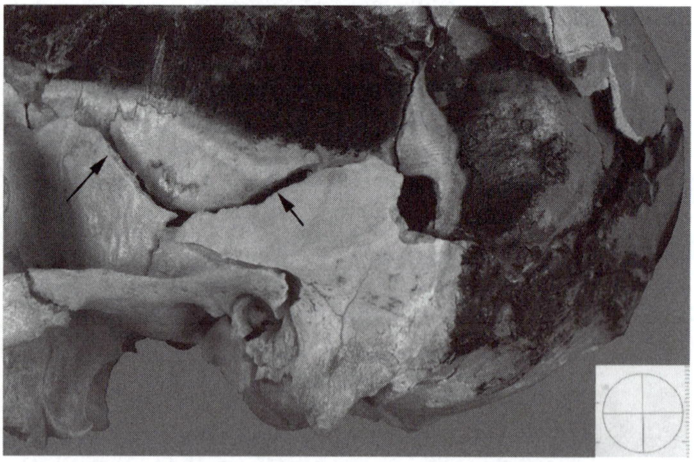

FIGURE 2.15 This figure is a close-up view of the left side of the skull documenting a gunshot wound just above the ear. The arrows illustrate a fracture that originates from the wound and travels forward, eventually terminating in the frontal bone.

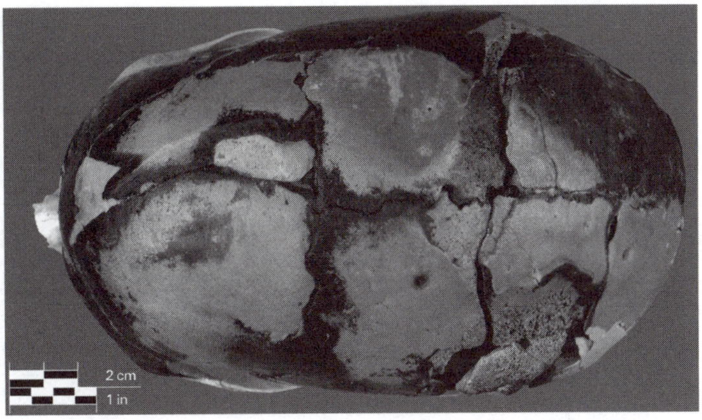

FIGURE 2.16 This figure represents the superior view of the burned car victim's cranium. Notice the gray calcined bone and the black charred bone.

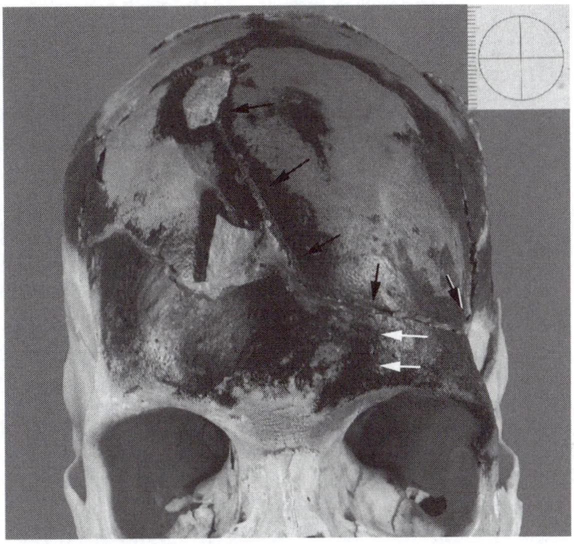

FIGURE 2.17 The front of the cranium illustrates differential burning, while the black arrows indicate a radiating fracture from the entrance of the gunshot wound that terminates in the coronal suture. The white arrows indicate a fracture produced by the bullet impacting the petrous portion of the temporal and producing secondary fractures.

affected less by the same intense heat; they are simply charred. This appears to be the result of oozing tissues and fluids shield against direct heat and slowing fire destruction in these regions. Figure 2.18 illustrates this with an inset revealing the same principles with cranial foraminae that also exhibit less burning. Burn patterns as exhibited in Figures 2.16 and 2.17 illustrate how color changes may indicate perimortem cranial fractures. Figure 2.17 also illustrates a secondary fracture (white arrows) originating from the bullet-impacted petrous portion. This fracture terminates into the original radiating gunshot entrance fracture. In this instance, correct noting and interpretation of color changes to bone allowed for recognizing perimortem fractures, sutures, suture disruption, and normal burn patterns on an otherwise complicated, partially destroyed burned cranium.

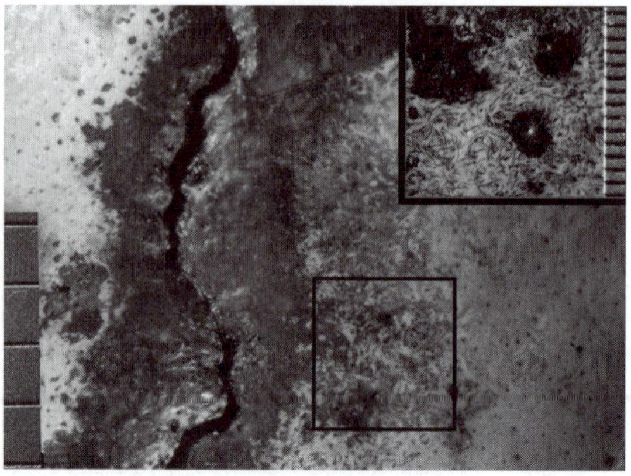

FIGURE 2.18 This figure is a close-up of the coronal suture. Notice the black (charred) discoloration that follows the suture border. The inset is an enlargement of the foramina in the cranial bone. Notice how they are also charred to a black color and indicate less burning destruction than areas of gray (calcined) bone.

Burned bone fracture biomechanics

Anthropologists have studied thermal fractures for more than a century. Numerous classification schemes have been suggested (Krogman, 1939, 1943a,b; Baby, 1954; Binford, 1963; Mayne, 1990; Symes *et al.*, 1996). These fracture classifications include the following:

1. Longitudinal. Longitudinal fractures to long bones are probably the most common of major burn fractures, occurring regularly and predictably. As a shaft heats to the point of evaporation and protein denaturalization, the bone matrix shrinks, facilitating structural failure. These longitudinal failures appear to originate in similar locations, commonly following the grain of bone, parallel to the osteon canals, although longitudinal fractures can also take a somewhat helical path down the long axis of the bone.
2. Step. These fractures are often associated with longitudinal fractures. A step fracture will extend from the margin of the longitudinal fracture transversely across the bone shaft, through the compact bone, fracturing the bone shaft at the intersection of another longitudinal fracture.
3. Transverse. Transverse fractures differ from longitudinal fractures in that they transect haversion canals. Transverse fractures are also common, since fire consumes most long bones transversely as it progresses up the shaft. Increasing tissue thickness and a pugilistic posture can hamper this progression on one side, while the exposed side consumes at a faster pace. These are very similar to or make up step fractures.
4. Patina. These superficial fractures, seemingly less destructive than other fracture types, appear as a fine mesh of uniformly patterned cracks similar to those seen in old china or an aged painting (Krogman, 1943a, b). This pattern is somewhat difficult to interpret but is often observed on flat areas of postcranial bones, and may be the result of a broad area receiving uniform amounts of heat,

compelling superficial cortical bone to shrink evenly over the surface. Others have suggested that patina patterns are due to the incineration of thin protective soft tissues. These are probably related ideas. Patina also appears on epiphyseal ends and cranial bones.

5. Splintering and delamination. These fractures are characterized by the splitting away of cortical bone layers from cancellous bone, the separation of the inner and outer tables of cranial bone, or the exposure of cancellous bone on epiphyses.
6. Burn line fractures. These fractures follow the burn borderline, seen clearly in reconstructions; it separates burned and unburned bone.
7. Curved transverse. The classic curved transverse fracture is the result of bone heating, then cracking as protective soft tissues and periosteum shrink, pulling the brittle surface of the thermally altered bone (thus also called 'muscle shrinkage lines') (Figures 2.19 and 2.20). A less common manifestation of curved transverse fractures may also form as 'concentric rings.' Concentric rings typically occur in fossae or areas of concentrated tissues, such as the popliteal region of the femur. They are a consequence of bone cortical thickness, shape, articulation, and soft tissue obstruction, but are not necessarily the byproduct of elastic muscle fibers shrinking. The curved transverse fracture also commonly results in 'coning,' where the fractured diaphysis appears arched at the fracture margin (see Figure 2.10A).

A common place for curved transverse fractures is on the shaft of the femur. In a typical fire, the first place to burn on a femur is the anterior knee area. As the fire consumes and separates the knee from the rest of the bone, tissue thickness increases dramatically in the fire destruction path along the diaphysis due to the powerful lower limb muscles, and thus the limb becomes progressively more resistant to heat destruction. In this way, due to tissue protection,

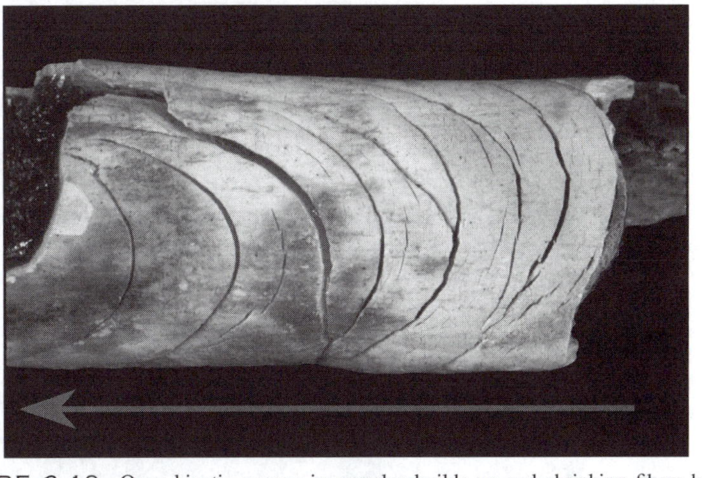

FIGURE 2.19 Once kinetic energy in muscles builds up and shrinking fibers begin to break free of connective restraints, the bone is systematically exposed. This jerky movement occurs in this case up the femur shaft as indicated by the residual curved transverse fractures. Extended exposure creates a heat shrinkage fracture on a femur at the point where tissue borders the exposed bone. These fractures are generally concave on the tissue side, as indicated by the arrow.

the last area of the femoral shaft to burn is its proximal third. Figure 2.20 illustrates a distal posterior femur. This is a case where the original examiners (including the first author) mistakenly thought that this pattern was a 'hot spot.' However, with our current knowledge of curved transverse fractures, we can now interpret this as the opposite. The 'bull's-eye' pattern is the last place to burn on a posterior knee, where pugilistic posture protects this area of the distal femur. This is essentially a cold (protected) spot of the distal femur.

As fire consumes the knee, all muscle insertions undergo destruction. With tendons and muscle ends released, the fibers can shrink unrestrained by insertions. Muscles shrink into bundled masses and slowly recede up the shaft as fire destroys the limb. The muscle fibers continue until they break loose from all attachments, whereupon movement accelerates. This acceleration is inhibited by connective tissues that restrain the receding muscle fibers. Like rubber bands storing kinetic energy, muscles override connective tissue resistance as they shrink, spurring rapid movement up the shaft until they reduce the stress with their short burst of kinetic energy. As shrinkage occurs, muscles eventually counteract any attachment resistance.

Muscles shrink on the shaft of the bone when the distal muscle bundles receive full exposure to fire, as heat destroys tissues. When muscle fibers shrink enough to separate from attachments, the remaining bundle of fibers,

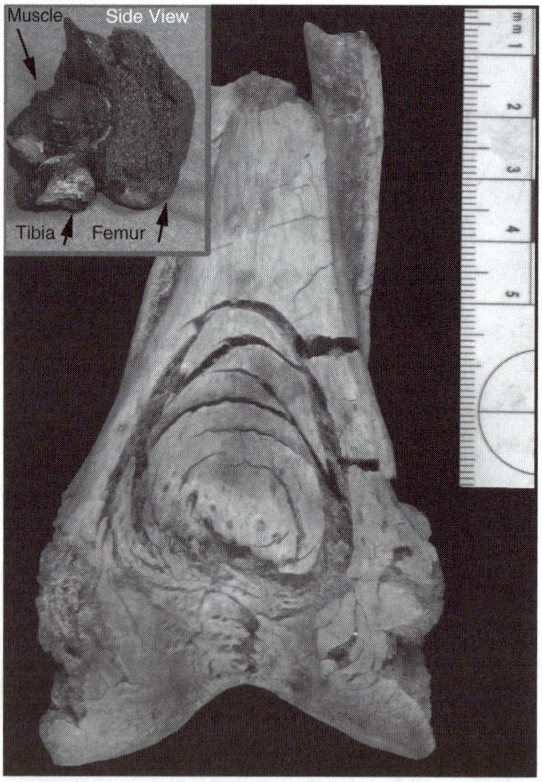

FIGURE 2.20 This figure is a posterior view of a distal femur. Here the heat has calcined the bone leaving a recognizable pattern. The 'bull's-eye pattern' is a reflection of the tissues shrinking. The inset shows a distal femur that has not been completely consumed by fire. Notice the existing muscle that is slowly shrinking due to destruction. This process is creating a shrinkage curved transverse pattern.

still present, temporarily protect the bone. While the muscle fibers are shrinking, yet restrained from movement by connective tissues, burn lines occur on the border of these tissues, where bone is exposed. With more shrinkage, muscle fibers break free of all attachments and contract rapidly until kinetic energy is lost. Once again, enough kinetic energy must build up to break free of connective restraints. This jerky movement occurs up to the femur shaft as indicated by the residual curved transverse fractures, where extended exposure creates a heat shrinkage fracture on a femur at the point where tissue borders exposed bone. Since the muscle is shrinking to its origin or insertion, the remaining bundle of fibers protecting the bone is forced into bundles (Figure 2.21). Incremental shrinking leads to predictable patterns. This pattern is also observable in other areas of the skeleton where muscles do not form a bundle, but form a sheath of soft tissues. In areas such as the cranial vault, you see quick shrinkage of protective tissues across the bone surface, a fracture line may then appear, and then repeated retreats of tissues occur forming more fracture lines. Figure 2.22 demonstrates this on a close-up of the nuchal muscle lines of a burned skull, where the soft tissue is pulled back on the head due to the strong neck muscles. The creeping of the soft tissues registers in repetitive tissue line fractures. Since the nuchal muscles cover a wide area of the skull, there is no opportunity for the muscle to form a narrow bundle, and no curvature occurs.

A second form of curved transverse fracture occurs near joints (epiphyses and metaphases) or where cortical bone is thin. A large joint is often slowly impinged upon by intense heat from all directions. For example, the tibial/talus joint is rotated (pedal inversion) in the pugilistic posture. This creates increased exposure to the lateral aspect of the joint that begins to burn

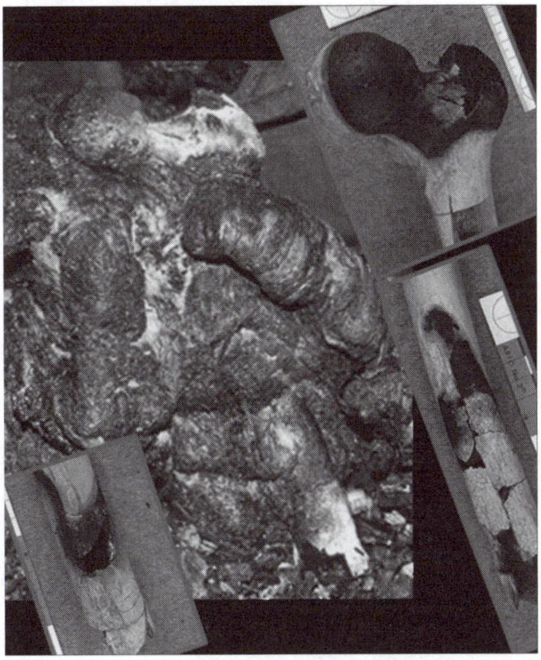

FIGURE 2.21 As tissue burns, muscles contract somewhat predictably to form patterned destruction of the bone. This figure illustrates an upper leg before and after (insets) processing. (see Plate 5)

FIGURE 2.22 This illustration demonstrates tissue shrinkage of the nuchal muscle lines of a burned occipital bone on the back of a skull. The photograph exhibits how soft tissues are pulled back on the head due to strong neck muscles. The creeping of the soft tissues registers in repetitive tissue line fractures. (see Plate 6)

early as heat encroaches into it producing a fine curved transverse pattern. The fire consumes the joint by slowly encroaching to the middle of the joint. A similar example is the posterior distal femur (see Figure 2.20).

Curved transverse cross-sectional fractures also leave diagnostic patterns. The fracture is curved, beveled, and is usually combined with or ends in steep stepped fractures. The examination of burned torsos in the morgue usually exhibits numerous limb ends shaped in a convex and stepped pattern. This pattern is suggestive of limbs being destroyed distally, and burning toward the torso. When examining the pattern in Memphis, we (Berryman, Smith, and Symes) originally coined the term 'coning' for this burned bone contour. Curved transverse fractures perpendicular to long bone shafts reveal an arched contour. The cross section of this fracture may also be arched with the inner cortical bone extending beyond the outer cortical bone. This is a function of the outer cortical bone being heated initially, with the inner bone protected. So coning fractures are often beveled and arched in cross section (Symes *et al.*, 1999a, b).

CASE 4 (HOMICIDE IN ARSON STRUCTURE FIRE)

Firefighters foraging through the remnants of an abandoned nightclub fire in Buffalo, NY, discover the body of an adult female at the base of a stairwell, faceup, with drug paraphernalia in her possession (see Symes *et al.*, 2005a,b for more information on this case). Although the fire was quickly contained, the victim had incurred severe heat damage, and perimortem cranial trauma was evident upon external exam. In addition, investigators observed blood at the back entrance and found evidence of accelerants, leading them to suspect the fire had been purposefully set in an attempt to conceal what was clearly a homicide. An astute medical examiner in autopsy enlisted the expertise of anthropologists after noticing the burn pattern was inconsistent with normal burn patterns and failed to manifest the expected pugilistic position.

The skull and the mandible illustrate a pattern of trauma indicative of massive force. The skull is fractured severely on the left side with fractures radiating from a center point immediately above the ear, and concentric fractures creating circular (caving-in) fractures. The fracturing pattern extends to the entire left face, including the zygomatic arch and the mandible. Certainly, the head trauma answers the cause and manner of death, but are we, as experts, not compelled to interpret the complete story behind this felonious act? Do taphonomic events mask an accurate interpretation? An accurate cause of death entails a complete list of traumatized tissues.

During the postautopsy exam, pathologists and anthropologists agreed that the right hand burned abnormally, and all commented that it did not fit the classic pugilistic posture. Rather than flexing in the pugilistic fashion, the victim's right hand exhibited slightly hyperextended proximal phalanges, exposing the fingertips and allowing uncharacteristic burn patterns, as illustrated in Figure 2.23 (after much of the soft tissue had been removed). The victim has an unusual pattern on the right arm that requires close investigation.

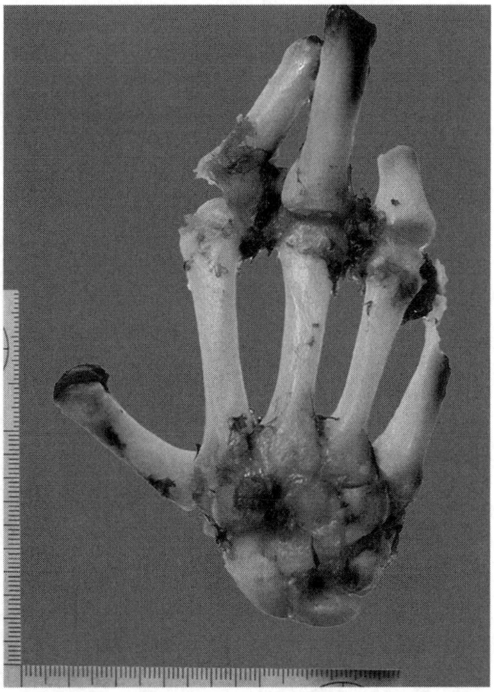

FIGURE 2.23 Rather than flexing into the pugilistic position, the suspected homicide victim's right hand exhibited slightly hyperextended proximal phalanges. This exposed the victim's fingertips and allowed for uncharacteristic burn patterns, indicating an abnormal pattern.

Examination of the right radius and the ulna indicates heat damage and fracture at the distal third of the shafts (Figure 2.24, inset). The ulna is burned at all ends, and no conclusions other than 'fire damage' can be drawn. A conservative interpretation of thermal destruction of the radius should take precedence, and therefore all injuries of the right forearm should be considered fire-related unless more information can be gathered. However, close examination of fracture shape may be the best evidence for perimortem trauma in this case. A consistent fracture pattern traverses burned and unburned bone uniformly in the distal radius (see Figure 2.24). All evidence appears consistent with wet bone. Upon reconstruction, additional microfractures indicate a typical fresh bone tension/compression fracture with the formation of the butterfly fracture (Figure 2.25). A brittle, burned bone could not form this smooth, shearing fracture pattern. Figure 2.26 is an unrelated forearm bone that illustrates a fracture, partially burned and unburned. The figure shows how the fracture contour changes from wet to dry bone. The fracture occurred after the arm burned.

Misgivings created from an outstretched hand of a burned homicide victim, and critical examination of the same forearm, confirm that there are more indicators of violence in this victim's death. The right radius indicates bending blunt force trauma of the shaft with the forearm shaft bending anterior (compression) to posterior (tension), the opposite

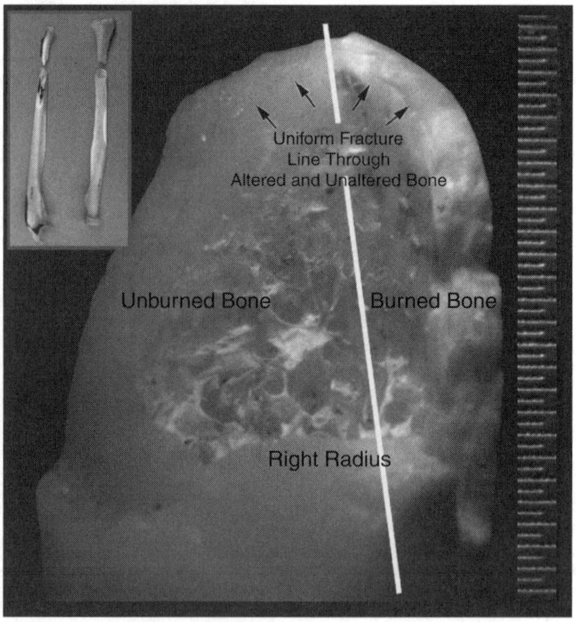

FIGURE 2.24 The distal radius demonstrates a consistent fracture pattern traversing burned and unburned bone uniformly (line separates the two). All evidence appears consistent with a wet bone fracture. (see Plate 7)

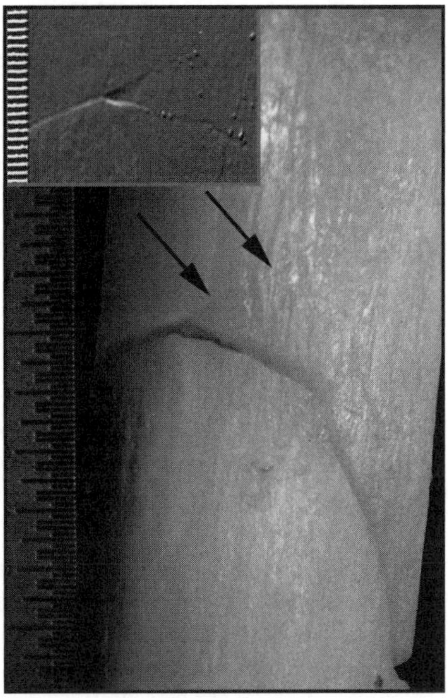

FIGURE 2.25 Upon reconstruction of the distal radius, additional microfractures indicate a typical fresh bone tension/compression fracture (arrows) with the formation of the butterfly fracture. The inset reveals microfracture embossed for visibility. All evidence appears consistent with a wet bone fracture.

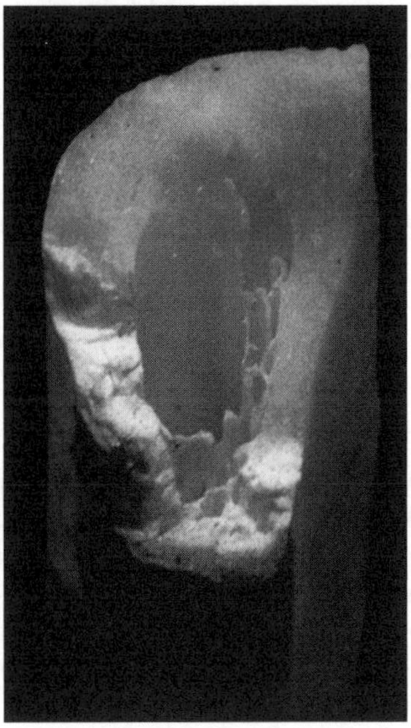

FIGURE 2.26 This figure illustrates a bone fracture of a partially burned ulna. The figure shows how the fracture contour changes from wet to dry bone. This pattern indicates that the fracture occurred after the arm burned. (see Plate 8)

direction of a typical Colles' fracture that occurs when a person falls on their wrist. This fracture appears to be perimortem, compromising the forearm and disallowing the pugilistic posture to occur in the right hand, since the fulcrum for the powerful wrist flexors in the forearm is compromised. A more accurate cause of death in this case is multiple bone fractures to the head and the right arm due to blunt force trauma.

SUMMARY

In the rapidly changing world of forensic science, more specifically forensic anthropology, there is a necessity to understand and be able to interpret fire modification of human remains. With the enormous potential for fire and heat alterations to inhibit scientists' abilities to interpret patterns to human remains, burn trauma analysis is inconsistent and slow to mature in many disciplines. Destructive forces of fire often significantly alter, damage, or even destroy many recognizable patterns, characteristics, and evidence that we normally depend upon. This likely contributes to the existence and persistence of old and untested theories concerning burned bone where inaccurate interpretations, such as exploding skulls, persist for decades (see Preface, this volume), and terminology is inconsistent.

As forensic anthropologists, we must examine burned bodies with the same procedural approach we would utilize with any taphonomical influence. We must work backward from the postmortem condition, understand how heat alters tissues, and eventually work to put the flesh back on the victim. The reversal requires an understanding of burning process signatures. Figures 2.7 and 2.8 chart initial, secondary, and final burning areas on the human skeleton. This burn pattern is predictable.

The current research suggests that three areas contribute significantly to burning destruction interpretation:

1. Body position and tissue shielding in bone,
2. Color change in thermally altered bone, and
3. Burned bone fracture biomechanics.

Burn patterns, when using these three groups of characteristics, become increasingly recognizable and manageable. As the case studies cited above demonstrate, deviations from normal patterns hoist a red flag to the examiner, and more information is likely needed to accurately interpret the cause and manner of death. In fatal fire cases, forensic anthropologists are responsible primarily for separating perimortem trauma from heat-induced fractures and for assigning temporal and sequential designations to trauma when possible. These findings may contribute to the determination of cause and manner of death, time of death, and perpetrator behavior.

While fire inhibits the recognition of perimortem trauma, an understanding of 'normal' burn patterns may indicate when something is awry. Whether it is simply an unsolved homicide from Buffalo, or one of thousands of examined victims of human rights atrocities, the medicolegal system requires more than *manner of death*. Cause of death requires an accurate analysis of *ALL* traumata. Understanding process signatures in burned bodies is one approach to contributing to this arena.

ACKNOWLEDGMENTS

I (S.A.S.) witnessed and photodocumented my first burned body in 1979. Dr. William M. Bass continued to take me to fire scenes in Eastern Tennessee, and that is my first recollection of seeing patterns of burning. Without Dr. Bass' natural curiosity and methodological approach, I may never have had the curiosity to write this chapter.

The Mercyhurst Archaeological Institute has a large role in this current work. The list of characters includes Luis Cabo-Perez, the editor, Dennis C. Dirkmaat for his support and direction, and James M. Adovasio for enabling capabilities. Kyra Stull, a graduate student from the Department of Applied Forensic Sciences, also contributed hands-on and editing skills. And finally, a thank you to Ranald Donaldson for his medical perspective. Partial funding was offered through the National Forensic Academy (R011007087).

REFERENCES

Baby, R.S. (1954). *Hopewell cremation practices*. The Ohio Historical Society Papers in Archaeology **1**, pp. 1–7.

Bass, W.M. (1984). Is it possible to consume a body completely in a fire? In *Human Identification: Case Studies in Forensic Anthropology* (T.A. Rathbun and J.E. Buikstra, eds.), pp. 159–167. Charles C. Thomas, Springfield, IL.

Bass, W.M. and Jantz, R.J. (2004). Cremation weights in East Tennessee. *Journal of Forensic Sciences* **49**(5), 1–4.

Bennett, J. and Benedix, D.C. (1999). Positive identification of cremains recovered from an automobile based on the presence of an internal fixation device. *Journal of Forensic Sciences* **44**(6), 1296–1298.

Berryman, H. and Berryman, C.A. (2007). Considerations in differentiating negligence from deliberate misconduct – lessons learned from tri-state crematorium. *Proceedings of the American Academy of Forensic Sciences* **13**, 385–386.

Binford, L.R. (1963). An analysis of cremations from three Michigan sites. *Wisconsin Archaeologist* **44**(2), 98–110.

Binford, L.R. (1972). An analysis of cremations from three Michigan sites. In *An Archaeological Perspective* (S. Struever, ed.), pp. 373–382. Seminar Press, New York.

Bonucci, E. and Graziani, G. (1975). Comparative thermogravimetric, X-ray diffraction and electron microscope investigations of burnt bones from recent, ancient, and prehistoric age. *Atti Memorie Accademia Nazionale die Lincei Scienze, Fisiche, Matematiche Naturali* **Series 8**(59), 517–534.

Bradtmiller, B. and Buikstra, J.E. (1984). Effects of burning on human bone microstructure: a preliminary study. *Journal of Forensic Sciences* **29**(2), 535–540.

Buikstra, J.E. and Swegle, M. (1989). Bone modification due to burning: experimental evidence. In *Bone Modification* (R.B. Bonnichsen and M.H. Sorg, eds.), pp. 247–258. Center for the Study of the First Americans, Orono, ME.

Civjan, S, Selting, W.J., de Simon, L.B., *et al.* (1971). Characterization of osseous tissues by thermogravimetric and physical techniques. *Journal of Dental Research* **51**(2), 539–542.

Crow, D.M. and Glassman, R.M. (1996). Standardization model for describing the extent of burn injury to human remains. *Journal of Forensic Sciences* **41**(1), 152–154.

DeHaan, J.D. (1997). Destruction of the body. In *Kirk's Fire Investigation*, pp. 370–376. Prentice Hall, Englewood Cliffs, NJ.

DeHaan, J.D. and Nurbakhsh, S. (2001). Sustained combustion of an animal carcass and its implications for the consumption of human bodies in fires. *Journal of Forensic Sciences* **46**(5), 1076–1081.

DeHaan, J.D., Campbell, S.J., and Nurbakhsh, S. (1999). Combustion of animal fat and its implications for the consumption of human bodies in fires. *Science & Justice* **39**(1), 27–38.

Delattre, V.F. (2000). Burned beyond recognition: systematic approach to the dental identification of charred human remains. *Journal of Forensic Sciences* **45**(3), 589–596.

Dirkmaat, D.C. (2002). Recovery and interpretation of the fatal fire victim: the role of forensic anthropology. In *Advances in Forensic Taphonomy: Method, Theory, and Archaeological Perspectives* (W.D. Haglund and M.H. Sorg, eds.), pp. 451–472. CRC Press, Boca Raton, FL.

Dirkmaat, D.C. and Adovasio, J. (1997). The role of archaeology in the recovery and interpretation of human remains. In *Forensic Taphonomy: The Postmortem Fate of Human Remains* (W.D. Haglund and M.H. Sorg, eds.), pp. 39–64. CRC Press, Boca Raton.

Dunlop, J. (1978). Traffic light discoloration in cremated bones. *Medicine, Science and the Law* **18**(3), 163–173.

Eckert, W.G. (1981). The medicolegal and forensic aspects of fires. *American Journal of Forensic Medicine and Pathology* **2**(4), 347–357.

Eckert, W.G., James, S., and Katchis, S. (1988). Investigation of cremations and severely burned bodies. *American Journal of Forensic Medicine and Pathology* **9**(3), 188–200.

Fairgrieve, S., and Molto, E. (1994). Burning point Canadian case studies of intentionally cremated human remains. In *Strength in Diversity, A Reader in Physical Anthropology* (A. Herring and L. Chan, eds.), pp. 385–402. Scholars Press Inc., Toronto.

Forbes, G. (1941). The effects of heat on histological structure of bone. *Police Journal* **14**(1), 50–60.

Gejvall, N.G. (1969). Cremations. In *Science in Archaeology* (D. Brothwell and E. Higgs, eds.), pp. 468–479. Praeger, New York.

Grevin, G., Bailet, P., Quaterhomme, G., and Olliver, A. (1998). Anatomical reconstruction of fragments of burned human bones: a necessary means for forensic identification. *Forensic Science International* **96**(2–3), 129–134.

de Gruchy, S. and Rodgers, T.L. (2002). Identifying chop marks on cremated bone: a preliminary study. *Journal of Forensic Sciences* **47**(5), 933–936.

Haglund, W.D. and M.H. Sorg (1997). Introduction to Forensic Taphonomy. In Forensic Taphonomy: The Postmortem Fate of Human Remains (W.D. Haglund and M.H. Sorg, eds.), pp. 1–9. CRC Press, Boca Raton.

Heglar, R. (1984). Burned remains. In *Human Identification. Case Studies in Forensic Anthropology* (T. Rathbun and J. Buikstra, eds.), pp. 148–158. Charles C. Thomas Press, Springfield, IL.

Herrmann, B. (1970). Anthropologische bearbeitung der leichenbränden von Berlin-Rudow. *Ausgrabungen in Berlin* **1**, 61–71.

Herrmann, B. (1977). On histological investigations of cremated human remains. *Journal of Human Evolution* **6**, 101–103.

Herrmann, N.P. and Bennett, J.L. (1999). The differentiation of traumatic and heat-related fractures in burned bone. *Journal of Forensic Sciences* **44**(3), 461–469.

Hiller, J.C., Thompson, T.J.U., Evison, M.P., *et al.* (2003). Bone mineral change during experimental heating: an X-ray scattering investigation. *Biomaterials* **24**(28), 5091–5097.

Holager, J. (1970). Thermogravimetric examination on enamel and dentin. *Journal of Dental Research* **49**(3), 546–548.

Holland, T.D. (1989). Use of the cranial base in the identification of fire victims. *Journal of Forensic Sciences* **34**(2), 458–460.

Icove, D.J. and DeHaan, J.D. (2004). *Forensic Fire Scene Reconstruction*. Prentice Hall, New York.

Işcan, M.Y. and Kennedy, K.A.R. (1989). Reconstruction of life from the skeleton: an introduction. In *Reconstruction of Life from the Skeleton* (M.Y. Iscan and K.A.R. Kennedy, eds.), pp. 1–10. Alan R. Liss, Inc., New York.

Kerley, E.R. (1980). Forensic anthropology. In *Forensic Sciences* (W.G. Eckert, ed.), pp. 102–113. The C.V. Mosby Company, St. Louis, MO.

King, C.G. and King, S. (1989). The archaeology of fire investigation. *Fire Engineering* **142**(6), 70–74.

Krogman, W.M. (1939). A guide to the identification of human skeletal material. *FBI Law Enforcement Journal* **8**(8), 1–29.

Krogman, W.M. (1943a). Role of the physical anthropologist in the identification of human skeletal remains, Part I. *FBI Law Enforcement Bulletin* **12**(4), 17–40.

Krogman, W.M. (1943b). Role of the physical anthropologist in the identification of human skeletal remains, Part II. *FBI Law Enforcement Bulletin* **12**(5), 12–28.

Krogman, W.M. (1949). The human skeleton in legal medicine. *Symposium on Medicolegal Problems* (S.A. Levinson, ed.), pp. 1–92. J.B. Lippincott, Philadelphia, PA.

Krogman, W. and Iscan, Y. (1986). *The Human Skeleton in Forensic Medicine*. Charles C. Thomas, Springfield, IL.

Le Fort, R. (1901). Experimental study of fractures of the upper jaw: part III. *Rev Chir Pris* **23**, 479–507. Reprinted (1972) in *Plastic and Reconstructive Surgery* **50**, 600–605.

Mayne, P.M. (1990). *The Identification of Precremation Trauma in Cremated Remains*. M.A. Thesis. University of Alberta. Edmonton, AB.

Mayne Correia, P.M., (1997). Fire modification of bone: a review of the literature. In *Forensic Taphonomy: The Postmortem Fate of Human Remains* (W.D. Haglund and M.H. Sorg, eds.), pp. 275–293. CRC Press, Boca Raton, FL.

Mayne Correia, P.M. and Beattie, O. (2002). A critical look at methods for recovering, evaluating, and interpreting cremated human remains. In *Advances in Forensic Taphonomy: Method, Theory, and Archaeological Perspectives* (W.D. Haglund and M.H. Sorg, eds.), pp. 435–450. CRC Press, Boca Raton, FL.

McKinley, J.I. and Bond, J.M. (2001). Cremated bone. In *Handbook of Archaeological Sciences* (D.R. Brothwell and A.M. Pollar, eds.), pp. 281–292. John Wiley and Sons, Chichester.

Murray, K.A. and Rose, J.C. (1993). The analysis of cremains: a case study involving the inappropriate disposal of mortuary remains. *Journal of Forensic Sciences* **38(1)**, 98–103.

Murray, E.A., Bonell, H.J., Huckaby, B.L., and Perzigian, A.J. (1991). *Cremation in a chicken coop: self-incineration or fowl play?* Paper presented at the 43rd Annual Meeting of the American Academy of Forensic Sciences, Anaheim, CA.

Nelson, R. (1992). A microscopic comparison of fresh and burned bone. *Journal of Forensic Sciences* **37**(4), 1055–1060.

Owsley, D.W. (1993). Identification of the fragmentary, burned remains of two U.S. journalists seven years after their disappearance in Guatemala. *Journal of Forensic Sciences* **38**(6), 1372–1382.

Piper, A. (2007). Unpublished Manuscript. *Identification of Translucency in Heat Modified Bone: Utility and Implications for Process Signature Interpretation*. M.S. thesis, to be submitted to Mercyhurst College.

Pope, E.J. (2007). Burned human remains: myths in forensic science. *Proceedings of the American Academy of Forensic Sciences* **13**, 379–380.

Pope, E.J. and Smith, O.C. (2004). Identification of traumatic injury in burned cranial bone: an experimental approach. *Journal of Forensic Sciences* **49**(3), 431–440.

Pope, E.J., Symes, S.A., Smith, O.C., and Gardner, C.D. (2001). Burning observations II: analysis of color patterns. *Proceedings of the American Academy of Forensic Sciences* **7**, 277–278.

Richards, N.F. (1977). Fire investigation – destruction of corpses. *Medicine, Science and the Law* **17**(2), 79–82.

Rootare, H.M. and Craig, R.G. (1977). Vapor phase adsorption of water on hydroxyapatite. *Journal of Dental Research* **56**, 1437–1448.

Schultz, M. (1986). Hitzeiduzierte veränderungen des knochens. Die mikroskopische untersuchung prähistorischer skekefunde. *Archaeologie un museum* **1**, 116–129.

Shipman, P., Foster, G., and Schoeninger, M. (1984). Burnt bones and teeth: an experimental study of color, morphology, crystal structure and shrinkage. *Journal of Archaeological Science* **11**, 307–325.

Sigler-Eisenburg, B. (1985). Forensic research: expanding the concept of applied archaeology. *American Antiquity* **50**(3), 650–655.

Sledzik, P.S., *et al.* (2007). *The fourth era of forensic anthropology: examining the future of the discipline*. Paper presented at the 59th Annual Meeting of the American Academy of Forensic Sciences, San Antonio, TX.

Stewart, T.D. (1979). Burned bones. In *Essentials of Forensic Anthropology, Especially as Developed in the United States*, pp. 59–68. Charles C Thomas, Springfield, IL.

Stiner, M., Kuhn, S., Weiner, S., and Bar-Yosef, O. (1995). Differential burning, recrystalization, and fragmentation of archaeological bone. *Journal of Archaeological Science***22**, 223–237.

Symes, S.A., Smith, O.C., Berryman, H., *et al.* (1996). Bones: bullets, burns, bludgeons, blunders, and why (Workshop). *Proceedings of the American Academy of Forensic Sciences* **2**, 10–11.

Symes, S.A., Smith, O.C., Berryman, H., and Pope, E.J. (1999a). *Patterned thermal destruction of human remains*. Paper presented to the 30th Anniversary of the T.D. Stewart Personal Identification in Mass Disasters, Central Identification Laboratory, Hawaii.

Symes, S.A., Smith, O.C., Gardner, C.D., *et al.* (1999b). Anthropological and pathological analyses of sharp trauma in autopsy. *Proceedings of the American Academy of Forensic Sciences* **5**, 177–178.

Symes, S.A., Williams, J.A., Murray, E.A., *et al.* (2002). Taphonomical context of sharp-force trauma in suspected cases of human mutilation and dismemberment. In *Advances in Forensic Taphonomy: Method, Theory, and Archaeological Perspectives* (W.D. Haglund and M.H. Sorg, eds.), pp. 403–434. CRC Press, Boca Raton.

Symes, S.A., Kroman, A.M., Rainwater, C.W., and Piper, A.L. (2005a). *Bone biomechanical considerations in perimortem vs. postmortem thermal bone fractures: fracture analyses on victims of suspicious fire scenes*. Poster presented to the Annual Meeting of the American Association of Physical Anthropology, Milwaukie, OR.

Symes, S.A., Dirkmaat, D.C., Woytash, J.J., *et al.* (2005b). Perimortem bone fracture distinguished from postmortem fire trauma: a case study with mixed signals. *Proceedings of the American Academy of Forensic Sciences*, **11**, 288–289.

Thompson, T.J. (2004). Recent advances in the study of burned bone and their implications for forensic anthropology. *Forensic Science International* **146S**, 203–205.

Thompson, T.J. (2005). Heat-induced dimensional changes in bone and their consequences for forensic anthropology. *Journal of Forensic Sciences* **50**(5), 1008–1015.

Thurman, M. and Willmore, L.J. (1980). A replicative cremation experiment. *North American Archaeologist* **2**(4), 275–283.

Trotter, M. and Peterson, R.R. (1995). Ash weight of human skeletons in per cent of their dry, fat-free weight. *Anatomical Record* **123**, 341–358.

Ubelaker, D., Owsley, D., Houck, M., *et al.* (1995). The role of forensic anthropology in the recovery and analysis of Branch Davidian compound victims: recovery procedures and characteristics of the victims. *Journal of Forensic Sciences* **40**(3), 335–340.

Van Vark, G. N. (1970). *Some Statistical Procedures for the Investigation of Prehistoric Skeletal Material*. Unpublished Ph.D. dissertation, Rijksuniversiteit de Groningen.

Van Vark, G.N. (1975). The investigation of human cremated skeletal material by multivariate statistical methods II. Measures. *OSSA*. **2**, 47–68.

Warren, M.W. and Maples, W.R. (1997). The anthropometry of contemporary commercial cremation. *Journal of Forensic Sciences* **42**(3), 417–423.

Warren, M.W. and Schultz, J.J. (2002). Post-cremation taphonomy and artifact preservation. *Journal of Forensic Sciences* **47**(3), 656–659.

Webb, W. and Snow, C.E. (1945). *The Adena People*. Publications of the Department of Anthropology No. 6, pp. 166–199. University of Kentucky, Lexington.

Wells, C. (1960). A study of cremation. *Antiquity* **34**, 29–47.

3

The Recovery and Study of Burned Human Teeth

Christopher W. Schmidt, Ph.D.

Department of Anthropology, Indiana Prehistory Laboratory, University of Indianapolis, Indianapolis, IN

Introduction

Teeth are among the most useful structures in the body for establishing a decedent's age, sex, ancestry, and skeletal idiosyncrasies (i.e., biological profile). In forensic contexts, positive identifications are commonly made via dental morphology, patterns of restoration, and pulp chamber DNA (Cook, 1984; Duffy *et al.*, 1991; Sweet and Sweet, 1995; Ubelaker *et al.*, 2002; Adams, 2003a,b; Gaytmenn and Sweet, 2003; Lee *et al.*, 2004; Shiroma *et al.*, 2004). In archeological contexts, where positive identification is rarely a major concern, teeth aid in determining an individual's biological profile, diet, and health. They also indicate certain cultural habits (e.g., betel nut chewing and pipe smoking) and types of personal adornment (e.g., tooth filing and ablation).

Recent and ancient human remains are found in various states of preservation that do not always correlate with time since death. For example, a body that was placed in a shallow grave 3 years ago may be much easier to identify than a body that was intentionally burned in a 55-gallon drum only a week ago. Cultural practices like burning a body can rapidly degrade bone and teeth, making identification difficult. For those with experience, however, it is possible to recover burned dental remains and to reconstruct the dentition for either forensic identification or bioarcheological interpretation.

Purpose and Caveats

This chapter is written as a guideline to help professional and student archeologists, biological anthropologists, forensic dentists, coroners and medical examiners, and crime scene technicians develop strategies for recovering and studying burned human teeth. Even though these guidelines derive from years of cremation study, it is impossible to cite every circumstance one might face when recovering burned human teeth. Likewise, it is probably best to remember that for every 'rule' there is an exception. Some things seen frequently by one observer may not appear in the burned bodies that another encounters, especially if the observers work with different culture groups or in very different physical environments. Finally, it is difficult at best to find all dental fragments that might be associated with a cremation or a burnt victim. Some teeth will fragment beyond recognition and reconstruction; taphonomic forces such as plants, animals, and water may make dental fragments unrecoverable. Thus, it is best to have a sound strategy going into the recovery so that confusion, mistakes, and surprises are kept at a minimum.

FIELD RECOVERY

The suggestions made here to recover burned teeth are a summary meant to aid those developing their skills in the archeological recovery of human remains. They are modifications to standard archeological procedures intended to benefit nonarcheology personnel and archeologists not yet familiar with the recovery of burned bodies. This chapter is only one component of the preparation necessary to lead such an excavation. There are many other aspects of archeological training (e.g., osteology/odontology, archeological method and theory, soil science, and surveying) that, likewise, are requisite for the proper recovery of burned remains.

Investigators find burn victims in all types of environments, indoors and outdoors. The discussion here, however, addressed only those remains found outdoors on the surface because they are, perhaps, the most susceptible to disturbance that may complicate the recovery. Animals such as rodents gnaw on bone while moles and insects often burrow through the ground, making tunnels into which burned fragments can fall. Rainwater can wash away tooth fragments, sometimes transporting burned remains meters from their points of origin (see Bontrager and Nawrocki, Chapter 13, this volume). Subsurface deposits can suffer disturbance from plants and animals, but they are generally better contained. In cases of buried cremated remains, one should follow standard archaeological excavation procedures until the cremation is located; then, follow the guidelines provided here.

FIELD PROCEDURES

(1) People should not conduct forensic recoveries unless they are fully prepared to take all proper biohazard precautions, for themselves and the people working with them, required by law and necessitated by common sense. A minimum recommendation is that fieldworkers wear gloves anytime forensic remains are handled, even in cases of significant burning. If the body is only partly burned and soft tissues are identifiable, one should consider a complete biohazard suit, and coverings for the face, shoes, and hands. One should use properly labeled biohazard bags for the remains and follow appropriate transport procedures once the field recovery is complete.

(2) Despite the current climate of intense popular interest in the forensic sciences, most states and cities strictly govern the recovery of human remains.

In Indiana (where I work) each county's coroner oversees all forensic cases, and the Office of the State Archaeologist oversees all archaeological human remain cases. One must know and follow all appropriate laws before being involved in the recovery of human remains.

(3) The best way to recover cremated teeth is to proceed archaeologically through the process. The first step is conducting a controlled-interval surface survey to find the extent of the cremation deposit. The remains may be in a small pile, or spread out over a large area. Next, one establishes a primary reference point, or datum, and constructs a grid from the datum over the area or areas of bone concentration. Usually, grid sizes used for the recovery of human remains vary from 1×1 to 5×5 m, but they can be of any size, and reduced or enlarged as needed. The grid facilitates mapping the burned elements; mapping is a critical step because scene photos do not provide the detailed location information necessary for each fragment. One can make the grid with pins and strings or electronically via the use of a surveying tool such as a total station. In either event, it is important to know where each bone and tooth is, so that one can recreate the circumstances of the cremation. Having established the grid, one carefully maps, photographs, and collects the surface elements (i.e., bones, teeth, artifacts, etc.).

(4) Once everything is removed from the surface, the investigator uses a trowel to gently scrape the sediment underneath the bone concentrations and wooden tools, such as chopsticks and skewers, to expose any buried bones and teeth. Troweling down through the sediment continues only as far as it produces burned fragments. One should not take a shovel and remove several centimeters of sediment below the surface concentrations since this will undoubtedly destroy important information needed to reconstruct the cremation event. It is better to work slowly and cautiously than to cause any harm. Trowel away from the burned bone concentration until no more bone or teeth fragments are found. One should pay close attention to small drainage ways and animal burrows that may course through the area, and always expect some fragments to work their way down slope.

(5) In the field, all sediments surrounding the burned bones should be poured through a 1/8-in. (roughly 3 mm) screen. This is a finer mesh than is typically used in archeology, but the usual 1/4-in. mesh is too large to recover isolated teeth and tooth fragments. If the sediment that passes through the screen includes tooth fragments, collect it and take it to the lab. There, use finer mesh screens to recover the fragments (if the field screen is used over a tarp, it is easy to collect the sediment that was poured through it). Generally speaking, reconstructing fragments smaller than 1/8 in. is extremely difficult, but it can be done and may be necessary for the resolution of the case.

(6) Dental fragments found at the scene or in the screen should be put in labeled plastic bags that are placed into labeled glass or plastic vials. This offers them the needed protection because burned teeth can be very fragile.

LAB PROCEDURES

(7) In the lab, clean the teeth with tap water and a soft-bristled toothbrush. During the cleaning process, lab personnel should periodically document characteristics of the teeth (such as restorations, nonburn-related fractures, and

pathological conditions) that could be obscured if a tooth were accidentally dropped or damaged during the cleaning process. Keep a fine-mesh screen over the drain when washing fragments in a sink.

(8) Reconstruct cleaned teeth with glue designed for porous material, such as Duco® or a similar acetone-based adhesive. Mincer *et al.* (1990) report that clear acrylic spray paint is a good stabilizing agent because it is readily available and affordable. Do not use consolidants excessively, however, and it is best to use a material that removes easily (i.e., is either water or alcohol soluble), because some can interfere with studies such as dental microwear. Well-known consolidants, for example, Acrysol® and polyvinyl acetate (PVA), are available in water- and alcohol-soluble forms, respectively. Avoid using consolidants in the field, especially if the sediment is adhering to the teeth. Sometimes there is no other option but to use a preservative on friable remains in order to recover them. However, such consolidation should be the last option; when possible it is better to take the remains and sediments to the lab "en masse" and to carefully remove sediments there.

(9) Reconstructed teeth are sorted by tooth type using standard dental identification procedures (e.g., Bass, 1995; Hillson, 1996; White, 2000; Hillson, 2005) and are viewed for taphonomic changes, including thermally induced ones. Rebuilding the teeth necessitates a sound knowledge of dental anatomy, thus my earlier admonition that one consult a dental expert for the analysis. At this point, the teeth are ready for detailed study.

THERMALLY INDUCED CHANGES IN COLOR, TEXTURE, AND SIZE

In general, thermal changes in teeth are similar to those made in bone (see Beach *et al.*, Chapter 8, this volume). Teeth that are exposed to lower temperatures and/or durations of heat tend to be dark black or brown in color. As temperature and/or duration increases, teeth turn blue-gray, then stark and chalky white; a condition known as calcination (Figure 3.1). At this stage, most of the protein and water have burned and evaporated and all that remains are the inorganic materials. However, because teeth consist of tissues of different mineral content, they do not change uniformly. An isolated tooth may have enamel fragments that seem unburned, while the dentin looks black and

FIGURE 3.1 Calcined tooth with large patch of dark gray. (see Plate 9)

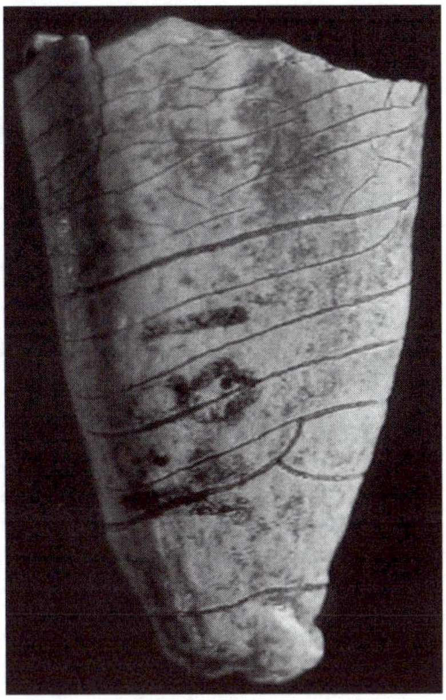

FIGURE 3.2 Transverse fracturing on root fragment. (see Plate 10)

gray. This difference occurs because dentin has a higher content of nonmineralized material that combusts and brings about changes in color. Enamel is already almost 100% mineral, so at first its changes are subtler.

The texture, or fracture pattern, found in burned teeth is also similar to that in bone, although certain fracture types are more common in dental remains (the texture terminology used herein follows guidelines presented by Symes *et al.*, Chapter 2, this volume). Most often, roots fracture transversely (Figure 3.2). Crowns tend to fracture along cusp margins where they are the thinnest. It is possible, however, for the enamel to fall off a tooth almost intact as the underlying dentin shrinks in the heat. The following discussion explores color and texture changes in detail for enamel and dentin. Because cementum is not always present on burned teeth, I do not discuss it here.

Burned teeth change noticeably in size. Buikstra and Swegle (1989) and Shipman *et al.* (1984) report marked losses in tooth size ranging, on average, about 10–15%. Dentin seems to exhibit greater shrinkage than enamel, likely due to its greater organic content. The remaining inorganic crystals collapse into the spaces left by the organic material, reducing the overall size of the tooth. Interestingly, if the fire gets hot enough (over 800°C), the crystals fuse to each other, which inhibits further heat-related shrinkage and fracturing (Shipman *et al.*, 1984; Buikstra and Swegle, 1989).

ENAMEL

The erupted and developmentally mature incisor and canine enamel may not change noticeably if it experiences only modest heat. As temperature and/or duration increases, the enamel starts to turn blue or gray (Muller *et al.*, 1998). This darkening can be exaggerated while the enamel is still connected to the dentin. The enamel is somewhat transparent and the underlying dentin, which darkens as a result of the heat, may show through. Later in the heating process, it is possible for the enamel to split along central, lingual, and labial grooves as well as along the cemento-enamel junction (Beach *et al.*, Chapter 8). Eventually, the crown becomes brittle and either falls off as a complete entity or shatters into many pieces (Muller *et al.*, 1998).

In certain instances, the enamel of the anterior teeth may be less thermally affected than that of the posterior teeth. This occurs especially when the soft tissues of the face have decayed prior to the burning; the anterior teeth easily fall out once the gingiva and periodontal ligaments are gone. The loss of anterior teeth presumably happens as one prepares the body for burning and/or stokes the fire. Once free of the jaws, the anterior teeth are protected from the heat because they are likely on the ground, either below or away from the fire. However, some researchers have found the enamel of the anterior teeth to be more affected by heat than the molar enamel; this seems to occur in incompletely burned bodies (like those recovered from car accidents), where the lips burn away quickly and expose the incisors, while the cheek teeth are still protected by the orofacial tissues (Delattre, 2000: Fig. 1; Sakoda *et al.*, 2000).

Premolar and molar crowns change color in much the same way as in the anterior teeth. The fragmentation of these teeth, however, is somewhat more predictable. Because the various cusps adjoin along deep grooves, it is these areas that fail first during thermal exposures. The crowns split longitudinally or transversely, sometimes splitting into quadrants, at other times into halves. Prolonged exposures will make these crowns chalky and brittle.

The process of molar fracture has been an area rife with speculation. In discussions among colleagues, some have argued for a 'popcorn model,' which states that molar crowns explode as moisture in the pulp cavity heats and expands, similar to the controversy surrounding the exploding skull theory (see Chapter 2). The result is a violent fracturing of the surrounding dentin and a subsequent destruction of the crown. Others suggest that enamel caps simply fall off as the dentin contracts. In their separate tooth-burning experiments, Muller *et al.* (1998) and Beach *et al.* (Chapter 8) report no evidence of exploding teeth. Certainly, the burning process is mechanically violent, and large cracks can literally pop bones open within the first few minutes of a fire. As of yet, the same has not been documented in teeth. If it is eventually determined that teeth can pop or explode in certain circumstances, such knowledge will affect our strategies for finding teeth and our explanations for the dispersal of dental fragments.

Crowns of Immature and Deciduous Teeth

In human jaws, enamel forms from an initial matrix of minerals and proteins. Before eruption, the enamel matures by losing nearly all of its protein and becoming almost completely mineralized (Ten Cate, 1994). Prior to maturation, the teeth have a much higher organic content than they have once they are completely formed. The higher percentage of organics means that immature crowns are more likely to go through the heat-related color changes evident in bones. There is an important implication here. First, when one encounters a complete, isolated crown with a dark-blue or gray coloring, it may indicate that the tooth has not yet fully formed (Figure 3.3). The cervical

FIGURE 3.3 Darkly colored crown of an unerupted premolar. (see Plate 11)

margin of the crown holds clues for this identification. A smooth cervical margin indicates that the crown is still forming; a fractured cervical margin indicates that the crown may have broken off the root. Another indicator that a tooth once had a root is the presence of occlusal wear. If masticatory (or interstitial) facets have formed on the cusps, one can assume that the tooth had a root and had erupted. If the tooth was resorbing its root (i.e., the tooth was getting ready to be shed), it would have at least a spur of dentin (or a fracture if the dentin broke away in the fire) near the labial aspect of the cervix since this remains after expectoration (Ten Cate, 1994).

Enamel completes its mineralization (i.e., maturation), from the cusp tip to the cemento-enamel junction; a crown with incompletely mature enamel has virtually no root formation. The extent of maturation is evidenced by a linear demarcation between very glossy enamel on the crown apex side of the line and coarser enamel where the crown is yet to mature. Recognizing crown maturation is especially important for determining the age of crown fragments, and it helps to explain stark color differences on a single crown.

It is possible to recover teeth that have not yet begun to mature. These teeth may be no more than 30% mineral and fragile, yet burned immature deciduous premolar crowns have been recovered from ancient archeological sites completely intact. Schmidt *et al.* (Chapter 14, this volume) include such teeth in their dental inventory from a 7000–9000-year-old cemetery. Immature crowns are relatively easy to spot because they are chalky to the touch and their surfaces are rough in appearance (Figure 3.4).

Mature, erupted deciduous crowns often show color changes more vividly than adult teeth, and fracture rather easily. Because deciduous teeth are used for such short periods of time, it is rare that the crowns are worn completely flat. This means that even if the enamel is absent it is quite possible that the underlying dentin will still provide an excellent indication of the tooth's crown anatomy. For example, lower deciduous premolars have a large eminence on their mesiobuccal aspect called the tubercle of Zuckerkandl. They also bear a very conical mesiolingual cusp. Both of these are visible on the dentin component on the crown even if all of the enamel is gone (Figure 3.5).

Frequently, immature deciduous teeth are still in their crypts at the time of cremation. As in adult teeth, it is important to view the cervical margin of

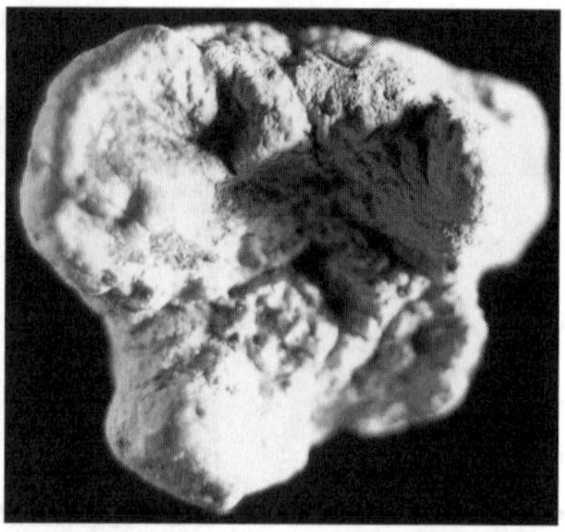

FIGURE 3.4 Chalky appearance of a calcined, immature crown. (see Plate 12)

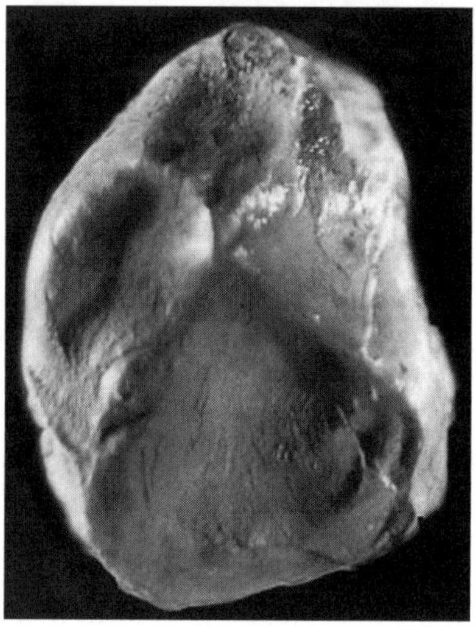

FIGURE 3.5 Distinguishing crown characteristics are evident in the dentin of a burned, enamel-less deciduous lower first molar crown.

deciduous crowns. Important age information may be gleaned if one is able to distinguish between a crown that is, for example, quarter formed and one that is completely formed, but fractured in the fire.

Unerupted Crowns

Tooth crowns still in their crypts at the time of the burning or the cremation may be adult or deciduous, immature or mature, and may have either no root present or partially complete roots. The bones around the crypts in the maxilla and the mandible can significantly reduce the thermal exposure of unerupted teeth. Specifically, unerupted mandibular teeth tend to protect nicely because of the density of the mandible itself. Thus, unerupted crowns may be darkened or slightly bluish when the surrounding mandibular bone is almost white. Teeth in maxillary crypts, especially those in the anterior part of the mouth, are slightly less protected as a result of the relatively thin maxillary bone just beneath the nasal aperture. One might ask, why have a discussion about teeth that are hidden away in their crypts? The fact is, despite the protection offered by the jawbones, unerupted teeth are frequently found in isolation where conflagrations are severe enough to fracture the jaws.

Moreover, unerupted teeth may be compromised because of small openings in the jaws, the gubernacular canals. These canals are the paths the unerupted teeth follow as they erupt and can allow heat directly into crypts. For the anterior deciduous and adult teeth, these canals appear as pinprick-sized openings. For the molars, though, the gubernacular canals may be several millimeters wide. The protection offered by the jaws, therefore, can be short-lived in intense fires and eventually even crypt-protected teeth can be heat-affected.

Soft Tissue Adherence

In some circumstances, oral and other soft tissues fuse to the crown surface. This material is generally charred in appearance and may have sediment imbedded in it if the body was placed outside on the ground. The soft tissue tends to protect the teeth if the fire loses its intensity relatively quickly (at least around the head). In these cases the enamel may look completely unaltered by heat despite being dotted with patches of charred flesh. All soft tissues should be removed in the lab and curated and/or disposed of appropriately. Another possibility is that teeth will fall or be forced from their crypts and become imbedded in the soft tissues of the head. If teeth are missing from a partially burned body, X-rays should be taken of soft-tissue masses, especially around the neck and the braincase, to locate them.

ROOTS

When remains are extensively burned, roots and root fragments are far more commonly recovered than crown fragments, despite being less mineralized. This is due, in part, to the fact that cheek teeth have multiple roots and so have a greater likelihood of being found simply because of their great frequency. However, there may be another variable that helps to keep roots around long enough to be recovered virtually intact: unlike enamel crowns, roots are always imbedded in alveoli, even after the tooth has erupted, which means they are constantly protected. As noted earlier, anterior teeth are single-rooted and are susceptible to falling out once the soft tissues have either decayed or burned away. The molars, by virtue of being multi-rooted, are firmly placed in their alveoli and rarely fall out until the roots themselves fracture. Usually the exception to this rule is the third molar because its roots often fuse and create what is essentially a large, single cone-like root.

Root Color and Texture

Heat-related color changes in roots are distinct. Often, it is even possible to determine how deeply a tooth sat in its socket during the fire; that part of the root that was protected will be black to dark blue or gray, while the exposed root will be gray or white. If a root falls out of its socket onto the ground, usually the groundside is darker in color than the more heat-exposed side that often becomes calcined. Because roots are protected for at least some of the burning period, individual roots are frequently multi-colored.

As roots lose their organic material and turn white in color, they become brittle and cracked. Roots have a tendency to fracture transversely as well as express small areas of checking (minute longitudinal and transverse fractures intersecting like squares on a checkerboard). The transverse fractures can extend completely through the root and even break the root into ring-shaped cross-sections of a few millimeters or so thick. Macroscopically, one distinguishes these from similar bone fragments because dental pulp cavities are usually quite small compared to the overall size of the fragment. By contrast, bones that are small enough to be confused with teeth tend to have thin cortices and large medullary cavities often bearing at least some traces of trabecular bone.

Distinguishing Molar Roots from Anterior Tooth Roots

One of the most difficult tasks in reassembling the dentition is distinguishing anterior roots from molar roots. At first consideration this seems easy, but heavily burned molar roots will separate and each of the two or three individual molar roots (radicals) will fall free looking very much like single anterior roots. Fortunately, in many instances, there is a telltale notch or spur on molar radicals that is absent on anterior teeth (Figure 3.6), which marks where the radical was once connected to another at the tooth's neck. In fact, it is possible to reconnect individual radicals by matching corresponding spurs and notches. For the most part, root morphology remains even after calcination, allowing one to distinguish mesial, distal, and lingual roots on upper molars and mesial from distal roots on lower molars. The notches and spurs are on the distal aspects of the mesial roots and on the mesial aspect of the distal roots. On maxillary lingual roots, the spur is on the radical's buccal aspect (Schmidt *et al.*, 2005).

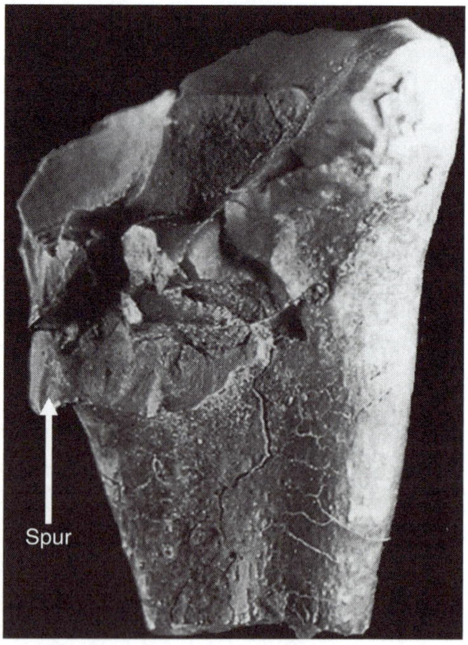

FIGURE 3.6 Spur projecting from single molar radical, not created in burned single-rooted teeth. (see Plate 13)

Distinguishing Wear-Related Crown Loss from Fire-Related Crown Loss

Finding burned tooth fragments is difficult enough without spending time looking for something that does not exist. Not all teeth have crowns remaining at the time of burning. Sometimes, especially in archeological contexts, the teeth are so worn that the enamel is completely gone. In these instances, the occlusal aspect of the root is smooth and may show a dark spot, or spots, where the reparative dentin protected the pulp cavity from the heavy wear. In other instances, the wear is extensive enough to expose the pulp chamber.

If the enamel cap is lost because of the heat, the underlying dentin will retain its coronal relief. For example, maxillary molars have four cusps and the underlying dentin has four corresponding conical points beneath these four cusps. The underlying dentin replicates the enamel surface so well that it is possible to identify and side the tooth almost as if the enamel crowns were still present, making it is possible to recover only the dentin part of the crown and still achieve a dental identification.

MICROSCOPY

Although it is possible to completely recover and identify cremated remains without the use of a microscope, there are instances where microscopy is very beneficial, especially when only a very small fragment has been found or when the fragment is found in isolation, outside the context of a mass of cremated bone. A scanning electron microscope (SEM) uses a beam of electrons fired down a vacuum chamber. An electron detector in the SEM picks up electrons from the specimen and converts them into an image one views on a computer screen. There is no need for thin-sectioning the specimen with an SEM, but some models require coating the specimen with a thin layer of metal to improve imaging.

Under the SEM, the external surface of the unburned mature enamel looks smooth. On anterior teeth the labial surface will have minute transverse lines that are artifacts of enamel layering during its formation. At magnifications around 1000×, the surfaces of mature teeth look smooth, although one will likely see small scratches and pits created during normal mastication. An immature enamel, however, has its surfaces covered with small pits created by a projection from the cells that secrete the enamel. These pits are smooth-edged and shallow and appear anywhere the outermost (aprismatic) layer of enamel has not yet formed.

When mature enamel is exposed to intense heat, the aprismatic layer can break down, exposing the underlying pits so that the surface takes on a dimpled appearance (see Yamamoto *et al.*, 1990; Muller *et al.*, 1998 for details on enamel breakdown). As the temperature and/or the duration of the fire increases, the margins of the pits gradually degrade until finally they are almost amorphous (Shipman *et al.*, 1984). The enamel microstructure begins to crack and eventually sizeable pieces may fall away, but the overall pattern of the microstructure remains intact to the point that human enamel fragments can be distinguished from those of other animals (Yamamoto *et al.*, 1990).

The outer surface of unburned dentin is covered with a layer that encapsulates the root (mantle dentin). Below this is another layer that is filled with numerous tubules coursing from near the pulp chamber to the outer layer.

As dentin heats up, it shrinks noticeably and takes on a roughened appearance microscopically. Eventually, the tubules change shape from round to oval and then begin to break down entirely. The matrix between the tubules takes on a globular or 'frothy' appearance with areas of glassy texture. Finally, the globules fuse into nodular spikes (Shipman *et al.*, 1984: 313).

In addition to visual inspection, the SEM can be used to determine a specimen's composition. The electron beam generates X-rays as it interacts with the sample; a special detector in the SEM collects these X-rays and determines the elements present. This process is called energy dispersive X-ray spectroscopy (EDS) and it, too, has been used for studying burned human teeth. At times, fragmentation is so severe that it is difficult to tell if a burned element is a bone or a tooth or something else entirely. Ubelaker *et al.* (2002) have employed the SEM/EDS to determine a specimen's elemental composition; bones and teeth have diagnostic proportions of calcium and phosphorous, which can be used to distinguish them from most other materials.

Like to the aforementioned SEM studies, several researchers have found heat-related histological changes in teeth and bones with a standard compound microscope (e.g., Herrmann, 1977; Nelson, 1992; Cattaneo, 1999). For this type of study, thin sections are mounted onto slides and light is transmitted through the specimen. Magnification ranges with standard microscopes are far less than those attainable with the SEM, but hard tissue structures are easily identifiable with these devices. Unfortunately, most histological studies of burned remains to date have focused on bones and have addressed teeth in only a cursory manner (e.g., Herrmann, 1977). These studies indicate that normally organized bone structure maintains much of its usual appearance even after heat exposure, but that there is shrinkage of bone microstructure (Nelson, 1992). The promise of histological analysis in cases of burned bone is great, especially when it is unclear whether or not a bone is even from a human. Cattaneo *et al.* (1999) found that histology could be used to distinguish human from nonhuman burned bone successfully, and that it was more efficacious than the analysis of either the DNA or the albumin. Given the value of burned bone histology, perhaps more studies will be conducted specifically on burned dental tissues.

DENTAL RESTORATIONS AND ORTHODONTICS

Earlier in this chapter anatomical idiosyncrasies were included in the definition of the biological profile: the physical characteristics that are unique to each person. It is important to determine the age, sex, and ancestry of a person, but these traits alone do not indicate exactly who someone is. By and large, anatomical idiosyncrasies are used to make positive identifications – from the unique loops and whorls of fingerprints to distinctive dental patterns. In instances of burned human remains, it is likely that the skin and its diagnostic characters, including the friction ridges on the hands and feet, are burned away or rendered unusable, a condition that makes the teeth particularly valuable for identification (e.g., Johanson and Saldeen, 1969; Fairgrieve, 1994; Andersen *et al.*, 1995; Chapenoire *et al.*, 1998). Dental idiosyncrasies can include (but are not limited to) the teeth that have erupted, the teeth that are present at the time of death (Adams, 2003a), dental morphology (external and internal), alignment, patterns of occlusion, and expression of pathological conditions. Furthermore, cultural practices that affect teeth can be very idiosyncratic, and none is more important than the practice of dentistry. Although dental treatments are documented well into antiquity, it is usually only with recent people that we concern ourselves with positive identification; therefore, the discussion here addresses only modern dentistry.

The following is a brief introduction to the dental materials and devices one might encounter in association with a burned body. The range of materials used in dentistry over the past 50 years is truly overwhelming and labs are constantly developing new products. It is impractical to list all dental materials here, but the reader should be aware of some of the more common dentistry artifacts and of what they are made. The range of materials used in dental restorations and prosthetics includes precious metals such as gold, amalgams, acrylic, composite resins (and an ever-growing number of plastics), ceramics and porcelain, vinyl, various steels including stainless, as well as a myriad of dental adhesives and even solder (Phillips, 1982). At the current rate of research into dental materials, the task of learning everything on the market is daunting. However, one need not have *a priori* knowledge of all dental materials; a general knowledge of dental appliances and procedures provides the impetus to look for and recover trace evidence of dental work. Even if the recovered appliance is unidentifiable because of significant thermal alteration, it can be tested chemically for diagnostic signatures of dental materials.

MATERIALS AND DEVICES

Fillings are direct restorations designed to patch areas of decay and are generally of three varieties: amalgam, resin, and gold. Amalgam fillings are silver or an alloy of silver, tin, and copper dissolved in mercury. These fillings appear metallic and are very durable. Resin fillings are composite materials used as alternatives to amalgams because they have a more natural color (Phillips, 1982). Fillings can be found on any tooth, but are most common on the molars; they can be very sizeable, taking up much of the crown or extend the length of the pulp chamber (i.e., a root canal). Sealants are composite materials usually used to fill grooves in the permanent molars of children. They are more prophylactic than restorative.

Inlays and onlays are indirect restorations where ceramic or gold (or another metal) is bonded to a particular location or, in the case of onlays, along the entire occlusal surface of the tooth. Fillings, inlays, and onlays all require the use of a dental bur (a.k.a. drill) to remove decay. However, inlays and onlays are lab-manufactured pieces that require further modification of the surface by the dentist to make certain the restoration fits properly. A crown, or cap, replaces almost the entire anatomical crown of the original tooth with a prosthesis, usually made of porcelain. Veneers are generally cosmetic porcelain facings attached to dental labial surfaces, although the vast array of crown and veneer products available today somewhat blurs the line between these two means of restoration.

Implants consist of a titanium post surgically placed into the jaw to support dental prosthetics ranging from single crowns to complete dentitions. Bridges fill in gaps in the tooth row by anchoring prosthetic teeth to natural teeth; they are built out of ceramics and porcelain fused to metal. Partial and full dentures are made from acrylic, but can have metal components (Rosenstiel *et al.*, 1995). They rest directly on the gingiva, although in some instances existing teeth or posts can support them. Splints buttress existing teeth that are no longer well enough seated in their alveoli to be used for mastication. They are made of metal and are often located on the lingual aspect of the teeth they support. Sometimes existing teeth require a splint before they can be used to anchor an appliance such as a bridge (Smith, 1990).

Orthodontic devices are used to correct malocclusion and include (but are not limited to) removable retainers and bruxism guards, as well as braces, brackets, and fixed retainers cemented onto the teeth. Like restorations, orthodontic technology has expanded markedly over the years and the number of available products is wide-ranging and ever changing. For example, braces use to be almost exclusively stainless steel, but today ceramic and clear proprietary materials are increasingly commonplace.

THERMAL PROPERTIES

Amalgam fillings will melt at a temperature of about 965°C (Norrlander, 1997). Resins can survive a burning event, but they will shrink and perhaps fall out (Rossouw *et al.*, 1999). Displaced amalgam and resin fillings remain radiopaque even after great heat exposure (Rossouw *et al.*, 1999); therefore, radiographs of partially burned bodies can locate lost fillings that have fallen from the mouth into other parts of the body. Gold may be used as a filling material, although it is used in other restorations as well (Bowers, 2004). It melts at just over 1000°C, but dental gold is usually an alloy whose melting point depends on the percentage of gold present. Porcelain's melting point is 1232°C. Most house fires tend to burn closer to 650°C (Norrlander, 1997), thus crowns and veneers are capable of surviving a fire.

Dentures may survive a fire [acrylic's melting point is 600°C (Norrlander, 1997)], but they will not last as long as other restorations (Jacob and Shalla, 1987). In the United States, 21 states require that dentists place identifying marks on dentures, usually the patient's name, but some include the social security number or driver's license number (Collins, 2004). Partial or complete edentulousness of the jaws may be an indicator of dentures (Jacob and Shalla, 1987). Thus, it is prudent to sort through burned plastics at the scene if there is any suspicion of denture use by the decedent (Borrman *et al.*, 1999; Marella and Rossi, 1999).

Evidence of dental appliances might be directly or indirectly evident among the burned remains. Finding a tooth that still contains an amalgam filling or a group of teeth bound by wire braces provides direct evidence of dental work. However, even when restorative and orthodontic devices are lost, there may be indirect evidence that they were present on a fire victim. Diagnostic grooves are created by the drilling and acid etching required to prepare teeth for restoration (Phillips, 1982). The drilling process is actually the use of a bur that grinds away decayed dental hard tissues. Steel and diamond burs, however, leave behind very different marks on the enamel; scratches created by steel burs are somewhat uniform and organized, while diamond tips create scratches that are far more irregular (Phillips, 1982). Fairgrieve (1994) used an SEM to locate the grooves from a filling, even though the amalgam had melted; this effort has led to positive identification of a burned victim. Usually, bridges and splints are worn for long periods of time and the wearer cannot remove them (Smith, 1990). Since these appliances connect to anchor teeth, they are likely to leave telltale grooves or marks. Braces and similar orthodontic devices may leave behind evidences of their presence in the form of adherence halos on the enamel. In instances where prosthetic crowns are lost to the flames, it is possible for implant posts to survive because the bone in which they sit protects them.

CASE STUDY: BAUMEISTER HOMICIDES

The case described in Chapter 13 of this volume involves one of the most notorious serial killers in Indiana's history. Over the span of a few years in the mid-1990s, at least 11 young adult males were allegedly sexually asphyxiated by Herbert Baumeister who placed the bodies in a wooded area behind his house. At a later point in time, he collected some of the bodies and burned them in a small drainage channel, near to where he originally placed them. The bodies were thoroughly burned and fragmented, and a sizable percentage of the bone and tooth fragments washed downslope as rainwater periodically filled the drainage channel.

In addition to thousands of bone fragments, forensic archeologists from the University of Indianapolis recovered approximately 100 tooth fragments, including crowns and roots, which ranged in color from unburned to calcined. None of the burned teeth retained its crown. Transverse fractures dominated, although longitudinal fractures also occurred. For the most part, the molar roots had separated and the enamel-less cervical dentin was generally in poor condition. Most of the fragments were dark blue, suggesting that the fire was not hot enough and/or long enough to fully calcinate all of the dental tissues. A majority of the teeth (and bones, for that matter) were highly fragmented, indicating that mechanical disturbances, such as stoking the fire, had broken up many of the elements. It is possible that fragmentation continued after the burning event via a variety of taphonomic forces, including plants, animals, and water.

Along with the burned dental fragments were a number of nearly completely unburned anterior teeth. Presumably these single-rooted teeth fell out of the partly decomposed jaws during their moving and preparation for burning. The unburned teeth were limited to maxillary and mandibular central and lateral incisors, all found just under the ground surface below the burned bone concentrations.

I used several morphological, pathological, and restorative traits to create three partial dentitions (Figure 3.7). One individual had a prominent deposit of calculus on the lingual aspect of his lower incisors and distinctively dark roots. Another had long, yellowish roots, and the third had robust roots and some interproximal resin fillings. The teeth that seemed to match up were placed into clay arches, which were then occluded to compare maxillary and mandibular wear facets. I started with the better-preserved anterior teeth and

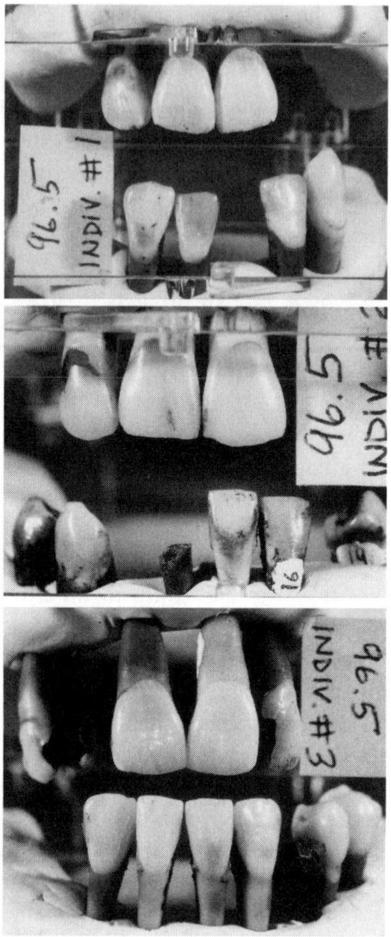

FIGURE 3.7 Three dentitions reconstructed from the Baumeister serial homicide case (photos by S.P. Nawrocki). (see Plate 14)

worked distally to add fire-damaged teeth to each dentition. This was done by careful comparison of the morphology of both the burned and unburned teeth. Some molar roots were reattached by fitting together spurs and notches, and a few burned canine roots were matched to the incisors by their shape and size.

One partly burned lower premolar crown had an amalgam filling, but it did not have an interproximal contact facet that matched any of the burned canine crowns. Its contact facets were not placed on its mesial- and distal-most points, indicating that the tooth erupted slightly rotated. Another premolar crown fragment matched up with the restored premolar, but it was difficult to determine which set of unburned incisors matched these premolars because no canine crown was similar enough to both the incisors and premolars. Eventually, I realized it was a blackened canine root fragment that belonged to these teeth. Similarities in enamel color and calculus thickness supported the connection between these particular premolars and anterior teeth, and together they yielded the third partial dentition.

Of the three reconstructed dentitions, a consulting forensic dentist positively identified two of them. The person with the restored premolar actually had that tooth misaligned – not rotated – in life, but making the effort to include that tooth in the dentition aided in the identification. Without these reconstructions, despite their difficulties, two victims might have gone unidentified.

CONCLUSION

It is vital to thoroughly recover and properly reconstruct burned dentitions despite the obvious difficulty and tedium. The archeological methods recommended to recover burned dental fragments are necessary because of significant fragmentation that can occur to the body in general, and to the teeth in particular, during a fire. Analyzing dental remains demands a detailed knowledge of dental anatomy and the ways in which heat affects teeth in order to reconstruct the dentition for positive identification. In total, dental analysts face numerous concerns when confronted with burned teeth, and all must receive due consideration (Table 3.1).

Biological anthropologists spend their lives examining and evaluating the complex nature of human teeth. The amount of detail and information gained

TABLE 3.1 Summary of Concerns and Considerations Germane to the Recovery of Burned Human Teeth

Concern	Consideration
How does fire affect teeth?	It changes their color to black, blue/gray or white; it fractures them mostly transversely; and it shrinks them
What affects the dispersal of burned bones and teeth?	Animals (including people), plants, and physical elements like water
What special care should a person take when recovering burned bones?	Take all biohazard precautions and know all pertinent laws regarding the recovery of human remains
How should burned human remains be collected?	Follow strict archeological guidelines so that the removal is as scientific as possible
Are there special techniques required to find burned teeth?	Because they are so fragmented, it is recommended that the archeological screening be done with a 1/8-in. or finer wire mesh screen
What happens to enamel as it burns?	Mature enamels will not change color at first. Then they will darken and fracture. Immature enamels start to change faster than the mature one
Can deciduous teeth survive a fire?	Yes, especially if the enamel is mature
What happens to unerupted teeth?	They are protected by the surrounding bone, but will fall out if the bone breaks open. They could show less color change and fracturing than the rest of the skeleton
How can broken molar roots and anterior tooth roots be distinguished?	Molar roots will have a small spur or notch near their cervical margins noting where they once were attached to other radicals at the cervix

(*Continues*)

TABLE 3.1 *(Continued)*

Concern	Consideration
How can incompletely formed and fractured teeth be distinguished?	Incompletely formed teeth often have a smooth inferior margin and no occlusal wear
Can occlusal wear affect burned teeth?	If a tooth is worn down to its neck, there is no point in looking for the crown. Worn teeth will have a smooth coronal aspect and evidence of reparative dentin
What technology can help?	Scanning electron and compound microscopes can be used to distinguish human from nonhuman teeth or to confirm if a tissue has been burned
How are teeth put back together?	In the lab, have a dental expert to oversee the reconstruction of the teeth and the dental arches. Use consolidants only when necessary
Can dental restorations and orthodontics survive a fire?	Yes, although not all to the same extent. Care should be taken to find direct and indirect evidences of dental work to facilitate a positive identification

from a dentition cannot be taken lightly and warrants serious dedication. I hope that continued study and collaboration results in better strategies for the recovery and study of burned human teeth so that increasingly effective procedures develop.

ACKNOWLEDGMENTS

Thanks are given to the people who offered much needed constructive criticism of this chapter, especially Saundra Cox, Molly Hill Schmidt, and Steve Symes. Moreover, this manuscript benefited greatly from conversations with Matthew Hill, D.D.S., who provided great insight into dental procedures, materials, and devices. Gregory Reinhardt deserves special acknowledgment for his tremendous assistance in editing. Any errors are solely the responsibility of the author. Stephen Nawrocki took the photos in Figure 3.7.

LITERATURE CITED

Adams, B.J. (2003a). Establishing personal identification based on specific patterns of missing, filled, and unrestored teeth. *Journal of Forensic Sciences* **48**, 487–496.

Adams, B.J. (2003b). The diversity of adult dental patterns in the United States and the implications for personal identification. *Journal of Forensic Sciences* **48**, 497–503.

Andersen, L., Juhl, M., Solheim J.M., and Borrman, H. (1995). Odontological identification of fire victims – potentialities and limitations. *International Journal of Legal Medicine* **107**, 229–234.

Bass, W. (1995). *Human Osteology: A Laboratory and Field Manual*, 4th edn. Missouri Archaeological Society, Special Publication No. 2, 361 pp.

Borrman, H.I., DiZinno, J.A., Wasen, J., and Rene, N. (1999). On denture marking. *Journal of Forensic Odontostomatology* **17**, 20–26.

Bowers, C.M. (2004). *Forensic Dental Evidence*. Elsevier, Amsterdam, 211 pp.

Buikstra, J.E. and Swegle, M. (1989). Bone modification due to burning: experimental evidence. In *Bone Modification* (R. Bonnichsen and M.H. Sorg, eds.), pp. 247–258. Center for the Study of the First Americans, University of Maine, Orono, ME.

Cattaneo, C., DiMartino, S., Scali, S., Criag, O.E., Grandi, M., Sokol, R.J. (1999). Determining the human origin of fragments of burnt bone: a comparative study of histological, immunological, and DNA techniques. *Forensic Science International* **102**, 181–191.

Chapenoire, S., Schuliar, Y., and Corvisier, J.M. (1998). Rapid, efficient dental identification of 92% of 13 train passengers carbonized during a collision with a petrol tanker. *American Journal of Forensic Medical Patholology* **19**, 352–355.

Collins, D. (2004). *The dentist's role in forensic identification: the release of dental records and radiographs, and denture labeling*. Council on Dental Practice, American Dental Association. www.ADA.com.

Cook, D.C. (1984). Adult age determination from the dentition. In *Human Identification* (T.A. Rathbun and J.E. Buikstra, eds), pp. 307–328. Charles C. Thomas, Springfield, IL.

Delattre, V.F. (2000). Burned beyond recognition: systematic approach to the dental identification of charred human remains. *Journal of Forensic Sciences* **45**, 589–596.

Duffy, J.B., Waterfield, J.D., and Skinner M.F. (1991). Isolation of tooth pulp cells for sex chromatin studies in experimental dehydrated and cremated remains. *Forensic Sciences International* **49**, 127–141.

Fairgrieve, S.I. (1994). SEM analysis of incinerated teeth as an aid to positive identification. *Journal of Forensic Sciences* **39**, 557–565.

Gaytmenn, R., and Sweet, D. (2003). Quantification of forensic DNA from various regions of human teeth. *Journal of Forensic Sciences* **48**, 622–625.

Herrmann, B. (1977). On histological investigations of cremated human remains. *Journal of Human Evolution* **6**, 101–103.

Hillson, S.W. (1996). *Dental Anthropology*. Cambridge University Press, Cambridge, 373 pp.

Hillson, S.W. (2005). *Teeth*, 2nd ed. Cambridge University Press, Cambridge, 388 pp.

Jacob, R.F. and Shalla, C.L. (1987). Postmortem identification of the edentulous deceased: denture tissue surface anatomy. *Journal of Forensic Sciences* **32**, 698–702.

Johanson, G. and Saldeen, T. (1969). Identification of burnt victims with the aid of tooth and bone fragments. *Journal of Forensic Medicine* **16**, 16–25.

Lee, S.S., Choi, J.H., Yoon, C.L., *et al.* (2004). The diversity of dental patterns in the orthopantomography and its significance in human identification. *Journal of Forensic Sciences* **49**, 784–786.

Marella, G.L. and Rossi, P. (1999). An approach to identification by means of dental prostheses in a burnt corpse. *Journal of Forensic Odontostomatology* **17**, 16–19.

Mincer, H.H., Berryman, H.E., Murray, G.A., and Dickens, R.L. (1990). Methods for physical stabilization of ashed teeth in incinerated remains. *Journal of Forensic Sciences* **35**, 971–974.

Muller, M., Berytrand, M.F., Quatrehomme, G., *et al.* (1998). Macroscopic and microscopic aspects of incinerated teeth. *Journal of Forensic Odontostomatology* **16**, 1–7.

Nelson, R. (1992). A microscopic comparison of fresh and burned bone. *Journal of Forensic Sciences* **37**, 1055–1060.

Norrlander, A. (1997). Burned and incinerated remains. In *Manual of Forensic Odontology*, 3rd edn. (C.M. Bowers and G.L. Bell, eds), pp. 16–18. American Society of Forensic Odontology, Eureka, MT.

Phillips, R.W. (1982). *Skinner's Science of Dental Materials*, 8th edn. W.B. Saunders, Philadelphia, PA, 646 pp.

Rosenstiel, S.F., Land, M.F., and Fujimoto, J. (1995). *Contemporary Fixed Prosthodontics*. Mosby, St. Louis, MO, 672 pp.

Rossouw, R.J., Grobler, S.R., Phillips, V.M., and van W. Kotze, T.J. (1999). The effect of extreme temperatures on composite, compomer, and ionomer restorations. *Journal of Forensic Odontostomatology* **17**, 1–4.

Sakoda, S., Zhu, B.L., Ishida, K., *et al.* (2000). Dental identification in routine forensic casework: clinical and postmortem investigations. *Legal Medicine (Tokyo)* **2**, 7–14.

Schmidt, C.W., Reinhardt, G.A., Nawrocki, S.P., and Hill, M.K. (2005) Analysis of burned dental remains. *American Journal of Physical Anthropology*, Suppl. 40, p. 183 (abstract).

Shipman, P., Foster, G., and Schoeninger, M. (1984). Burnt bones and teeth: an experimental study of color, morphology, crystal structure and shrinkage. *Journal of Archaeological Science* **11**, 307–325.

Shiroma, C.Y., Fielding, C.G., Lewis, J.A., *et al.* (2004). A minimally destructive technique for sampling dentin powder for mitochondrial DNA testing. *Journal of Forensic Sciences* **49**, 791–795.

Smith, B.G.N. (1990). *Dental Crowns and Bridges: Design and Preparation*. Year Book Medical Publishers, Inc., Chicago, IL, 254 pp.

Sweet, D.J. and Sweet, C.H. (1995). DNA analysis of dental pulp to link incinerated remains of a homicide victim to a crime scene. *Journal of Forensic Sciences* **40**, 310–314.

Ten Cate, A.R. (1994). *Dental Histology*. Mosby, St. Louis, MO, 532 pp.

Ubelaker, D.H., Ward, D.C., Braz, V.S., and Stewart, J. (2002). The use of SEM/EDS analysis to distinguish dental and osseus tissues from other materials. *Journal of Forensic Sciences* **47**, 940–943.

White, T.D. (2000). *Human Osteology*. Academic Press, San Diego, CA, 563 pp.

Yamamoto, K., Ohtani, S., Kato, S., *et al.* (1990). Morphological changes in human and animal enamel rods with heating – especially limits in temperature allowing discrimination between human and animal teeth. *Bulletin of the Kanagawa Dental College* **18**, 55–61.

4

Analysis of Human Cremains: Gross and Chemical Methods

John J. Schultz, Ph.D.[1], Michael W. Warren, Ph.D.[2], and John S. Krigbaum, Ph.D.[2]

[1]*University of Central Florida, Department of Anthropology, Orlando, FL*
[2]*University of Florida, Department of Anthropology, Gainesville, FL*

Introduction

Cremated human remains, or cremains, have long held the interest of anthropologists. Early scientific studies of cremains focused on interpreting prehistoric cremation practices and the role of this funerary practice in the lifeways of ancient cultures (e.g., Baby, 1954; Wells, 1960; Binford, 1963; Merbs, 1967). While scientific studies of cremated remains in the archaeological records continue, physical anthropologists in the United States have been increasingly called upon to examine commercially cremated remains that have become the subject of litigation involving issues of disputed identity, excessive commingling, or negligent cremation practices. Biological anthropologists and odontologists have been tasked as the logical choice among professionals for these analyses since cremains consist primarily of calcined bones and teeth. In instances where the anthropologist is serving as the primary expert, an odontologist should be consulted if evidence is found for dental restorations, orthodontics, prosthodontics, or maxillofacial surgery.

The primary reason for increasing litigation is in part due to the ever escalating numbers of people that are choosing cremation over burial. The number of people choosing cremation over burial is expected to increase significantly over the next 20 years. The Cremation Association of North America (CANA) reported that in 2004 nearly 30% of the people who died that year in the United States were cremated. CANA projects that by 2010 this number will increase to nearly 36%, and by 2025 greater than 45% of all the deceased in the United States will be cremated.

A second reason for increasing litigation is significant media coverage of incidents involving crematoria where unethical procedures have been alleged, such as the highly publicized Tri-State Crematory in Georgia and the Bayview Crematory in New Hampshire. A number of books have discussed the expanding role of forensic anthropologists analyzing cremains, citing some of these highly publicized incidents (Maples and Browning, 1994; Iverson, 2001; Bass and Jefferson, 2003).

As a result of the increased involvement of forensic anthropologists analyzing contemporary cremations, there has been a concomitant growth in the body of literature by forensic anthropologists that is directed toward new methods and procedures in cremains analysis. Cremation research by physical anthropologists was initially geared toward identifying cremains based on recognizable bone and nonosseous artifacts (Murray and Rose, 1993; Huxley, 1994; Kennedy, 1996; Murad, 1998; Warren and Schultz, 2002), cremation weights (Warren and Maples, 1997; Bass and Jantz, 2004), and descriptions of the commercial cremation process (Eckert *et al.*, 1988; Murad, 1998; Warren and Schultz, 2002). However, due to the efficiency of the newest processing methods, traditional gross methods of cremation analysis may be of limited value when trying to determine the identity and composition of purported cremains. Advances in newer processing methods by the cremation industry have made the forensic analysis of cremains a more challenging task (Murad, 1998; Warren and Schultz, 2002). In many instances, it is difficult or impossible to identify the decedent based solely on the analysis of osseous fragments and cremation weights. As a result, cremation research is now focusing on using chemical methods to determine the elemental composition of cremains to answer some basic questions that cannot be answered from more traditional methods (Warren *et al.*, 2002; Bodkin *et al.*, 2005).

In this chapter, we provide a detailed overview of the contemporary cremation process that begins with a human body and ends with a volume of inorganic matter that can fit in a small box or an urn. Next, we discuss various methods that cremation analysts have traditionally used to analyze cremated remains. Lastly, we explore the potential of using chemical methods that help to answer some basic questions about the elemental make-up and basic properties of cremated bones and teeth.

CREMATION PROCESS

Contemporary cremation is a two-step taphonomic process that reduces a body to small bone fragments and ash. The remains are first cremated in a retort (crematorium oven) to reduce the body down to the inorganic fraction of bone. After the cremation cycle is completed, the cremated remains are removed from the retort and processed, or pulverized, in order to reduce the overall volume for either inurnment or scattering.

CREMATION PHASE

The cremation process begins by placing the body into a cremation container that will incinerate completely. The container is most often a cardboard box, but a specially designed wooden coffin lacking metal hardware may occasionally be used. The cremationist places the container into the retort with the body oriented feet first. A common method of placing the body into the retort is by using a wheeled table with a hydraulic lift. The top of the table is constructed with rollers that allow the cremationist to easily push the cremation container into the retort.

Prior to placing the body in the retort, the body is normally checked for a pacemaker. If a pacemaker battery is in place, it is removed from the body by the cremationist because it can explode during the cremation process and damage the retort (Huxley, 1994; Murad, 1998). A cremation tag, less than 3 cm in diameter, is normally placed in the retort with the body during the cremation process, although we have observed it being placed just inside one of the front corners of the retort by a cremationist (Figure 4.1). Alternatively, the tag may be taped to the outside of the retort during cremation, in which case it would appear unburned. Cremation tags can be made out of a variety of materials (stainless steel, brass, and aluminum), but tags made of aluminum are not placed in the retort with the body because they have a low melting temperature of 1200°F (649°C). After the body has been cremated, the cremation tag is often added to the processed cremains. Each cremation tag is etched with its own identification number and sometimes the crematorium name is included on the tag. The identification number on the tag will be noted in the associated paperwork by the cremationist, and the cremation tags ultimately serve as a means for identifying the cremains to prevent them from being mistakenly substituted for another.

The inside of a modern retort consists of two chambers: the main chamber where the body is placed is called the cremation chamber, while a secondary chamber, also called an after-chamber or afterburner chamber, is usually

FIGURE 4.1 View inside a retort showing the numerous bone fragments that survive the cremation process and the cremation tag (inset) that was placed at the front of the retort (arrow) during the cremation.

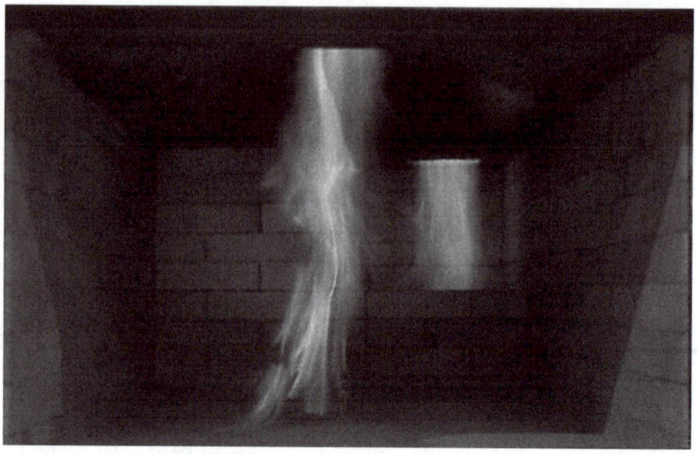

FIGURE 4.2 View inside the retort showing the location of the cremation burner (front) in the cremation chamber and the afterburner (back) located in a section of the after-chamber. (see Plate 15)

located at the back of the cremation chamber and under the floor of the cremation chamber. The retort contains two burners in the ceiling (Figure 4.2). The cremation burner is located in the cremation chamber and the second burner, called the afterburner, is located in the afterburner chamber at the back of the cremation chamber. The cremation chamber serves to incinerate the body and the cremation container. Because of clean air standards, all of the smoke and odor that are created from the cremation process must be burned off before leaving the retort; the afterburner system allows complete combustion of smoke and odor with minimal particulate emissions. Retorts are designed to route the combustion gases from the cremation chamber into the afterburner chamber which has been superheated by the cremation burners, in order to ensure complete combustion of all materials and to reduce visible flue emissions. The gases are drawn through the chambers by a draft inducer, located at the bottom of the stack, which injects high velocity air upward into the stack creating a suction that draws gases through the chambers and up the stack out of the retort.

Both chambers are lined with two types of refractory fire brick (Dale Waters, personal communication). The standard brick, called hard brick, is very dense and makes up the majority of the brick lining of both chambers. The more porous brick, called insulating or soft brick, lines the middle of the walls of the cremation chamber and a few places in the second chamber. The insulating brick is used in combination with the hard brick for several reasons: the insulating brick does not take long to heat up so operating temperatures can be reached sooner; the insulating brick has better insulting value than the hard brick thereby retaining more heat in the chamber and reducing fuel consumption; and the combination of the bricks decreases the cooldown time period of the retort.

The average time for a cremation cycle of newer retorts is around 2 h or less, not counting the preheating of the chamber or the cooldown. Larger more powerful retorts can cremate a body in less than 2 h. Most modern retorts are fueled with natural gas. Temperatures can range from 1400°F to

1800°F (760°–982°C) in the cremation chamber, with the highest temperatures occurring in the middle of the cycle when the body and the container have been fully ignited. Obese decedents, whose bodies contain more fuel in the form of high caloric content lipids, will often burn at temperatures greater than the peak setting programmed by the cremationists. Therefore, the temperature of some cremations may climb to 2000°F (1093°C) or more. The minimum temperature for the after-chamber according to air quality regulations in Florida for newer retorts is 1600°F (871°C) (Dale Waters, personal communication).

PROCESSING PHASE

Once the cremation is completed, there are many large diagnostic bone fragments that survive the combustion process (Figures 4.1 and 4.3). Bones can display various colors that include white, pale yellow, yellow, reddish-brown, very dark gray-brown, dark gray, reddish-yellow, black, medium blue, gray-white, gray, and light gray (Shipman *et al.*, 1984; also see Symes *et al.*, this volume). The various colors are the result of different temperatures attained by bones, the length of time that bones were subjected to fire, and the proximity of bones to the fire (Heglar, 1984; Shipman *et al.*, 1984). The color of commercially cremated bones normally ranges from chalk white to blue-gray. The chalk white to blue-gray colors represent the final stage of calcination where there is a complete loss of the organic fraction of bones. The fusion of the bone salts results in a china-like quality (Ubelaker, 1978; Correia, 1997). If the overall color of commercial cremains varies significantly from the normal chalk white to bluish-grey, a number of interpretations can be made. It may suggest that the temperatures in the retort were not high enough or that the temperatures varied throughout the retort. There also could have been metals introduced into the cremation environment (Dunlop, 1978), or the purported cremains may represent a foreign substance substituted for actual cremated human remains.

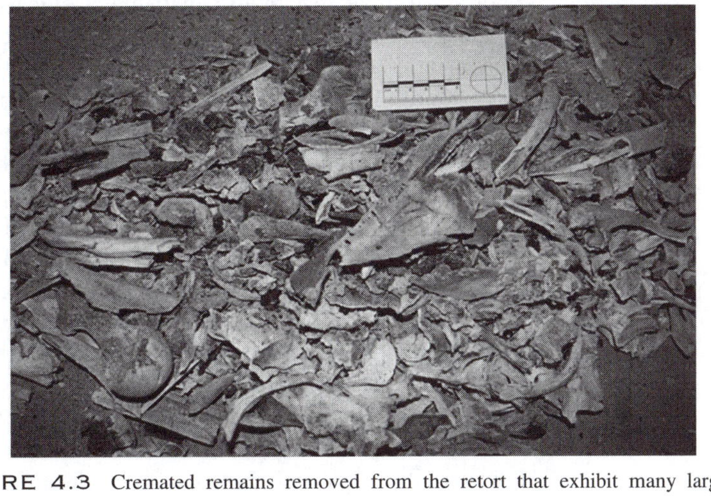

FIGURE 4.3 Cremated remains removed from the retort that exhibit many large and diagnostic bone fragments.

FIGURE 4.4 Cremated remains that are being swept to the front of the retort with a brush where they are directed though the hopper into an external collection pan.

After the retort has cooled down to a safe enough temperature to remove the cremains, the cremationist uses a wire-bristled brush or a hoe to sweep the remains to the front of the retort for removal (Figure 4.4). When using an older style retort, the cremated remains can be swept directly into a collection bin that is temporarily attached to the front of the retort when the door is opened. Cremated remains are removed from newer retorts by sweeping them into an opening in the floor at the front of the retort that leads to an external cooling and collection hopper (Figure 4.4). The remains are collected in an external collection pan at the bottom of the hopper where they can cool as another body is placed into the retort. The collection pan is then removed from the retort to process the cremated bone. Prior to processing, the cremationist will remove large medical (e.g., joint replacements and plates used for internal fixation and stabilization of fractures) and dental prosthetics. A magnet is used to remove magnetic objects, such as staples, nails, and screws. It is important to note that many small metal objects are not removed from the cremains and will survive processing. In particular, it is common to find small metal surgical objects that are nonmagnetic and paramagnetic within the cremains. These artifacts are usually small and take on the same color as the bone fragments. All of the larger material that is removed is discarded. We have seen this material being placed into a bucket or a bin at a number of crematories. Even though this material is sterile after being subjected to intense heat, it can be considered 'biohazardous material' in some jurisdictions and should be handled and disposed of by a licensed waste handler. We have been involved in one case in Florida where over 200 orthopedic appliances (Figure 4.5) were inappropriately discarded in a river. As outlined by Ubelaker and Jacobs (1995), the serial number and manufacturer information that is etched into certain classes of orthopedic hardware can be helpful for the purpose of establishing identity. In this case, the cremation process obliterated the vendor logos, casting, lot and serial numbers, making them illegible. However, the logo and the lot number were preserved on a single tibial component (Figure 4.5) from a total knee arthroplasty. The manufacturer was

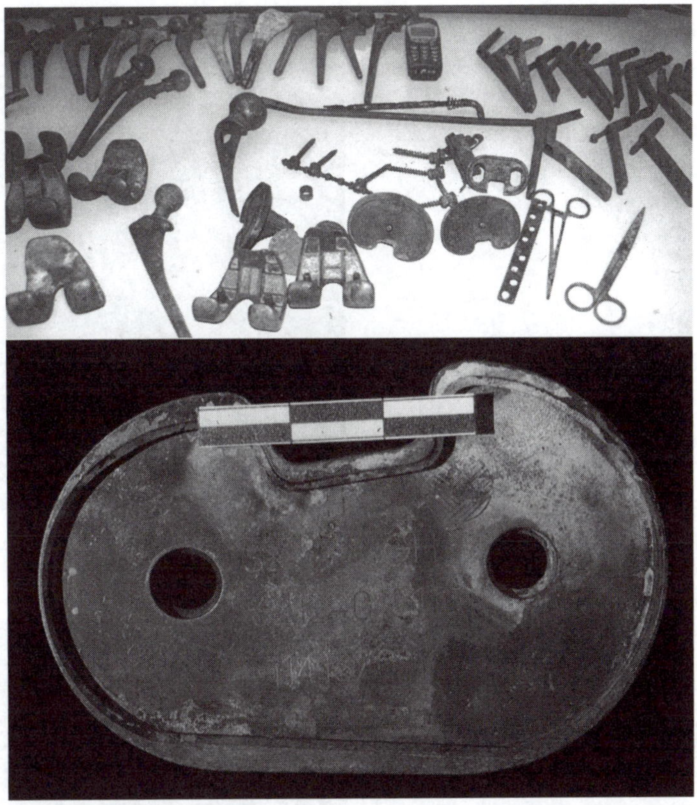

FIGURE 4.5 A subset of the various orthopedic appliances (top picture) that were removed from cremations at a crematorium and illegally dumped in a river. The information that was etched into one tibial component (barley discernable in the bottom picture) was used to implicate the crematorium that was cited for the illegal dumping violation.

able to provide a short list of patients that received devices with the identified lot number. After a short investigation, the crematorium was cited by the State of Florida's Environmental Protection Agency for an illegal dumping violation.

Cremated remains are reduced to a smaller volume for either inurnment and/or scattering by means of a processor, a pulverizer, or a cremulator. There are a variety of processing methods that have been regularly employed in the United States such as hand processing, ball/hammer mill processing, and the newest method called the rotary hinge blade processor. The method of processing determines the size of the bone and tooth fragments, and as a result, it is possible to determine the specific processing method that was used to reduce cremated remains based on the particulate size (Warren and Schultz, 2002). Hand processing consists of using a blunt object such as a piece of wood or a cremation magnet to pulverize cremated remains. We have observed cremations from Europe that were processed using this method, and in the United States, cremated remains from neonates, infants, and small children are sometimes hand processed to preserve sufficient volume for memorialization. Hand processing results in complete bones and large diagnostic bone and tooth fragments that are easily identifiable (Figure 4.6). Cremated remains

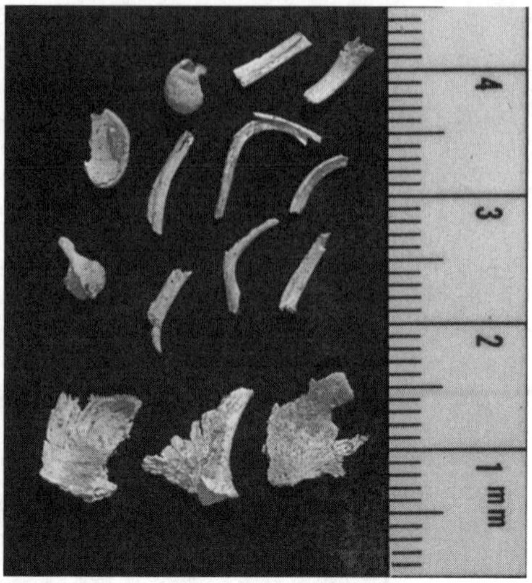

FIGURE 4.6 Identifiable bone fragments (ribs, scapula, and cranial) from a cremated fetus that were hand-processed.

FIGURE 4.7 Cremated remains were processed in the older ball or hammer mill processor by adding the remains to a rotating steel drum (inset) with perforations that contained a number of metal balls or cylinders that would pulverize the fragments so that they would fall through the perforations.

from adults are less frequently hand processed in the United States because the resultant volume generally exceeds the capacity of a standard size urn.

Older ball or hammer mill processors are mechanical processors that were the standard for many years (Figure 4.7). Cremated remains are placed in a perforated steel drum that contains a number of metal balls or cylinders. As the drum rotates, the hammers pulverize the cremains into smaller and smaller fragments inside the drum until they are small enough to fall through the perforations (approximately 4 mm in diameter) into a collection bin at the bottom of the machine. This process continues until all of the material is small enough to fit through the perforations. This reduction method results in diagnostic tooth and bone fragments and complete ear ossicles are often preserved. In addition, there is excellent survivability of the small nonosseous artifacts. It is important to note that damage to the perforations can result in larger openings in the drum that would allow preservation of larger fragments. In addition, bone fragments and nonosseous materials can become lodged in the perforations and then dislodged during successive cremations resulting in commingling.

A recurring problem in the industry is the occasional situation in which the volume of the remains exceeded the capacity of the industry standard-sized urn when hammer mill processors were used. Funeral directors were burdened with the ethical decision to inform the family about the additional cremains. The majority of crematories in the United States and Canada now use what has been coined a rotary hinge blade processor, first introduced in 1987 by an industrial engineering firm in Florida (Warren and Schultz, 2002). According to the manufacturer, this processor was created to reduce human cremains to a small enough volume to fit in the industry standard-sized urn of 200 $in.^3$ The processor looks very similar to a food processor for a reason – the prototype was based on an industrial-grade food processor. Cremated remains are placed into a metal pot with a metal blade at the bottom that is hinged on both sides with ends that are angled upward to provide lift to facilitate mixing of the remains (Figure 4.8). There is a timer on the machine and the manufacturer asserts that only 30-s cycles are required to process the remains.

FIGURE 4.8 Cremated remains are now processed in the newer rotary blade processor by placing the remains into a pot with a rotary blade (inset).

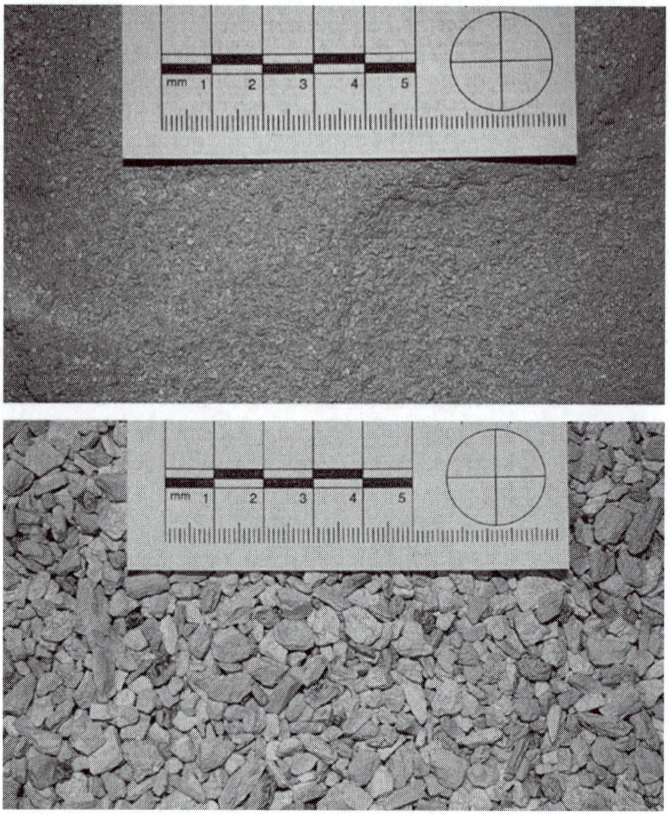

FIGURE 4.9 The majority of the processed bone from a rotary blade processor is ash (material less than 1 mm in the top picture) and nonidentifiable fragments (greater than 1 mm in the bottom picture).

While older processing methods reduced bones down to small fragments that were easily identifiable, the newest rotary blade processors reduced bones to primarily ash and nonidentifiable bone fragments (Figure 4.9) (Warren and Schultz, 2002). The reduction is time-dependent (i.e., the longer the processor is turned on, the smaller the fragments will become). As a result, it may be difficult, or impossible in some cases, to determine if the remains are human based on the osseous material alone. Although the majority of the cremains are nondiagnostic, it may be possible to retrieve diagnostic osseous fragments depending primarily on the condition of the blade and the length of the processing cycle (Warren and Schultz, 2002). As the blade wears in length, larger bone and tooth fragments will survive the processing cycle. While it may not be possible to make an identification based on the osseous material, the nonosseous materials (cremation artifacts) have excellent survivability with rotary blade processors and provide important diagnostic clues for identification (Warren and Schultz, 2002). Of course, the presence of artifacts is contingent on the decedent's life history (e.g., dental work or surgical procedures).

METHODS USED TO ANALYZE CREMAINS

The goals of a forensic cremains analysis are to (a) confirm that a specimen represents the cremated remains of a human, (b) confirm that the remains represent a single individual and are not significantly commingled, and (c) establish that the remains are most likely those of a specific individual. The identification in cremation cases is based on the preponderance of presumptive evidence available. In almost every case, no positive lines of evidence for identity exist. In other words, an examination may reveal that the cremains are *most likely* those of a given individual. In a case where multiple lines of evidence are consistent with the known information provided about the decedent, the examiner will provide an expert opinion that all available evidence either supports or does not support an identification.

A variety of methods are used to analyze cremated remains including cremation weights, microscopic examination of the morphology of osseous materials (bones and teeth), and microscopic examination of nonosseous materials (cremation artifacts). Although it may be possible to extract, amplify, and sequence DNA from *burned* bones (see Chapter 7), commercial cremation occurs at a temperature, and for a duration, that destroys all organic materials, including DNA. Therefore, using DNA as a method of identification is not possible for cases involving contemporary commercial cremation. More recently, chemical analyses have been used to determine the elemental composition of cremains in order to find signature elements of bone and/or elements consistent with the known life history of the decedent.

CREMATION WEIGHTS

Prior to removing the cremains from the bag in which they are wrapped, the overall weight of the cremains should be assessed. Cremation weights can provide limited baseline information. Weights exceeding the ranges established by researchers may suggest the presence of more than one individual. Weights less than expected may point to remains that are either incomplete (i.e., some portion of the remains have been removed prior to examination), or those of a juvenile or nonhuman. All things considered, the weight of the cremated remains should correspond with published values for the sex and general skeletal robusticity of the decedent (Warren and Maples, 1997; Bass and Jantz, 2004).

OSSEOUS MATERIAL

Identifiable bone and dental fragments can be located with the aid of magnification and radiography. In addition, standard sieves should be used to sift the fragments and particles into their respective sizes (e.g., 4, 2, 1, and <1 mm.). The recognition of diagnostic materials improves with the segregation of the cremains into different sizes. Bone and tooth fragments can be used to determine if the cremated remains are those of a human; to establish whether or not the decedent had a dentition; to reveal the developmental stage of the dentition and, therefore, the potential age of the individual; and finally, to help detect the presence of various age-related pathologies. Also, calcined

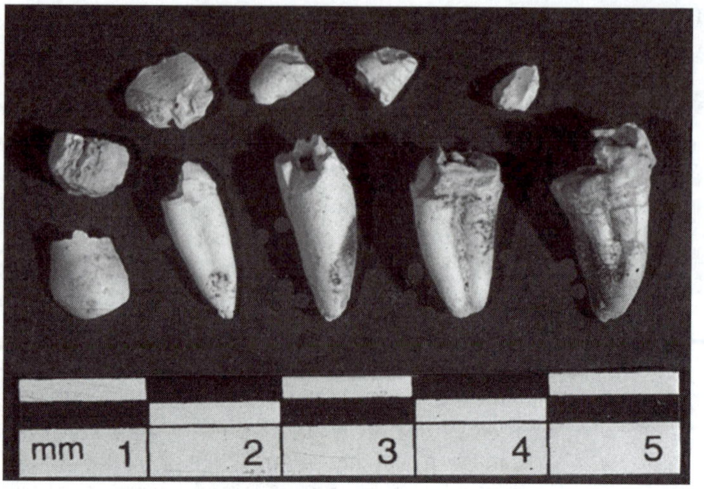

FIGURE 4.10 Identifiable tooth fragments (crown and root) recovered from a cremation that was processed by a rotary blade processor.

atherosclerotic blood vessels have been found among cremated remains that have been processed by the older hammer mill machines (Warren *et al.*, 1999). Since atherosclerosis is an age-related pathology, the presence of these calcined vessels provides a clue that the decedent was most likely a mature adult. While cremated bones that were processed using the older hammer mill machines result in large diagnostic bone and tooth fragments, including complete ear ossicles, we have examined cremains that have been processed with the rotary blade processor that exhibited significant amounts of diagnostic osseous materials (Figure 4.10).

CREMATION ARTIFACTS

The most important diagnostic material used to establish identity relies on cremation artifacts (nonosseous materials). We compared cremation artifacts with those in a known reference collection housed in the Forensic Anthropology Research Laboratory at the University of Florida. A small chemistry stir magnet was used to assist with the removal of the metallic material from the cremains that is magnetic. However, numerous medical artifacts are comprised of nonmagnetic alloys and cannot be removed using a magnet. Also, radiography and light microscopy using low magnification can be helpful to locate metal objects in the cremains.

For purposes of classification, cremation artifacts have been divided into five categories by Warren and Schultz (2002): medical, dental, mortuary, personal, and miscellaneous. Identifiable medical artifacts that may be recovered include surgical staples used for skin closure, vascular clips used to ligate blood vessels during surgery (Figure 4.11), complete and fragmentary sternotomy sutures, embolism filters, surgical wire, pacemaker leads (Figure 4.12), and fragmentary pacemaker pieces. Dental artifacts (Figure 4.13) include fragments from metallic crowns, posts, bridgework, and porcelain crowns and caps. Mortuary artifacts can relate the mortuary history of the decedent. Was the body embalmed? Was there a visitation? In what

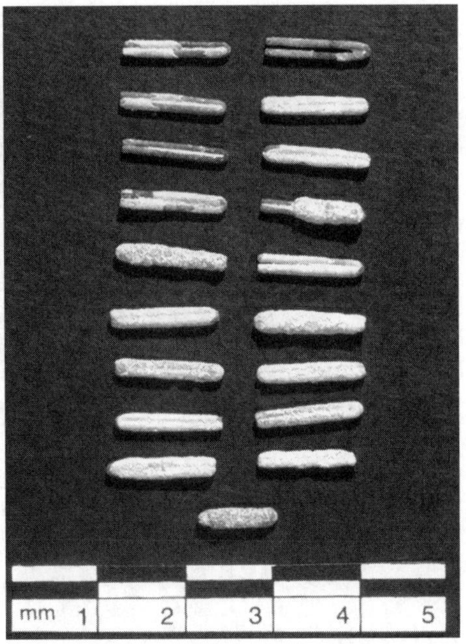

FIGURE 4.11 Nineteen vascular clips used to ligate blood vessels during surgery, which were recovered when analyzing one cremation.

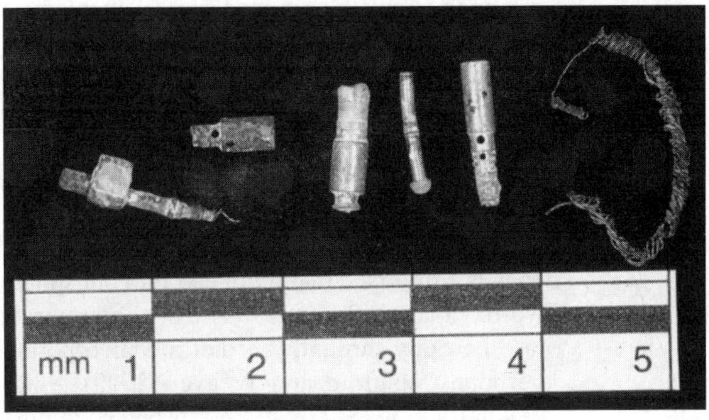

FIGURE 4.12 Pacemaker lead components that are commonly recovered from cremated remains.

type of container was the body cremated? These mortuary artifacts include injector needles that are inserted into the maxilla and the mandible to 'set the features' and to secure the mouth from opening and various staples that are used to construct the wood and cardboard cremation caskets in which the cremations take place. Personal artifacts are often noted in cremains, ranging from fragments of jewelry (chain fragments from a necklace or a bracelet are commonly recovered) to clothing such as bra clasps and zipper fragments. The clasps of hospital identification bracelets commonly survive the incineration process. It is relatively common to recover mementos that were added by the family to the cremated remains after the cremation process. These items

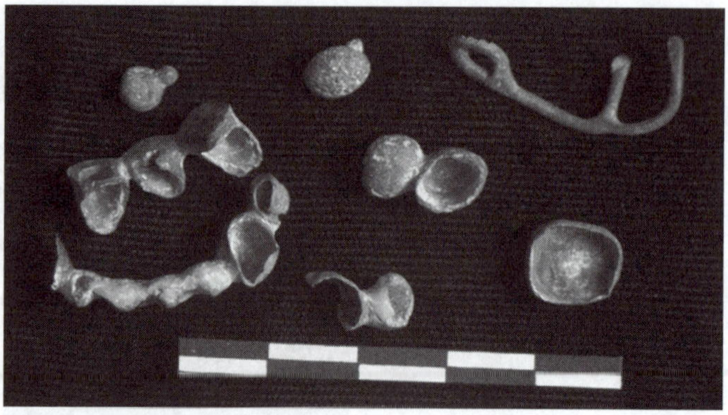

FIGURE 4.13 Various bridge and metallic crown material that was recovered from one cremation.

are easy to recognize because they are not burnt and signal the investigator that the urn has been opened prior to examination.

CHEMICAL METHODS

Because of the difficulty in identifying cremains processed with the rotary blade processor, chemical techniques have enabled us to answer some basic questions about the composition of cremains that cannot be answered from cremation weights and gross and microscopic methods. The elemental signature of questionable cremated remains can provide an additional signal for detecting contaminated, commingled, and inappropriately identified cremains. Bones are structurally composed of an inorganic mineral phase that accounts for 60–70% of its dry weight and the remainder consists of an organic phase (Nordin and Frankel, 2001). The organic material consists mainly of collagen and the inorganic material that gives bone its solid consistency is comprised mainly of calcium phosphate as a poorly formed crystalline hydroxyapatite (Sillen, 1989). Also, many trace elements that are introduced into the body through the diet are incorporated into the inorganic phase of the bone (Sandford and Weaver, 2000). Since the organic fraction of contemporary human cremations has been combusted, only the inorganic mineral component of the skeleton should be left for analyze. Therefore, variations of the expected calcium/phosphorous (Ca/P) ratio, elevated levels of minor elements normally found in bones, and the presence of rare trace elements can be used to distinguish between bones and foreign materials (Ubelaker *et al.*, 2002; Warren *et al.*, 2002). On the other hand, finding elevated levels of normally occurring minor elements or rare trace elements in bone tissues can be used to corroborate skeletal idiosyncrasies thought to have been acquired during the decedent's life. In such instances, the chemical analysis becomes an unparalleled means of identifying cremains.

A number of different analytical techniques used to determine the multielemental composition of cremains have recently been reported in the literature, which include proton-induced X-ray emission (PIXE), inductively coupled plasma mass spectrometry (ICP-MS), and inductively coupled plasma optical emission spectroscopy (ICP-OES). The elemental composition of cremains was first reported by Warren *et al.* (2002), when PIXE was used to compare the elemental signature of several cremains samples with the recorded standard for the elemental composition of bones and other materials to detect foreign materials introduced into, or substituted for, actual cremated remains. Proton-induced X-ray emission is a nondestructive method used for simultaneous trace multielement analysis that requires only a few milligrams of sample. This method bombards the sample with an impinging beam of low-energy (1–2.5 MeV) protons to induce innermost shell ionization of the target atoms. The ionization process is followed by a rearrangement of the electronic architecture of the excited atoms, resulting in the emission of characteristic X-rays that identify the chemical elements in the sample.

More recently, inductively coupled plasmas (ICP-OES and ICP-MS) have been used for the simultaneous trace multielement analysis of cremated samples to detect elevated levels of specific elements in cremains that may reflect elements acquired during the decedent's life. For example, Bodkin *et al.* (2005) used ICP-OES to detect high levels of lead in cremains of two individuals who suffered gunshot wounds and had carried the projectiles as a foreign body within their soft tissues. Both ICP-OES and ICP-MS measure most of the elements of the period table with the exception of H, O, N, F, Cl, Br, and I. When ICP-MS is used for the cremation analysis, only a very small sample is needed (0.05 g) from the original cremains. The small sample size is an advantage of this method because there is very limited destruction of the original sample. Samples are analyzed by decomposing the neutral elements in a high-temperature argon plasma or gas to atomize and ionize the elements. The elements are then identified by their mass-to-charge ratios and their concentration is determined as a proportion of the sample.

FORENSIC CASE STUDY USING ICP-MS

In this case, a man opened his deceased wife's cremains to add her remains with those of his mother and father. In doing so, he noticed that it was of different color than those of his parents. He sent the cremains to one of the authors (M.W.W.) and asked that they be analyzed. The remains were processed by a rotary blade processor, and although the bone structure was clearly present under light microscopy, the cremains were, in fact, different in color from any that the author had ever seen – almost a dark gun-metal gray color. Traditional methods of identification using gross and microscopic methods were not possible because of the processing method that was used. Therefore, we initiated a small pilot project to determine if the complainant received the correct cremains and to explore the use of ICP-MS in assessing the elemental composition of cremated remains. This case study involved analyzing the 'problem cremation' with three other cremated samples from 'known' contexts:

1. The 'problem cremation' detailed above.
2. A recent historical cremation from Europe.
3. The cremains of a donated anatomical specimen.
4. A known commingled cremation.

The second sample is from a recent historical cremation, performed in Germany, which dates to 1962. These cremains were hand-processed and have been stored in a copper urn since inurnment. The third sample is a known human cremation from a donated anatomical cadaver, and the fourth sample is from a cremation in which demonstrable commingling was detected. These three additional samples provide a snapshot of what the baseline elemental expectations should be for cremated samples.

Samples from each cremation were dissolved in an acid solution and then diluted for analysis using a Thermo-Finnigan Element 2 (ICP-MS) housed in the Department of Geological Sciences at the University of Florida. All four samples were first analyzed to determine which elements comprise cremated remains and to compare the elemental composition of the 'problem cremains' with the other three samples. The composition of the suite of analytes detected for the samples in parts per million (ppm) is shown in Figure 4.14. The major elements, Ca and P, and essential elements (e.g., Mg, Na, Zn, and Fe) mirror the composition of bones, as calcium phosphate is the chief constituent of the inorganic bone mineral, hydroxyapatite (Sandford and Weaver, 2000). Thus, this analysis confirms that all four of the cremation samples are comprised of the inorganic fraction of bones, and the elemental composition of the 'problem cremains' indicates that this set of cremains was not contaminated or substituted with a foreign material.

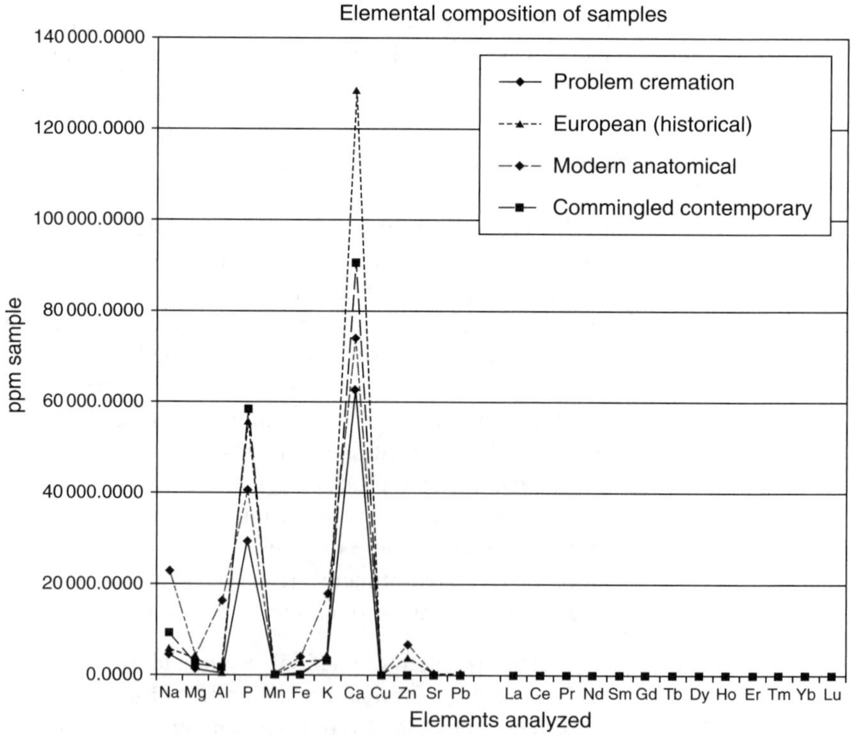

FIGURE 4.14 Elemental composition of the four cremation samples.

A Standard Reference Materials (SRM) is required when performing a chemical analysis to determine the elemental composition of a specific material using an analytical method such as ICP-MS. The SRM 1400 Bone Ash by NIST (National Institute of Standards and Technology, Gaithersburg, MD) was required to provide data calibration for the equipment and to provide baseline elemental data to interpret the cremation samples. The SRM 1400 Bone Ash contains the naturally occurring elements that are found in bones (Hinners *et al.*, 1998), and thus is used as a comparative reference when seeking to determine the specific major, minor, and trace elements in bone, or purported bone, samples. Furthermore, the detection of rare earth elements or increased levels of essential elements that differ from the SRM 1400 Bone Ash standard can reflect elements acquired during a decedent's life and may provide a means of identifying cremains.

The elemental composition of the four sets of cremains samples to the SRM 1400 Bone Ash standard is compared in Figure 4.15. The peaks in Figure 4.15 are substantially different from the standard, whereas valleys near zero conform to the SRM 1400 Bone Ash standard. All four cremation samples display variations of the detected elements compared to the SRM 1400 Bone Ash standard that includes elements normally found in bones and not normally found in bones. The one marked element of the historical cremation that is noteworthy is Cu. The increased Cu level in the European cremation was most likely the result of storing the cremains in a copper urn. The elemental composition of the commingled cremation is most similar to the SRM 1400 Bone Ash standard. Interestingly, the modern anatomical cremation sample deviated the most from the SRM 1400 Bone Ash standard

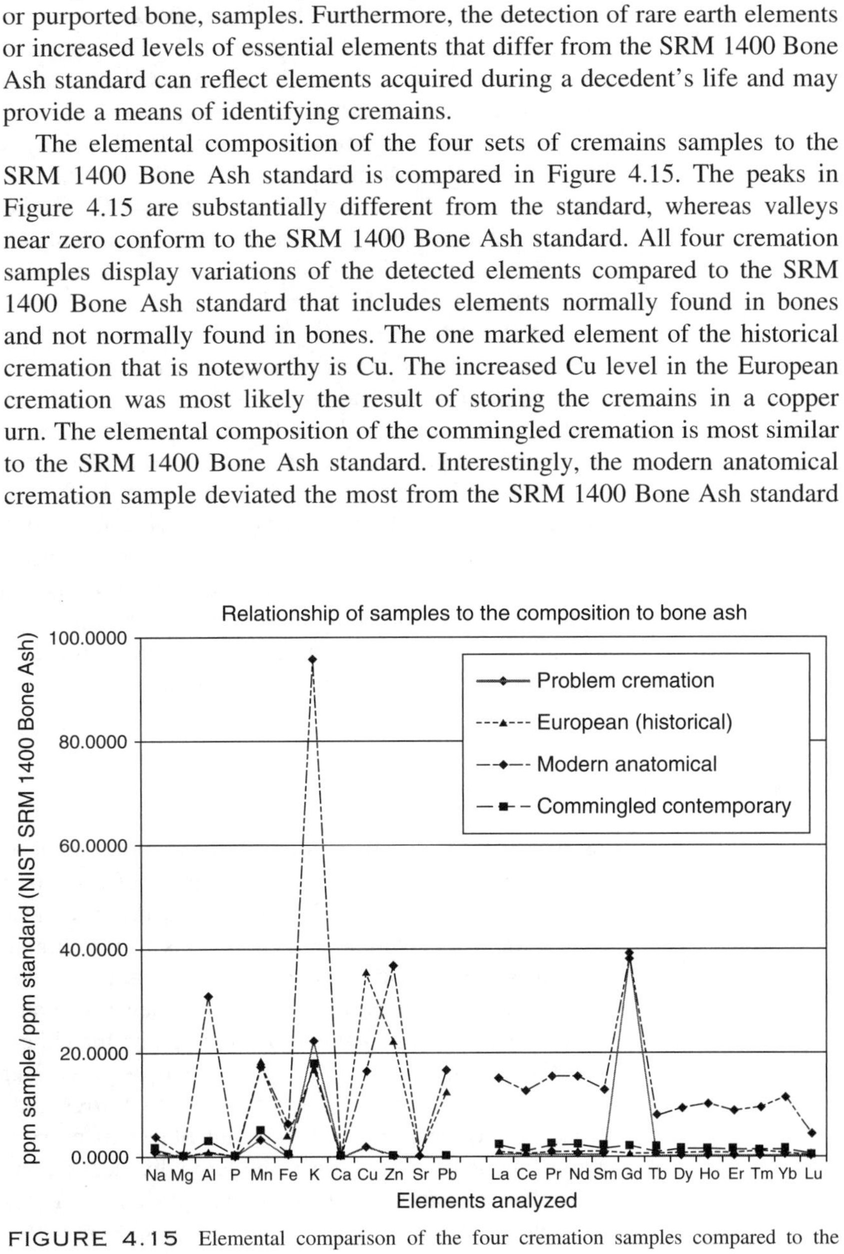

FIGURE 4.15 Elemental comparison of the four cremation samples compared to the reference NIST SRM 1400 Bone Ash standard.

with elevated levels of many elements including marked peaks of Al, K, Zn, and gadolinium (Gd). A number of these detected and elevated elements in the anatomical cremation sample may be due to the preservative fluids that were injected into the body so that the cadaver could be used for anatomical dissection. Finally, elevated levels of K and Gd were noted in the 'problem cremains' sample. While increased levels of K were noted in all four samples compared to the SRM 1400 Bone Ash standard, K is normally found in bones. However, marked levels of Gd were noted in the anatomical and 'problem cremation' samples.

Gadolinium can be introduced into the body as a contrast medium in radiological imaging procedures (Kirchin, 2003; Gibby *et al.*, 2004), and unlike metals with a known biological function, Gd does not have a known pathway for excretion from the body and can persist primarily within the liver and bones (Gibby *et al.*, 2004). While it is possible that the presence of Gd was the result of contrast mediums for both the anatomical and 'problem cremation' samples, we were able to confirm that Gd in the 'problem cremation' sample was most likely the result of the decedent having undergoing MRI and CT imaging studies incidental to diagnosis and treatment for renal failure. Although the color of the 'problem cremains' sample called the validity of this set of cremains into question, the elemental composition confirmed that they were cremated remains and provided supportive evidence that the complainant most likely had received the proper cremated remains of his wife. However, the strange color of the 'problem cremains' sample was most likely not due to the presence of Gd because the modern anatomical specimen that also contained Gd appeared normal in color. Unfortunately, the only way to further investigate the origin of the strange color would be to observe cremations taking place at the crematorium in question and possibly examine more cremations that were performed at this location. Overall, while the purpose of this forensic case study was primarily to assist with the identification of the 'problem cremains' sample, it demonstrates the utility of incorporating an elemental analysis when analyzing cremations.

CONCLUSIONS

As the number of people choosing cremation increases, we should continue to see a concomitant growth in the number of cases in which forensic anthropologists are consulted. It is essential for those forensic anthropologists who are examining contemporary cremations to be familiar with the entire cremation process. Contemporary cremation is a two-step taphonomic process that begins by cremating a body, and then processing the bone fragments to reduce the overall volume of the remains for either inurnment or scattering. Of forensic importance, the method of processing determines the size of recognizable bone and tooth fragments (Warren and Schultz, 2002). Hand processing consists of using a blunt object to pulverize cremated remains, and results in complete bones and large diagnostic bone and tooth fragments. Older ball or hammer mill processors mechanically pulverize the cremains through 4-mm perforations in a drum, which results in the preservation of diagnostic tooth and bone fragments and excellent survivability of the small nonosseous artifacts. Finally, the newest rotary blade processors reduce bones to primarily ash and nonidentifiable bone fragments, but there is excellent survivability of the nonosseous artifacts. Due to the efficiency of the newest processing methods, traditional gross methods of cremation analysis may be of limited value when trying to determine the identity and the composition of purported cremains. Although cremation artifacts have excellent survivability with the newest rotary blade processors, there may be instances when there are no medical or dental artifacts because of the deceased's medical and dental history.

Therefore, the use of chemical methods for elemental analysis should now be considered as a regular step when analyzing cremains. In particular, chemical techniques can answer some basic questions about the composition of cremains that cannot be answered from gross and microscopic methods alone. Chemical methods can be used to determine whether the disputed cremains are comprised of bones or a foreign material that was substituted for the cremains. In addition, chemical methods may provide the only supporting evidence for the identification of cremains by detecting elevated levels of normal minor elements found in bone tissue or the presence of rare trace elements that were acquired during the decedent's life.

ACKNOWLEDGMENTS

The authors wish to acknowledge Kenneth Robinson and Dale Walter of Industrial Equipment and Engineering, Inc. for their technical assistance. Also thanks to Chin-Hsin Liu for comments on the preparation of these specimens, as well as additional specimens being tested for a more comprehensive study. Special thanks to Prof. Paul Mueller and Dr. George Kamanov (Department of Geological Sciences, University of Florida) for their interest, assistance, and access to the Thermo-Finnigan Element 2 ICP-MS.

REFERENCES

Bass, W.M. and Jantz, R.L. (2004). Cremation weights in east Tennessee. *Journal of Forensic Science* **49**, 901–904.

Bass, W.M. and Jefferson, J. (2003). *Death's Acre: Inside the Legendary Forensic Lab, the Body Farm, Where the Dead do Tell Tales*. G.P. Putnam's Sons, New York.

Baby, R.S. (1954). Hopewell cremation practices. *Ohio Historical Society Papers in Archaeology* **1**, 1–17.

Binford, L.R. (1963). Analysis of cremations from three Michigan sites. *Wisconsin Archeologist* **44**, 98–110.

Bodkin, T., Potts, G., Brooks, T., and Shurtz, K. (2005). Elemental analysis of human cremains using inductively coupled plasma optical emissions spectroscopy (ICP-OES) to distinguish between legitimate and contaminated cremains. *Proceedings of the American Academy of Forensic Sciences* **11**, 307.

Correia, P.M.M. (1997). Fire modification of bone: a review of the literature. In *Forensic Taphonomy: The Postmortem Fate of Human Remains* (W.D. Haglund and M.H. Sorg, eds.), pp. 275–293. CRC Press, Boca Raton, FL.

Cremation Association of North America. Statistics. http://www.cremationassociation.org/docs/WebConfirmed.pdf

Dunlop, J. (1978). Traffic light discoloration in cremated bones. *Medicine Science and the Law* **18**, 163–173.

Eckert, W.G., James, S., Katchis, S. (1988). Investigation of cremations and severely burned bodies. *American Journal of Forensic Medicine and Pathology* **9**, 188–200.

Gibby, W.A., Gibby, K.A., and Gibby, W.A. (2004). Comparison of Gd DTPA-BMA (Omniscan) versus Gd HP_DO3A (ProHance) retention in human bone tissue by inductively coupled plasma atomic emission spectroscopy. *Investigative Radiology* **39**, 138–142.

Heglar, R. (1984). Burned remains. In *Human Identification: Case Studies in Forensic Anthropology* (T. Rathbun and J.E. Buikstra, eds.). Charles C. Thomas, Springfield, IL.

Hinners, T.A., Hughes, R., Outridge, P.M., *et al.* (1998). Interlaboratory comparison of mass spectrometric methods for lead isotopes and trace elements in NIST SRM 1400 Bone Ash. *Journal of Analytical Atomic Spectrometry* **13**, 963–970.

Huxley, A.K. (1994). Analysis of ceramic substrate found in cremains. *Journal of Forensic Science* **39**, 287–288.

Iverson, K.V. (2001). *Death to Dust: What Happens to Dead Bodies?* 2nd edn. Galen Press Ltd., Tucson, AZ.

Kennedy, K.A. (1996). The wrong urn: commingling of cremains in mortuary practices. *Journal of Forensic Science* **41**, 689–692.

Kirchin, M.A. (2003). Gadolinium contrast agents for MR imaging: safety update. *Highlights in MRI* **1**, 1–3.

Maples, W.R. and Browning M. (1994). *Dead Men do Tell Tales: The Strange and Fascinating Cases of a Forensic Anthropologist.* Broadway Books, New York.

Merbs, C. (1967). Cremated human remains from Point if Pines, Arizona. *American Antiquity* **32**, 498–506.

Murad, T. (1998). The growing popularity of cremation versus inhumation: some forensic implications. In *Forensic Osteology: Advances in the Identification of Human Remains* (K.J. Reichs, ed.), pp. 86–105. Charles C. Thomas, Springfield, IL.

Murray, K.A. and Rose, J.C. (1993). The analysis of cremains: a case study involving the inappropriate disposal of mortuary remains. *Journal of Forensic Science* **38**, 98–103.

Nordin, M. and Frankel, V.H. (2001). *Basic Biomechanics of the Musculoskeletal System*, 3rd ed. Lippincott Williams & Wilkins, Baltimore, MD.

Sandford, M.K. and Weaver, D.S. (2000). Trace element research in anthropology: new perspectives and challenges. In *Biological Anthropology of the Human Skeleton* (M.A. Katzenberg and S.R. Saunders, eds.), pp. 329–350. Wiley-Liss, Inc., New York.

Shipman, P., Foster, G., and Schoeninger, M. (1984). Burnt bones and teeth: an experimental study of color, morphology, crystal structure, and shrinkage. *Journal of Archaeological Science* **2**, 307–325.

Sillen, A. (1989). Diagenesis of the inorganic phase of cortical bone. In *The Chemistry of Prehistoric Bone* (T.D. Price, ed.), pp. 211–229. Cambridge University Press, Cambridge.

Ubelaker, D.H. (1978). *Human Skeletal Remains: Excavation, Analysis, Interpretation.* Taraxacum, Washington, DC.

Ubelaker, D.H. and Jacobs, C.H. (1995). Identification of orthopedic device manufacturer. *Journal of Forensic Science* **40**, 168–170.

Ubelaker, D.H., Ward, D.C., Braz, V.S., and Stewart, J. (2002). The use of SEM/EDS analysis to distinguish dental and osseous tissue from other materials. *Journal of Forensic Science* **47**, 940–943.

Warren, M.W. and Maples, W.R. (1997). The anthropometry of contemporary commercial cremation. *Journal of Forensic Science* **42**, 417–423.

Warren, M.W. and Schultz, J.J. (2002). Post-cremation taphonomy and artifact preservation. *Journal of Forensic Science* **47**, 656–659.

Warren, M.W., Falsetti, A.B., Hamilton, W.F., and Levine, L.J. (1999). Evidence of arteriosclerosis in cremated remains. *American Journal of Forensic Medicine and Pathology* **20**, 277–280.

Warren, M.W., Falsetti, A.B., Kravchenko, I.I., *et al.* (2002). Elemental analysis of bone: proton-induced X-ray emission testing in forensic cases. *Forensic Science International* **125**, 37–41.

Wells, C. (1960). The study of cremation. *Antiquity* **34**, 29–37.

5

Thermally Induced Changes in the Stable Carbon and Nitrogen Isotope Ratios of Charred Bones

Mark R. Schurr, Ph.D.[1], Robert G. Hayes, Ph.D.[2], and Della C. Cook, Ph.D.[3]

[1] *Department of Anthropology, University of Notre Dame, Notre Dame, IN*
[2] *Department of Chemistry and Biochemistry, University of Notre Dame, Notre Dame, IN*
[3] *Department of Anthropology, Indiana University, Bloomington, IN*

INTRODUCTION

For many cultures, cremation was a common or even dominant portion of the mortuary program. Historical examples include the Romans and modern Hindus (Barber, 1990). A complete bibliography of archaeological cremations would be extremely long and would span all of the world's continents and most of human prehistory. The earliest known cremation, from Lake Mungo, Australia, is dated at about 40 000 years ago (Bowler *et al.*, 2003). Cremations are extremely common in the European archaeological record, with examples ranging from the Mesolithic (Arts, 1987) through the Neolithic of Central and Western Europe (Whittle, 1985), Bronze Age (Coles and Harding, 1979: 263), to the Iron Age (Phillips, 1980: 52, 261–263), including Anglo-Saxon and Roman Britain (McKinley, 1994). In North America, cremations are known from the Hohokam and Mogollon cultures of the southwestern United States (Reinhard and Fink, 1994). Examples from eastern North America span the entire cultural sequence for the region, from one of the earliest Late Paleoindian burials known from the Great Lakes region (Mason, 1981: 117) through the Early Archaic of the Southeast (Chapman, 1973), Late Archaic (Perino, 1968), Ohio and Illinois Hopewell (Baby, 1954; Asch, 1976), the Late Woodland (Buikstra and Goldstein, 1973), and the Late Prehistoric period (Perino, 1971; Brown, 1981; Schurr, 1987).

Cremations, whether accidental or deliberate, are difficult to work with because they are usually fragmentary. However, cremation is rarely complete, and cremated burials often provide bone fragments suitable for many types of osteological investigations. Morphological analyses of cremated bones can provide information about sex, age, health, and cause of death. It is often assumed that cremation produces highly oxidized bone fragments that are little more than inorganic ash, but this is not the case for most prehistoric cremations. Many cremations are incompletely burned (Barber, 1990), and substantial portions of the skeleton, especially the most durable elements such as the skull and long bones, may remain largely intact. Bone fragments often display a gradient of thermal alteration, ranging from calcined through blackened to apparently unaltered, sometimes even on the same bone.

Burned bones can also be subjected to chemical studies, although these are rarely done. Earlier studies have generally been confined to quantifying chemical changes that occur when bone is burned (McCutcheon, 1992; McKinley, 1994; Reinhard and Fink, 1994; Taylor *et al.*, 1995) and distinguishing burned bones from bones blackened by other processes (Sillen and Hoering, 1993; Shahack-Gross and Bar-Yosef, 1997). Many cremations produce bones that still contain organic matter in the form of thermally altered organics or reduced carbon. This is especially true for prehistoric cremations, where the large amount of fuel necessary for a total cremation may not have been readily available to nonindustrial societies (Barber, 1990). In many cases, charred bones from archaeological contexts can be expected to contain more organic carbon than uncharred bones from the same context, because reduced carbon in charred bone is more resistant to diagenetic change than uncharred collagen and other organic fractions of bone. Cremated burials would provide an abundant and promising source of paleodietary data through isotope studies if correct techniques were available.

BONE CHARRING AND ITS EFFECT ON THE ISOTOPIC COMPOSITION OF BONE ORGANIC MATERIAL

Bones go through a well-defined sequence of chemical changes with increasing temperature. As summarized by McCutcheon (1992), temperatures between 20°C and 300–350°C lead to loss of unbound water and initial carbonization of the organic phase, while temperatures above 350°C result in complete combustion of the organic phase.

It has been reported that stable carbon and nitrogen isotope ratios cannot be used to reconstruct prehistoric diets or matter flows in prehistoric ecosystems using burnt bones because heating changes stable isotope ratios. In a series of laboratory experiments, DeNiro *et al.* (1985) found that heated bones showed pronounced shifts in isotopic composition. Shifts of up to 4‰ or 5‰ (expressed in the standard 'δ' notation) were observed, levels that would produce significant errors in isotopic reconstructions of prehistoric diet. Because of these thermal alterations, burnt human bones from cremated burials apparently cannot be used for subsistence studies. Thus, stable isotopes do not appear to be useful for paleodietary studies for much of the Late Bronze Age and Roman periods in much of Europe, where cremation was the primary disposal practice. In addition, when cremation may have been restricted to some members of the society, such as at Middle Mississippian chiefdoms where elites may have been partially cremated by burial in burned charnel houses or at northern European Bronze Age sites where both cremated and uncremated burials have been found, it is not possible to determine if differential mortuary treatments were associated with differential diet. Another example would be the reconstruction of diets for people who have been accidentally burned (e.g., in some sort of catastrophe).

CONTROLLED HEATING EXPERIMENTS TO EVALUATE ISOTOPIC CHANGES IN CHARRED BONES

CONTROL PREPARATION PROCEDURES

Modern controls were produced by heating small (0.1 g) pieces of cow bone that had been dried for 48 h at 70°C, a temperature empirically chosen as high enough to remove unbound water but low enough not to produce measurable chemical changes other than dehydration. Heating times ranged from 0.5 to 3 h at temperatures ranging from 150°C to 600°C. Three replicates were heated at each time and temperature for a total of 93 samples at 31 time and temperature combinations. The bone pieces were heated at the bottom of an open 150-ml beaker, with the pieces evenly spaced around the bottom. The beaker was placed into a small furnace at room temperature and then the temperature was increased to the set point. The heating time was measured from the moment that the oven attained the set point temperature. Weights were taken before and after heating to quantify the weight loss. The carbon and nitrogen contents of the burned bones were measured using a Costech elemental analyzer. Collagen and acid-insoluble charred organic materials (a mixture of thermally degraded collagen and insoluble char) were then extracted by slow demineralization in weak acid (0.25 M HCl) using procedures commonly used for isotopic studies of prehistoric bones (Moore *et al.*, 1989). Stable carbon and nitrogen isotope ratios of the extracted organic materials were determined with a Finnegan Delta Plus mass spectrometer equipped with a Carlo-Erba elemental analyzer and a Conflo II interface.

CONTROL RESULTS

Changes in weight loss, percentage of C and N of bones, and extraction yields were highly correlated and were more strongly controlled by temperature than by time. Stable carbon and nitrogen isotope ratios remained unaltered at and below 200°C. Figure 5.1 is a contour map showing how %N (the best measure of protein preservation) varies with time and temperature. The results were similar for the other variables. There was a relatively slight but consistent change at temperatures below 350°C. As the closely spaced contour lines show, an abrupt change occurred at 350°C, previously shown to be the temperature at which carbonization begins (McCutcheon, 1992).

Figure 5.2 shows $\delta^{13}C$ for the standard samples. The results are coded by heating time. As one sees, we find no dependence of $\delta^{13}C$ on extent of charring, although only very small amounts of extract were recovered from the samples that had been heated to higher temperatures, and the material resembles charcoal. In contrast to the earlier results of DeNiro *et al.* (1985),

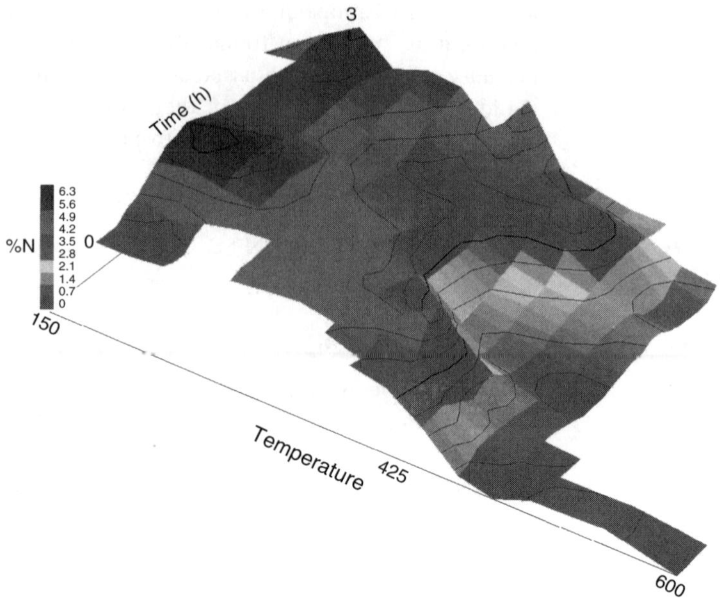

FIGURE 5.1 Changes in the percentage of nitrogen with heating time and temperature. (see Plate 16)

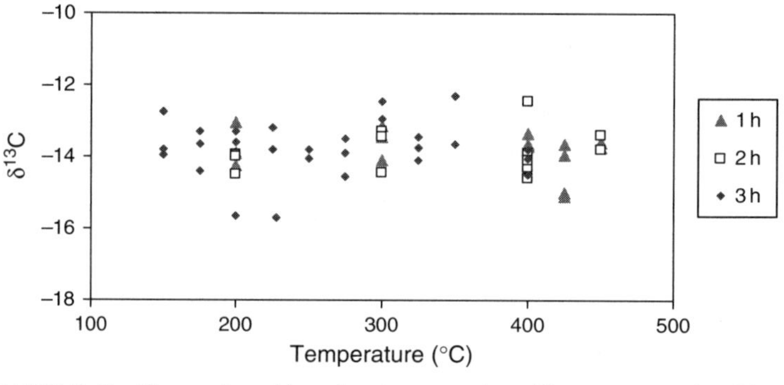

FIGURE 5.2 Changes in stable carbon isotope ratios with temperature. (see Plate 17)

we found that $\delta^{13}C$ values did not change systematically during heating although the isotope ratios show increased scatter around the mean value of $-13.2 \pm 0.1‰$ for unburned samples, compared to unburned controls. This suggests that burned bones from human cremations can provide reliable data for reconstructing aspects of prehistoric diet such as maize consumption or terrestrial/marine diets in the absence of C4 plants (because most marine foods and C4 plants both contain relatively large amounts of ^{13}C, it is not possible to distinguish the relative contributions of marine foods and C4 plants from diets that contain carbon from both sources), perhaps with some loss of precision over what could be expected from using well-preserved, uncharred bones.

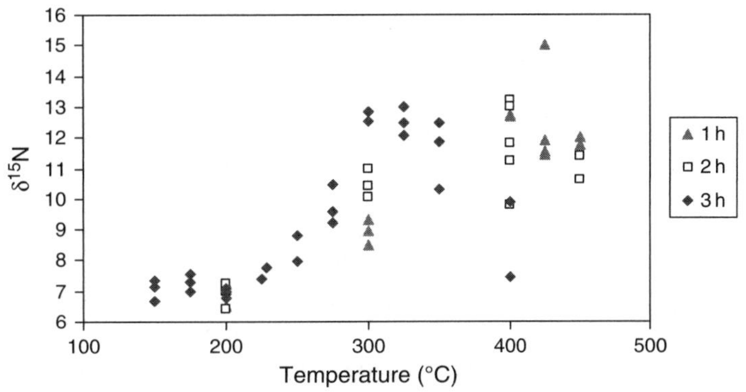

FIGURE 5.3 Changes in stable nitrogen isotope ratios with temperature. (see Plate 18)

Differing extraction procedures or heating conditions may explain why these results differ from an earlier study (discussed in more detail below).

Stable nitrogen isotope ratios, on the other hand, were significantly effected by heating (Figure 5.3). The $\delta^{15}N$ values of the charred controls increased with temperature and changed markedly at temperatures between 200°C and 300°C. Changes appeared to be largely independent of heating time, except around 300°C, where there appears to be a distinct time effect for samples heated for 1, 2, or 3 h. The maximum shift in $\delta^{15}N$ values occurred at 300°C for samples heated for 3 h, and abruptly at temperatures above 300°C for shorter heating times. The maximum enrichment in ^{15}N with heating was around 4‰–6‰ from the original value of 7.5 ± 0.3‰. From these results, it appears that burned bones would not be useful for studying topics like diet trophic level, terrestrial marine contributions, use of fresh water aquatic ecosystems, or weaning times.

EVALUATING PREHISTORIC CREMATIONS

Before the control results can be applied to prehistoric samples, it is necessary to determine whether prehistoric bones exhibit the same pattern of unaltered $\delta^{13}C$ values in increased $\delta^{15}N$ values. It is possible that diagenetic changes in prehistoric bones could produce different patterns of isotopic composition from modern laboratory controls. In order to evaluate the effect of diagenesis on prehistoric burned bones, we determined the stable isotope ratios of burned human bones from four prehistoric sites in eastern North America, ranging in age from Late Archaic (*ca.* 1500 BP) to Middle Mississippian (*ca.* AD 1400). Four different archaeological components from three sites were sampled. Three of the sites (Klunk, the Middle Mississippian Component at the Yokem site, and the Angel site) produced both burned and unburned bones. A Late Woodland cremation deposit at Yokem produced only burned bones, probably from a burned surface structure or a charnel house. If the control results are applicable to prehistoric samples, we should expect to see a relatively narrow range of $\delta^{13}C$ values and higher $\delta^{15}N$ values in charred samples compared to uncharred ones.

PETE KLUNK MOUND GROUP

The Pete Klunk mound group is located in the lower Illinois River valley. It provided the oldest prehistoric samples used here. The group contains Late Archaic, Middle Woodland, and Late Woodland burials (Perino, 1968). The Middle Woodland burials have received the most attention (Brown, 1981), but the Late Archaic burials are relevant to this project. Mound 7 contained over 35 Late Archaic adult burials. Seven of these were from crematory features and exhibit wide ranges of burning, from slightly blackened to calcined. Burials that were not placed in crematory basins include redeposited partial cremations, and unburned primary and secondary burials. In several cases, femora and other long bones indicate a burning gradient, ranging from apparently unburned on one end to heavily charred on the other. It is, therefore, possible to sample burned and unburned burials from Klunk Mound 7, and also to sample differentially burned portions of a single bone. Seventeen burned and unburned bone fragments from Klunk were analyzed, including three sets of paired samples showing differential burning along long bones (Table 5.1). The results from the Klunk Late Archaic component, dated prior to about 1500 BC, will be especially useful because these burials predate the appearance of maize agriculture. All individuals in the population would have consumed an exclusively C3 diet, and therefore should have stable carbon isotope ratios close to −20.5‰ (Van der Merwe and Vogel, 1978).

TABLE 5.1 Percentage of N and C, Extraction Yields, Stable Nitrogen and Carbon Isotope Ratios, and C/N Ratios of Demineralized Extracts for the Prehistoric Samples

Site	Burial No.	Appearance	N (%)	C (%)	Yield (%)	$\delta^{15}N$ (‰)	$\delta^{13}C$ (‰)	C/N ratio
Klunk	7-32	Uc	2.55	9.32	16.10	11.5	−20.2	3.22
Klunk	7-53	Uc	1.05	5.48	3.80	10.4	−21.0	4.11
Klunk	7-03-6	Uc	2.38	9.07	13.70	11.4	−20.4	3.31
Klunk	7-61	Uc	1.38	6.38	11.10	9.9	−19.7	3.52
Klunk	7-64	Uc	1.76	7.58	4.50	11	−21.1	
Klunk	7-57	Uc	1.20	6.56	6.30	9.1	−20.4	3.77
Klunk	7-01	C	0.72	6.71	5.70	17.3	−23.7	
Klunk	7-01	C	0.82	6.00	2.10	16.7	−23.7	7.12
Klunk	7-01	C	0.68	5.25	1.70	17.2	−23.3	6.45
Klunk	7-01	C	0.73	6.11	7.60	16.7	−23.7	
Klunk	7-01	C	1.08	6.82	9.30	17.9	−25.0	5.53
Klunk	7-01-07	C	1.09	6.28	3.30	9.9	−13.8	3.56
Klunk	7-01	C	0.85	6.19	2.30	15.1	−22.8	6.84
Klunk	7-05	C	0.87	5.92	3.70	17.4	−22.4	5.15
Klunk	7-03-6	C	0.94	6.34	4.50	13	−22.9	8.03
Klunk	7-01	C	1.00	8.20	0.00	15.4	−24.7	9.10
Klunk	7-01	C	0.88	6.67	5.20	15.6	−24.2	7.99
Klunk	7-01	C	1.33	7.91	9.10	17.3	−23.9	
Klunk	7-01	C	0.94	6.44	7.80	15.8	−23.1	5.53
Yokem Md 2	2-10	Uc	0.61	5.04	4.50	9.3	−12.9	3.49
Yokem Md 2	2-11	Uc	1.54	7.00	7.30	10.6	−15.4	3.68
Yokem Md 2	2-10	Uc	1.85	7.56	10.70	9.3	−13.3	3.41
Yokem Md 2	2-17	Uc	2.60	9.91	12.20	10.9	−14.3	4.22
Yokem Md 2	2-15	Uc	2.72	9.87	16.60	10.7	−13.5	3.31

(*Continues*)

TABLE 5.1 *(Continued)*

Site	Burial No.	Appearance	N (%)	C (%)	Yield (%)	$\delta^{15}N$ (‰)	$\delta^{13}C$ (‰)	C/N ratio
Yokem Md 2	2-13	Uc	2.70	9.53	17.60	10.4	−13.6	3.35
Yokem Md 2	2-17	Uc	3.46	11.72	19.90	10.4	−13.5	3.28
Yokem Md 2	2-10	Uc	0.39	4.60	1.40	15.3	−17.4	7.77
Yokem Md 2	2-10	C	0.00	3.19	0.50	12.8	−18.0	9.40
Yokem Md 2	2-11	C	0.00	3.08	0.80	11.8	−17.2	9.96
Yokem Md 2	2-17	C	0.40	4.70	3.00	15.5	−16	4.95
Yokem Md 2	2-10	C	0.50	4.87	0.30	11.6	−16.5	7.11
Yokem Md 2	2-10	C	0.99	6.28	0.40	11.7	−18.4	7.44
Yokem Md 2	2-10	C	1.02	6.20	2.70	15.3	−15.8	4.55
Yokem Md 2	2-10	C	1.03	6.51	4.30	15.0	−17.2	5.25
Angel	W11A-29	Uc	0.79	4.29	4.10	8.1	−9.4	3.65
Angel	W11A-20	Uc	1.62	6.74	10.10	8.2	−10.0	3.54
Angel	W11A-20	Uc	0.37	3.87	0.10	8.5	−9.0	3.51
Angel	W11A-29	C	0.85	5.32	3.50	12.1	−11.6	4.20
Angel	W11A-29	C	1.49	6.67	6.20	13.5	−9.9	4.18
Angel	W11A-20	C	0.78	4.66	4.00	14.3	−11.8	4.49
Yokem LW	Crem 6-7	C	0.51	5.04	8.00	13.9	−21.7	6.93
Yokem LW	Crem 6-7	C	0.89	5.53	4.30	13.8	−21.3	5.67
Yokem LW	Crem 6-7	C	0.74	5.12	5.50	12.0	−22.2	8.19
Yokem LW	Crem 6-7	C	1.26	7.02	3.10	13.2	−21.1	6.39
Yokem LW	Crem 6-7	C	1.34	6.87	8.60	15.3	−21.6	5.13
Yokem LW	Crem 6-7	C	0.81	4.25	5.30	12.9	−21.3	6.83
Yokem LW	Crem 6-7	C	0.24	3.47	1.30	15.3	−21.7	6.94
Yokem LW	Crem 6-7	C	1.05	4.41	7.20	11.1	−20.8	5.90

The expected carbon-stable isotope ratio has been confirmed by Schoeber (1998), who measured $\delta^{13}C$ and $\delta^{15}N$ values for 13 adult burials from the Archaic component at the site. She obtained an average $\delta^{13}C$ of −20.2‰ and an average $\delta^{15}N$ of 9.6‰, providing baseline values.

YOKEM UPPER MISSISSIPPIAN AND LATE WOODLAND

The Yokem site, located in the Mississippi River valley of West-Central Illinois, is a Middle Mississippian site with burials in burned charnel houses accompanied by many unburned burials (Perino, 1971). Three of the 10 mounds at the site represent Middle Mississippian mortuary activities dated between AD 1200 and 1300. Mound 1 supported an unburned charnel house surrounded by noncharnel house burials, a mortuary pattern that has been well-described for this region from the Schild site in the lower Illinois valley (Goldstein, 1980). Mound 2 was a Middle Mississippian platform mound that supported a charnel house that had been burned and then covered with earth. The burned charnel house contained eight burials, of which seven were partially burned with the charnel house. Twelve unburned burials were placed in the mound surrounding the charnel house. Because the burials in the charnel house were partially burned, it has been possible to sample both burned and unburned portions of the seven burned charnel house burials, along with the unburned charnel house burial. The isotope ratios for the charnel house burials can be compared to those from the 12 unburned burials in the same mound. At Yokem, both fleshed and unfleshed burials were in the charnel house when it was burned, providing more comparative data on the effects of burning primary and secondary burials. Rose (2001) measured stable carbon and nitrogen isotope ratios for 18 unburned adult burials from the Yokem site. The average stable isotope ratio of Rose's sample from Yokem ($\delta^{13}C = -14.3‰$, $\delta^{15}N = 10.7$) reflects significant maize consumption.

Seventeen samples were analyzed from the Yokem site (Table 5.1). All the sampled burials were located within a charnel house on Mound 2 that had been burned down. Three burials provided bones that were charred along with ones that did not appear charred, providing an opportunity to assess the effects of heating on isotope ratios of bones from single individuals. The other two burials provided samples that did not appear to have been heated.

YOKEM LATE WOODLAND CREMATORY

Along with Middle Mississippian burials, the Yokem site also contained several Late Woodland crematory deposits that were located between the Middle Mississippian mounds at the site. Little is known about these features because the results of their excavation have not been published. They contained only cremated and partially cremated bones, and probably date to sometime between AD 500 and 1000.

ANGEL UPPER MISSISSIPPIAN

The Angel site was occupied between AD 1300 and 1450, slightly later in time than Yokem, and was also inhabited by Middle Mississippian maize agriculturalists (Black, 1967). Some burials were partially cremated when fire was employed as part of the mortuary ritual, but at a low intensity, resulting in partially charred skeletons. At Angel site, both burned and unburned bones are available from two partially cremated skeletons (Table 5.1). In addition, 47 stable carbon and nitrogen isotope ratios have already been obtained from adults at the Angel site (Schurr, 1992; Schurr and Schoeninger, 1995), providing abundant baseline data on expected human isotope ratios at this site (mean $\delta^{13}C = -9.0‰$, $\delta^{15}N = 8.1‰$). The Angel stable carbon isotopes reflect the highest level of maize consumption for the sampled sites.

RESULTS

Unfortunately, the most reliable way of estimating thermal histories of prehistoric bones is the examination of the physical appearance. We can, therefore, examine only the isotope ratios of prehistoric bone samples to see if they are consistent with the behavior of the modern standards. All the three sites produce the expected pattern of increased $\delta^{15}N$ values when charred and uncharred samples were compared. As shown in Figures 5.4–5.6, charred bones have enriched $\delta^{15}N$ values compared to their uncharred counterparts at each site. The maximum enrichment is $\sim 5‰$, similar to the control bones. However, in contrast to the modern controls, $\delta^{13}C$ values of charred bones are slightly reduced compared to their unburned counterparts. The percentage

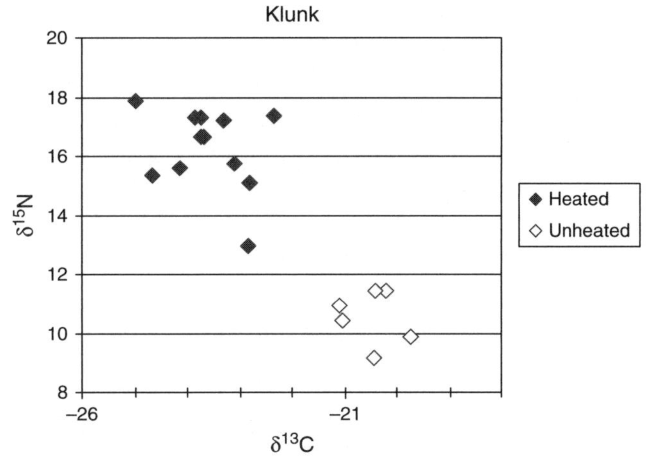

FIGURE 5.4 Stable isotope ratios from the Klunk Late Archaic burials.

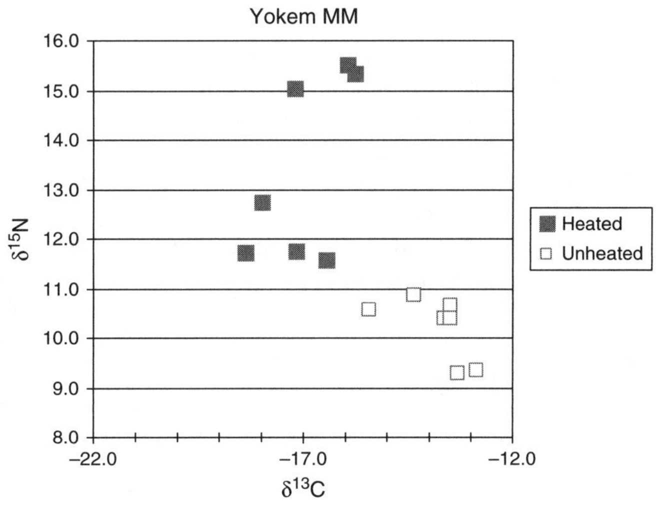

FIGURE 5.5 Stable isotope ratios from the Yokem Middle Mississippian burials.

of nitrogen in the bones are also generally reduced with charring, except at Angel where there was little difference between charred and uncharred bones. The percentage of carbon showed no consistent pattern.

Figure 5.4 shows data from burials in Klunk Mound 7, which contained burned and unburned Late Archaic burials, from the period around 1500 BC. These data cluster more than the others, with a sharp distinction between charred and uncharred bones in $\delta^{15}N$ and the expected shift of about 5‰. Stable carbon isotope ratios for visibly charred bones are reduced by 2.4‰ compared to those that appeared uncharred or came from uncremated burials, a difference that is not quite significant at a confidence level of 0.05 but that is probably real.

Figure 5.5 shows isotope ratio data from bones of Mound 2 at the Yokem site. As one sees, the uncharred bones of this sample have very similar $\delta^{15}N$

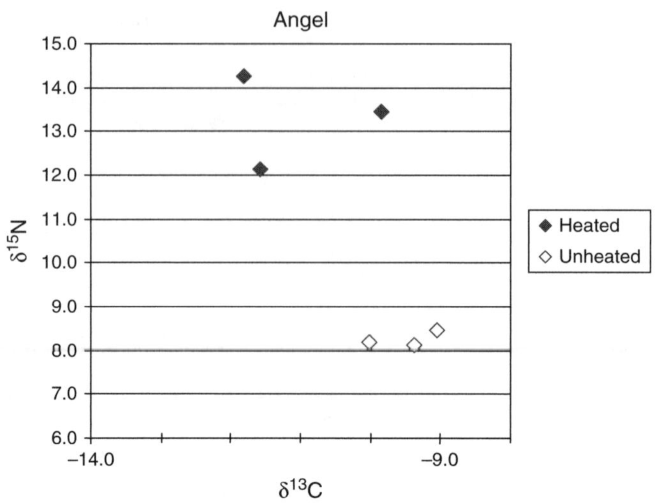

FIGURE 5.6 Stable isotope ratios from the Angel Middle Mississippian burials.

values. The mean stable carbon and nitrogen isotope ratios for the seven apparently uncharred samples from Mound 2 are very similar to the mean values of $-14.3‰$ and $10.7‰$ that Rose (2001) obtained from noncharnal house burials at the site. Once again, charred bones have $\delta^{13}C$ values that are depleted and $\delta^{15}N$ values that are enriched in the heavy isotopes relative to the uncharred bones.

The results from Angel (Figure 5.6) are very similar to those from Yokem. The uncharred samples produced results very similar to the overall average for 47 burials from the site (Schurr and Schoeninger, 1995). As before, nitrogen isotope ratios increased and carbon isotope ratios decreased for the charred samples.

Figure 5.7 shows isotope ratio data from the Late Woodland cremation feature at the Yokem site. This feature produced only burned bones. As one

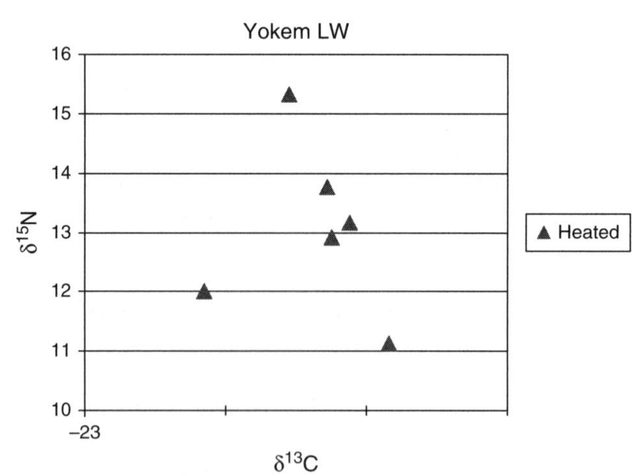

FIGURE 5.7 Stable isotope ratios from the Yokem Late Woodland burials.

sees, the $\delta^{15}N$ values are elevated and vary widely, but the $\delta^{13}C$ values fall, mostly, in a narrow range. No unheated bones were available from the Yokem Late Woodland cremation deposit, so we cannot compare heated and unheated bones. However, within the heated fragments, enrichment in ^{15}N correlates well with the degree of heating obtained from the overall appearance. The sample with the highest $\delta^{15}N$ value was the only calcined sample, and the sample with the lowest value appeared to have been only lightly heated to produce a slight browning.

One unexpected result that contradicts with those obtained for the control samples was that the charred samples from sites with uncharred controls all produced lower $\delta^{13}C$ values, perhaps as a result of diagenesis. The depletion in ^{13}C is especially pronounced at Klunk, but is present in the carbonized samples from all three sites.

DISCUSSION

Our data on heated modern bone controls (Figures 5.2 and 5.3) clearly show that $\delta^{13}C$ values do not change over the entire range of heating temperatures, and that $\delta^{15}N$ values increase by approximately 5‰ over a temperature range centered around 300°C. The lack of an easily identifiable dependence on heating time (except perhaps for samples heated at around 300°C) disagrees with earlier work (DeNiro *et al.*, 1985). In that work, shifts in both $\delta^{13}C$ values (of about 2.5‰) and $\delta^{15}N$ values (of as much as 6‰, but the shift profile with temperature was complex) were found for whole goat tali that were heated to maximum temperatures as low as 170°C for a total of 3 h (the heating temperature was not constant, but rose to a maximum after 1 h, then fell in the next 2 h). In a separate series of experiments in which powdered pig bone was heated at 200°C for various times, some of them quite long, small shifts in isotope ratios were seen for heating times as short as 2 h, and after 6 h, larger shifts were seen, but they varied widely from one data point to another.

It is not clear to us why our results differ from those of DeNiro *et al.*, but their experimental procedure was quite different from ours, and they used whole bone, whereas our samples (both controls and prehistoric) were almost exclusively cortical bones. Our samples were heated in an environment that allowed free exposure to air, while many of their samples were heated under conditions that restricted access to oxygen because they wanted to investigate the effects of conditions expected during cooking. Different extraction procedures may also play a role in the differing results, as they used a method that includes a rapid demineralization of the bone followed by hydrolysis in weak acid to extract the organic phase.

Our data on prehistoric bones support, in general, the idea that the results we have for modern bones can be extended to prehistoric bones, but only to some extent. We have no definite explanation for the shift in that is observed $\delta^{13}C$ data. It may represent a difference in the diagenetic trajectories of heated and unheated bones, but examination of this proposal will require more data. It is interesting to note that for the three sites that provided both charred and uncharred samples, the shift $\delta^{13}C$ is more pronounced with age, suggesting that diagenetic changes over time are indeed playing a role in the difference observed between the control samples and the prehistoric ones. Another possibility is that the samples are contaminated with humic acid, a C3 contaminant that would reduce stable carbon isotope ratios. For well-preserved samples, it is a common practice to soak the demineralized extracts in dilute sodium hydroxide solution for 24 h to remove any humic acid contaminants. This procedure is not effective for charred extracts because the extracts themselves are very soluble in basic solutions, probably because they contain many fragmented amino acids with free carboxyl groups. It would, therefore, be a difficult analytical problem to separate humic acid contaminants from charred extracts produced by demineralization. This theory is somewhat supported by the higher C/N ratios of the charred demineralized extracts compared to their uncharred counterparts (Table 5.1).

Clearly, for the extension of our results to prehistoric bones to be of any use, a method is needed to estimate the temperature to which prehistoric bones have been heated, or better, a way of estimating some measure of time at temperature. We have been exploring the use of electron spin resonance (ESR) for this purpose (Hayes and Schurr, 2002) and, more recently, of Fourier transform infrared spectroscopy (FTIR). However, it is still possible to obtain some generalized information about prehistoric diet from charred bones. For example, the $\delta^{13}C$ values of the Yokem Late Woodland samples are intermediate between those of the charred samples from the Klunk Late Archaic and Yokem Upper Mississippian, suggesting that at least some maize was consumed by the Late Woodland inhabitants of Yokem.

CONCLUSIONS

Under the set of procedures used here, the average carbon isotope ratios of heated bones do not change significantly, but nitrogen isotope ratios increase by up to approximately 6‰ for bones heated to 300°C and above. Similar results were obtained from bones from prehistoric sites from eastern North America, except for an age-dependent decrease in stable carbon isotope ratios for the prehistoric bones. Further analytical work will be needed before it is possible to use the stable isotopes of charred bones to precisely reconstruct prehistoric diets. However, the results from paired samples show that bone fragments that appear to be uncharred can provide reliable isotopic reconstructions, even when the fragments are in close proximity to heavily charred areas of the same bone. As many cremations are partial and show highly variable degrees of thermal alternation, this suggests that stable isotopes can be used to reconstruct paleodiets of cremated individuals if care is taken to select samples that appear to have escaped thermal alteration.

ACKNOWLEDGMENTS

This project is supported by a grant from the NSF Archaeometry Program Grant 0313592. Analytical facilities were provided by the Center for Environmental Science, University of Notre Dame. We are especially grateful to John Yellen for his advice on project design, and to Patricia

Maurice and Dennis Birdsell for access and assistance with the CEST facilities. Most of the lab work were performed by our undergraduate research team: Beth Bollwerk, Maria Campos, Laura Hammond, Melanie Irvine, Maggie Jordan, and Meghan Lawless. Many of the controls were prepared by Sean Bergin (ND 2003).

REFERENCES

Arts, N. (1987). Mesolithic settlement area with a human cremation grave at Oirschot V, municipality of Best, the Netherlands. *Helinium* **15–16**(1), 98–100.

Asch, D. (1976). *The Middle Woodland Populations of the Lower Illinois River Valley: A Study in Paleodemographic Methods*. Scientific Papers No. 1. Northwestern Archeological Program, Evanston, IL.

Baby, R.S. (1954). Hopewell cremation practices. *Ohio Historical Society Papers in Archaeology* **1**, 1–7.

Barber, P.T. (1990). Cremation. *Journal of Indo-European Studies* **18**(3–4), 379–388.

Black, G.A. (1967). *Angel Site: An Archaeological, Historical, and Ethnological Study*. Indiana Historical Society, Indianapolis, IN.

Bowler, J.M., Johnston, H., Olley, J.M., *et al.* (2003). New ages for human occupation and climatic change at Lake Mungo, Australia. *Nature* **421**, 837–840.

Brown, J.A. (1981). The search for rank in prehistoric burials. In *The Archaeology of Death* (R. Chapman, I. Kinnes, and K. Ransborg, eds.), pp. 25–52. Cambridge University Press, Cambridge.

Buikstra, J.E. and Goldstein, L. (1973). *The Perrins Ledge Crematory*. Reports of Investigations No. 28. Illinois State Museum, Springfield, IL.

Chapman, J. (1973). *The Icehouse Bottom Site 40MR23*. Department of Anthropology, University of Tennessee, Knoxville, TN.

Coles, J.M. and Harding, A.F. (1979). *The Bronze Age in Europe: An Introduction to the Prehistory of Europe c. 2000–700 BC*. St. Martin's Press, New York.

DeNiro, M.J., Schoeninger, M.J., and Hastorf, C.A. (1985). Effect of heating on the stable carbon and nitrogen isotope ratios of bone collagen. *Journal of Archaeological Science* **12**, 1–8.

Goldstein, L.G. (1980). *Mississippian Mortuary Practices: A Case Study of two Cemeteries in the Lower Illinois Valley*. Scientific Papers No. 4. Northwestern University Archaeological Program, Evanston, IL.

Hayes, R.G. and Schurr, M.R. (2002). Electron spin resonance studies to explore the thermal history of archaeological objects. In *Archaeological Chemistry. Materials, Methods and Meaning* (K.A. Jakes, ed.), pp. 151–168. American Chemical Society, Washington, DC.

Mason, R.J. (1981). *Great Lakes Archaeology*. Academic Press, New York.

McCutcheon, P.T. (1992). Burned archaeological bone. In *Deciphering a Shell Midden* (J.K. Stein, ed.), pp. 347–368. Academic Press, New York.

McKinley, J.I. (1994). Bone fragment size in British cremation burials and its implications for pyre technology and ritual. *Journal of Archaeological Science* **21**, 339–342.

Moore, K.M., Murray, M.L., and Schoeninger, M.J. (1989). Dietary reconstruction from bones treated with preservatives. *Journal of Archaeological Science* **16**, 437–446.

Perino, G. (1968). The Klunk Mound Group, Calhoun County, Illinois: the Archaic and Hopewell occupations. *Illinois Archaeological Survey Bulletin No. 6*, 9–128.

Perino, G. (1971). The Yokem site, Pike County, Illinois. *Illinois Archaeological Survey Bulletin No. 8*, 149–186.

Phillips, P. (1980). *The Prehistory of Europe*. Indiana University Press, Bloomington, IN.

Reinhard, K.J. and Fink, T.M. (1994). Cremation in southwestern North America: aspects of taphonomy that affect pathological analysis. *Journal of Archaeological Science* **21**, 597–605.

Rose, F. (2001). *A Fresh Perspective: Isotopic Evidence for Subsistence in the Mississippi River Valley of West-central Illinois, and Comparison with the Illinois River Valley*. Unpublished Ph.D. dissertation, Department of Anthropology, Rutgers University.

Schoeber, T.M. (1998). *Reinvestigation of Maize Introduction in West-Central Illinois*. Paper-in-lieu-of-thesis, Department of Anthropology, University of Illinois.

Schurr, M.R. (1987). A "Partial Cremation" from the Angel Site, Vanderburgh County, Indiana. *Proceedings of the Indiana Academy of Science* **96**, 91–93.

Schurr, M.R. (1992). Mortuary and isotopic variation in a Middle Mississippian population. *American Antiquity* **54**, 300–320.

Schurr, M.R. and Schoeninger, M.J. (1995). Associations between agricultural intensification and social complexity: an example from the prehistoric Ohio Valley. *Journal of Anthropological Archaeology* **14**, 315–339.

Shahack-Gross, R. and Bar-Yosef, O. (1997). Black-coloured bones in Hayonim Cave, Israel: differentiating between burning and oxide staining. *Journal of Archaeological Science* **24**, 439–446.

Sillen, A. and Hoering, T. (1993). Chemical characterization of burnt bones from Swartkrans. In *Swartkrans: A Cave's Chronicle of Early Man* (C.K. Brain, ed.), pp. 243–249. Transvaal Museum Monograph, No. 8. Transvaal Museum, Pretoria.

Taylor, R.E., Hare, P.E., and White, T.D. (1995). Geochemical criteria for the thermal alteration of bone. *Journal of Archaeological Science* **22**, 115–119.

Van der Merwe, N.J. and Vogel, J.C. (1978). 13C content of human collagen as a measure of prehistoric diet in woodland North America. *Nature* **276**, 815–816.

Whittle, A.W.R. (1985). *Neolithic Europe: A Survey*. University of Cambridge Press, Cambridge.

6

Bone Color as an Interpretive Tool of the Depositional History of Archaeological Cremains

Joanne B. Devlin Ph.D. and Nicholas P. Herrmann Ph.D

Department of Anthropology, University of Tennessee, Knoxville, TN

INTRODUCTION

Cremains present a difficult interpretative challenge to forensic anthropologists and bioarchaeologists. Often standard osteological measures and methods fail to accurately capture the variation and complexity of these samples. Color change, shrinkage, and fragmentation, in addition to fracturing and warping, are confounding components of thermally altered remains. These characteristics are observed in varying degrees resulting from the modification or destruction of organic and inorganic components of bone. At the most basic level, burned bone features result from a combination of heating temperature and duration of exposure. Although a specific and clearly defined relationship between thermal-induced changes and the intensity and duration of heating is not currently established, a general relationship and basic pattern is accepted. This enables the assessment of observable features toward a determination of the circumstances that produced the particular burned bone assemblage. The present study of bone color, a component of a larger GIS-based examination of the prehistoric Walker Noe collection, aids in the reconstruction and interpretation of the burial deposit.

BURNED BONE RESEARCH

Early examinations of burned bone were motivated by attempts to interpret skeletal collections from archaeologically derived cremation sites (Webb and Snow, 1945; Baby, 1954; Wells, 1960; Binford, 1963). Baby (1954) recognized a three-way classification of burned bone incorporating modification of element color and surface morphology: labeling bone as completely incinerated, incompletely incinerated or non-incinerated. This observable variation within Hopewellian mortuary collections was attributed to several factors including duration of exposure and proximity of bone to the fire (Baby, 1954). More formal examination of cremation was initiated by Wells (1960) who documented the effects of cremation and suggested that the degree of distortion was in part associated with firing temperature. Binford (1963) extended the study of cremations in his analysis of Late Archaic burial sites in Michigan, noting that it is possible to assess the preincineration condition by differentiating between degree and patterns of warping and fracturing. Binford (1963) noted that the resultant condition and appearance of bone is attributable to not only the duration and intensity of the fire but also the location and condition of the material prior to heat exposure. Although many of these early studies sought to understand aspects of site formation and focused upon variation in surface condition as potential indicators of preincineration condition, subsequent considerations focused on assessing the condition of bone as a predictor of heating parameters.

The recognition of varied surface conditions as an indicator of the degree of heat exposure has been a focal point of both laboratory and natural context examinations (Thurman and Wilmore, 1981; Shipman *et al.*, 1984; Gilchrist and Mytum, 1986; Buikstra and Swegle, 1989; Spenneman and Colley, 1989; McCutcheon, 1992; Nicholson, 1993). Research-driven examinations of the last several decades have sought to examine particulars of the impact of heat upon bone by investigating the association between exposure temperature and surface color (e.g., Shipman *et al.*, 1984) and the microscopic and structural impacts of heat upon skeletal material (e.g., Herrmann, 1977; Shipman *et al.*, 1984; McCutcheon, 1992; Nicholson, 1993; Thompson, 2005).

Pioneered by the work of Shipman *et al.* (1984), the assessment of surface color has been part of the primary protocol in the examination and interpretation of burned bone assemblages. Subsequently, the association between particular surface colors and exposure temperatures has become commonplace. However, Shipman *et al.* (1984: 320) state that the reliance upon color as an indicator of exposure temperature is 'an essentially imprecise criterion both because of individual differences in the ability to perceive fine color distinctions and because burnt bones may change color if they are buried.' Nonetheless, they note that changes in the hue, chroma, and value of surface color are attributable to changes in the chemical composition that result from exposure to heat and that recognition of color can be informative as to the range of temperatures to which a bone has been exposed. The duration of exposure to a heat source is an additional force that can impact the condition and appearance of bone as addressed experimentally (see, e.g., Degraff, 1961; Bennett, 1998; Pope, 2007; Walker *et al.*, chapter 7, this volume). Experimental work and collection analyses demonstrate the potential utility of regarding surface color as a basic indicator of the taphonomic processes to which bone has been exposed.

The observable changes in surface color and element integrity are outward reflections of the modification of the chemical composition of bone in the presence of heat. Bone, as an organic material, evidences predictable changes when exposed to heat and displays a basic chronology of changes. At the most basic level, heating leads to the alteration of the chemical compounds and subsequent dehydration of the tissue resulting in the development of a brittle material. Reflected in the observable modification of the overall structure, heated bone is fragile and exhibits warping, shrinkage, fracturing, and color change. Researchers have reported identifiable stages in the process of cremation (Bonucci and Graziani, 1975; see Correia, 1997; Thompson, 2005 for review). The first stage labeled dehydration is characterized by the breaking of hydroxyl bonds resulting in the loss of water. As reviewed by Correia (1997) these events are associated with exposure to temperatures up to 600°C. Stage 2, decomposition, is characterized by the removal of the organic components, associated with temperatures ranging from 500 to 800°C. Exposure to temperatures ranging from 700 to 1100°C causes the loss of carbonates, which characterizes the third stage, referred to as inversion. The final stage, referred to as fusion, is associated with temperatures in excess of 1600°C and is characterized by melting of the crystals (see Correia, 1997).

Recently, DeHaan and Nurbakhsh (2001) suggested that the observable features of burned bone are not simply attributable to high exposure temperatures, but instead reflect a complex interaction of factors. In particular, they note the potential of blood, marrow, moisture, and fat in specimens which can influence heating conditions. Obviously, the preincineration condition of bone dictates the varying levels to which these components are present and subsequently influences the postincineration condition. Further DeHaan and Nurbakhsh (2001) note that the variation in flame temperatures that occurs naturally must be appreciated and suggest that such tremendous fluctuation would have a less destructive effect on bone that is dry.

BONE COLOR

Anthropologists have incorporated the use of surface color into the description of bone affected by heat dating back to the early reports of Baby (1954). Described by Bonucci and Graziani (1975), progression through heating stages (e.g., dehydration, decomposition, and inversion) is associated with outwardly apparent modification, including change in the surface color of bone. With increasing exposure to heat, bone progresses through a sequence of colors, from unburned tan to shades of dark brown to black, progressing to blue and gray and finally to white. Bonucci and Graziani (1975: 531) claim that 'color itself can be used to some extent for deducing the approximate value of the temperature' correlating ochraceous and brownish colors with exposure to 200–300°C (392–572°F), black with temperatures of 300–350°C (572–662°F), grayish with temperatures of 550–600°C (1022–1112°F), and white for heating in excess of 650°C (1202°F). Black color and an increase in the fragility of bone involve combustion and progressive incineration of the organic materials of collagen and carbon. Continued exposure to heat will impact the crystalline structures and produce gray and white colors and overall shrinkage of the elements.

Substantiated several decades later in a systematic manner through the work of Shipman *et al.* (1984), surface appearance was referenced by using the Munsell Soil color charts. Shipman *et al.* (1984) standardized the description of surface colors on experimentally burned bone recognizing five distinct color stages associated with specific increases in temperature. Each stage is characterized by variation in hue, value, and chroma and is marked by dominant surface colors and multiple secondary colors. Shipman associated the dominant colors of pale yellow and very pale brown with exposure to temperatures less than 285°C (545°F). Temperatures from 285 to 525°C (545–977°F) were associated with dominant colors of pink and black and secondary colors of very dark grayish brown and brown; 'common colors are reddish brown, very dark gray-brown, neutral dark gray and reddish yellow' (Shipman *et al.*, 1984: 313). Shipman *et al.* (1984: 313) characterize stage three, delineated by temperatures up to 645°C (1193°F), by surface colors of 'neutral black, with medium blue and some reddish-yellow appearing' and identify Munsell colors of light gray and secondary colors of brown and light brownish gray. Bone heated to 940°C (1724°F) exhibits neutral white and light blue gray colors. Elements heated to temperatures in excess of 940°C were found to exhibit neutral white and medium gray colors.

McCutcheon (1992) and Nicholson (1993) echoed the association between heating temperature and surface colors, though variations in particular temperatures were noted. McCutcheon (1992) combined the assessment of surface color with microscopic criteria to recognize a three-level system for estimating exposure temperatures. He observed pale brown to black in specimens heated to 340°C (644°F), light brownish gray in elements heated to 600°C (1112°F), and white color-dominating bone heated in excess of 650°C (1202°F) (McCutcheon, 1992). Undoubtedly, surface color cannot be regarded as a sole indicator of the maximum exposure temperature; however. color can be interpreted in terms of its position along the color continuum as evidence of the extent of thermal exposure.

COLOR MODELS

Traditionally, anthropologists have assessed bone surface colors using Munsell Soil Color Charts. Although this approach requires little equipment and is easy to use, Munsell relies upon human perceived assessment of three dimensions of color; the three components include hue, chroma, and value. Hue refers to the apparent similarity with red, yellow, green, and blue (or a combination of two of them) components of a color. The symbol N designates a perceived color that is devoid of hue. The value component indicates the lightness while chroma suggests how rich the color is. The three components are incorporated into a Munsell score that consists of three symbols such that the first reflects hue (a number and letter designation), the second indicates value (ranging from 0, black to 1, white). The third component is chroma that reflects the strength of the color commonly scored from 0 (neutral) to 8 (strong) values. This overall score is supplemented with a text notation to most accurately describe a surface color (Munsell Color Company, 2000). Assignment of a Munsell color is dependent upon establishing a match to a color contained in the charts. This technique is susceptible to variation in lighting that can affect color interpretation and produce interobserver error.

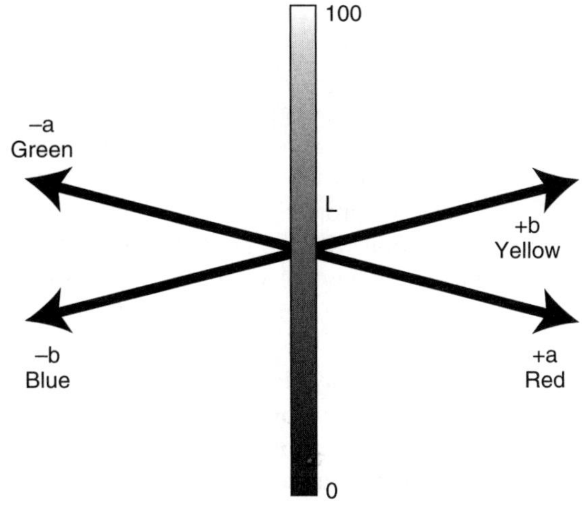

FIGURE 6.1 L*a*b* color space.

Further, comparative assessments of colors are prohibited as the Munsell representation of a color does not readily inform of the actual color, nor are differences in scores on a scale.

The present investigation incorporates the use of a standardized approach to the collection of bone surface color data relying upon the use of the CIE L*a*b (CIELAB) uniform color space (see Fairchild, 2005 for review). This system falls under guidelines established by the International Commission on Illumination (CIE) (see Ebner, 2007 for review). The CIELAB color system considers all colors to be combinations of surface and illuminant color. Colors are identified by three coordinates as illustrated in Figure 6.1. The L* axis reflects the lightness (values range from 0 to 100), while the a* axis reflects the red/green colors (positive values to negative values, red to green, respectively), and the b* axis indicates the yellow/blue colors (positive values to negative values, yellow to blue, respectively). As noted by Robertson (1990) in Fairchild (2007), the L* scale is comparable to the Munsell value scale. In the present examination, L*a*b color is recorded using a measurement device thereby allowing for precise and consistent color recognition. In addition, the three color dimensions, each of which are linear, enable mathematical modeling as colors can be plotted in a three-dimensional space and compared. Though both Munsell and CIELAB systems are three-dimensional, the latter enables specification and measurement of differences in color and as such permits statistical consideration of color data.

WALKER-NOE 15GD56

Cremated remains examined in this study were recovered from the Walker-Noe site (15GD56), a small Middle Woodland period mound (120 BC–200 AD; Pollack *et al.*, 2005) situated in the central Bluegrass Region of Kentucky. Fieldwork conducted during the fall of 2000 yielded over 18 kg of charred and calcined human bones. Ceramic and lithic artifacts associate activity at the mound with the Adena culture. Excavations at the site revealed evidence of in situ cremations indicated by substantial burned soils and burned black walnut suggesting the use of the mound as a crematory (Sharp *et al.*, 2003, Pollack *et al.*, 2005; Herrmann *et al.*, 2007). Excavation of the mound yielded more than 61 000 pieces of debitage, primarily Boyle chert. Several recovered projectile points demonstrate attributes that indicate contextual association with the cremation practices at the mound. Further, six points do not exhibit any use-wear and two additional points demonstrate heat alteration, suggesting their preparation and use as ceremonial objects during cremation activities.

Initial excavation involved the removal of the plow zone that yielded a 1.25 m^2 area of burned red clay loam to a depth of 5 cm (see Figure 6.2). This central region is surrounded by a region of dark reddish brown clay loam ranging in depth from 5–10 cm. A third region of dark brown silty clay loam expanded the area. Concentrations of cremated bone were identified directly above and around the central burned area. In addition, aggregations of burned bone were present in regions peripheral to the central burned zone.

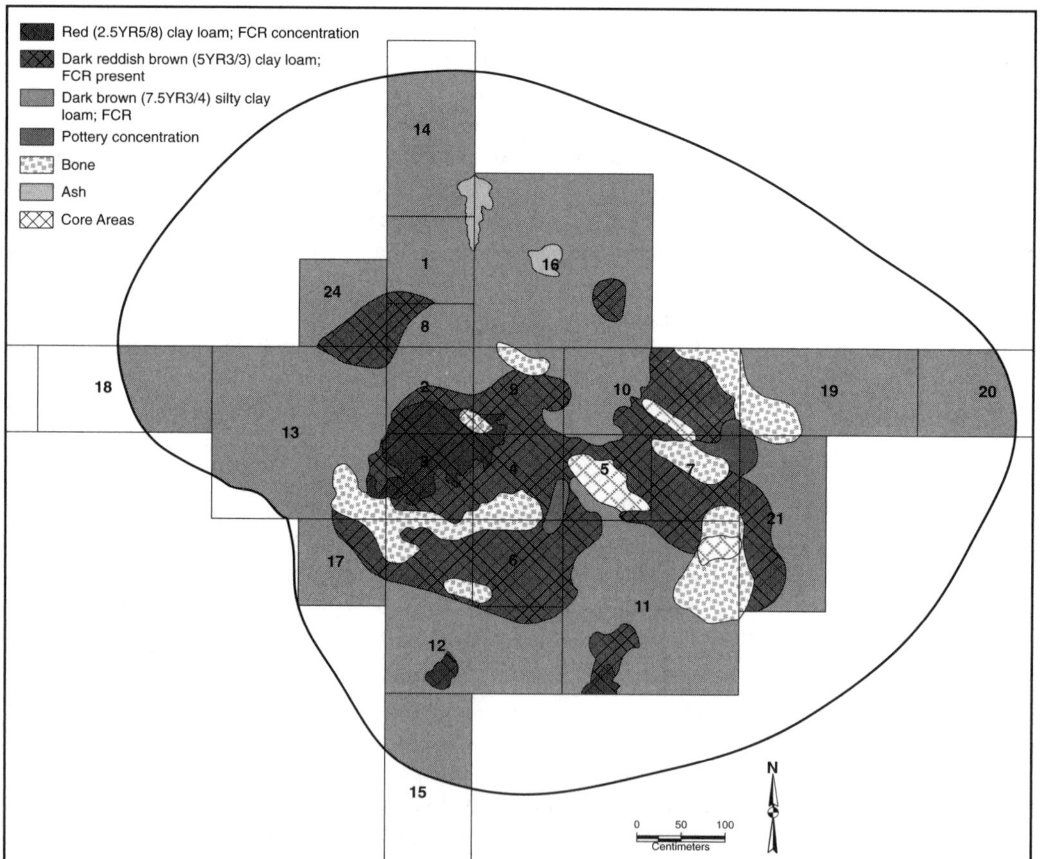

FIGURE 6.2 Walker-Noe site plan map (modified from Pollack *et al.* 2005). (see Plate 19)

All skeletal material was located in association with wood charcoal, ash, and burned soil. Excavation indicates that a central region served as the crematory with remains deposited in adjacent regions surrounding the central zone. The analyses of skeletal material were directed at illuminating particulars of crematory activities (i.e., number of individuals present, differential burning, and whether the site was used repeatedly).

WALKER-NOE 15GD56 CREMAINS

The bone fragments examined in this study represent the commingled remains of more than 20 individuals (Herrmann *et al.*, 2005; Herrmann *et al.*, 2007; Herrmann and Devlin, 2008; Bennett-Devlin *et al.*, 2006a). The assemblage is characterized by extreme fragmentation and a high degree of thermal alteration. The majority of fragments are less than 4 cm in diameter, with numerous specimens significantly smaller in size. The condition of the skeletal material from the Walker-Noe site reflects a thorough incineration process with all bone surfaces demonstrating evidence of advanced heat exposure. Surface colors associated with incomplete combustion of the organic components (i.e., browns and blacks) were not observed. All specimens exhibit evidence of calcination. In addition, fragments display extreme shrinkage and moderate degrees of warping, though the latter does vary across the assemblage. On the basis of the comparison to average mandibular measurements for the Kentucky Adena as reported by Webb and Snow (1945), shrinkage exceeds 13% (Bennett-Devlin *et al.*, 2006a,b). Surface cracking and fracturing is apparent on the majority of specimens, both endocranial and extocranial surfaces.

Bone fragments were initially sorted into skeletal division categories of cranial, appendicular, axial, and unidentifiable (UNID) when possible fragments were identified to a particular element. Fragments were macroscopically assessed in terms of surface colors, level of distortion, apparent degree of shrinkage, and overall fracture and cracking patterns (see Herrmann *et al.*, 2005; Bennett-Devlin *et al.*, 2006a,b; Herrmann *et al.*, 2007, Herrmann and Devlin, in press for additional analysis and discussion of the Walker-Noe cremains). Identifiable fragments were digitized using a modified version of the *BoneEntryGIS* software for ArcView 3.x (Marean *et al.*, 2001). Utilizing the software extension, skeletal elements were recorded by location, fracture pattern, external and internal color, and presence of other traditional human skeletal characteristics (pathology and discrete variants). Figure 6.3 illustrates the minimum number of elements (MNE) summation using the *BoneEntryGIS* application for the maxillae recovered from Walker-Noe. The software is unique in that it provides localized MNE based upon fragment overlap. Additionally, a critical aspect of the Walker-Noe analysis (or the examination of any crematory assemblage) must include assessment of heat-generated attributes such as shrinkage and bone color. In the present study, supplementing our GIS-based fragment analysis, bone specimens were subjected to color assessment and interpretation by skeletal division and archaeological context.

Surface colors were assessed on all specimens using an X-rite CA 22 spectrophotometer. This handheld instrument systematizes color evaluation by measuring areas on the sample greater than 1/4 in. Color data were sent

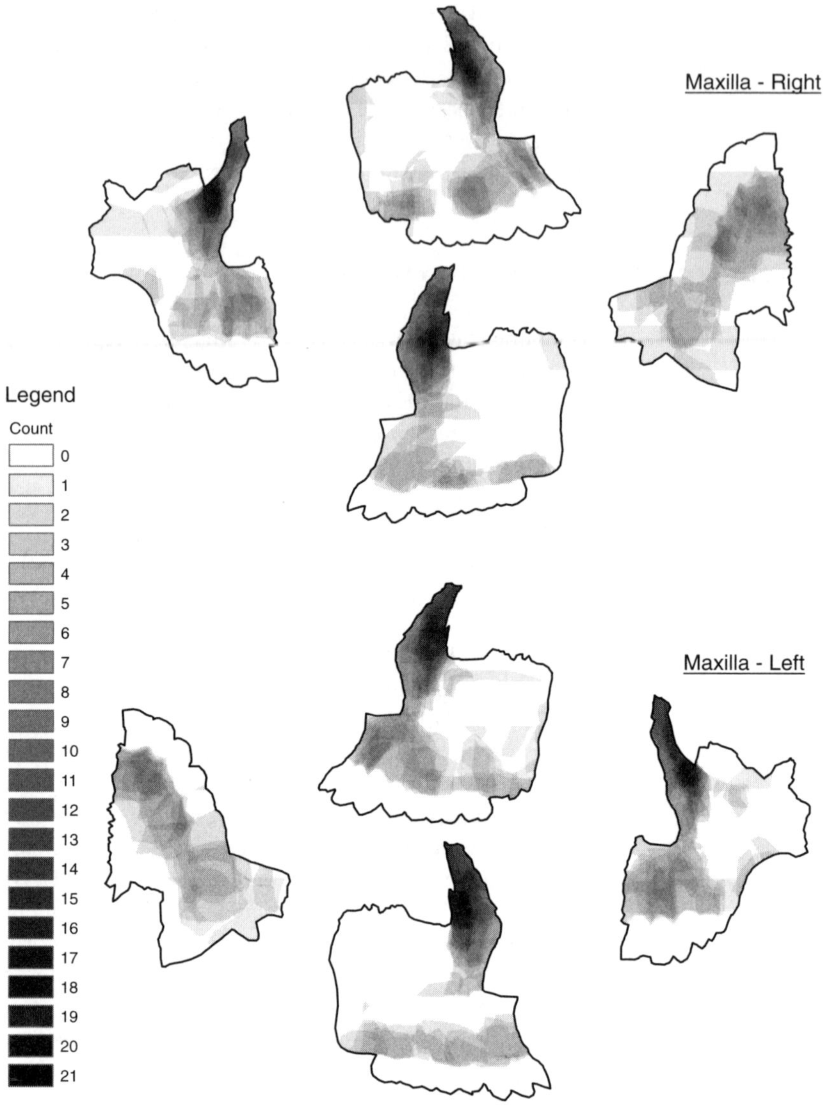

FIGURE 6.3 MNE Calculation from composite grid values.

to a host computer running Matchrite Color Designer software. Data were collected in CIELAB colors. The free software utility Munsell Conversion Program (version 7.0.1 available at http://livingstonmanor.net/Munsell/File Downloads.htm) allows the translation of Munsell colors into LAB colors. This enables comparison of experimentally derived bone colors identified by researchers (e.g., Shipman *et al.*, 1984; McCutcheon, 1992) with Walker-Noe specimens.

A total of 3843 fragments were assessed for primary surface color. See Table 6.1 for breakdown by site area and skeletal division. Analysis was performed on a total of 2589 fragments collected from the core area and 1254 fragments recovered from the peripheral units at the site. A total of

TABLE 6.1 Mean L*a*b* Color Values by Site Area and Skeletal Division

	L	(std)	a	(std)	b	(std)	n
Site area							
Core	58.02	(11.47)	4.94	(1.66)	16.31	(3.94)	2589
Periphery	58.69	(11.30)	4.62	(1.68)	15.47	(4.05)	1254
Skeletal division							
Appendicular	56.68	(10.56)	4.76	(1.64)	15.70	(3.89)	894
Axial	46.62	(10.86)	5.34	(1.26)	15.24	(3.05)	374
Cranial	64.00	(9.28)	4.72	(1.72)	17.10	(3.78)	1195
UNID	57.41	(10.80)	4.84	(1.74)	15.53	(4.27)	1380

1195 (31% of all fragments) cranial fragments were assessed; approximately 44% of these (516) were identifiable to specific element and further to exact anatomical location. Of the identifiable cranial fragments, 389 were recovered from the core of the site while 127 were collected from peripheral excavation units. Of the remaining fragments assessed for surface color, 894 (23% of total sample) were identified as appendicular, 374 (10%) were classified as axial elements, and 1380 (36%) were categorized as UNID. Although both external and internal colors were recorded, the focus of this study is upon the interpretation of ectocranial and external primary color for each fragment with respect to depositional location and skeletal division.

RESULTS

The three coordinate components of surface colors (L*, a*, b*, respectively) were considered both in terms of site location and skeletal division. Figure 6.4 illustrates the variation by site context for density values of each of the three color components. A comparison of the site areas in terms of the three color components demonstrates that fragments from both the core and the periphery of the crematory exhibit similar color distributions for L*, a*, and b* axes (see also Table 6.1). The L* values show minimal, nonsignificant variation in lightness (whiteness) of the fragments between areas ($t = 1.71$, d.f. = 3841, $p = 0.09$). Although the means are not significantly different at the 0.05 level, the L* plot demonstrates distributional variation in modes between areas with a higher peak density for fragments from the periphery of the site. Fragments exhibit much less variation across the sample in terms of a* and b* color axes compared to the L* axis. The means for a* values are significantly different between the core fragments and the periphery fragments ($t = 5.581$, d.f. = 3841, $p = 0$), with the higher mean value for core fragments indicative of more red coloration for these specimens. In addition, significant differences exist between b* values based on the site location ($t = 6.1403$, d.f. = 3841, $p = 0$). The higher mean value for core fragments is associated with colors that are more yellow in appearance.

Density values for the three color axes plotted by skeletal division are illustrated in Figure 6.5 and represented in Table 6.1. The L* axis values are significantly different between the four skeletal divisions. Table 6.2 presents

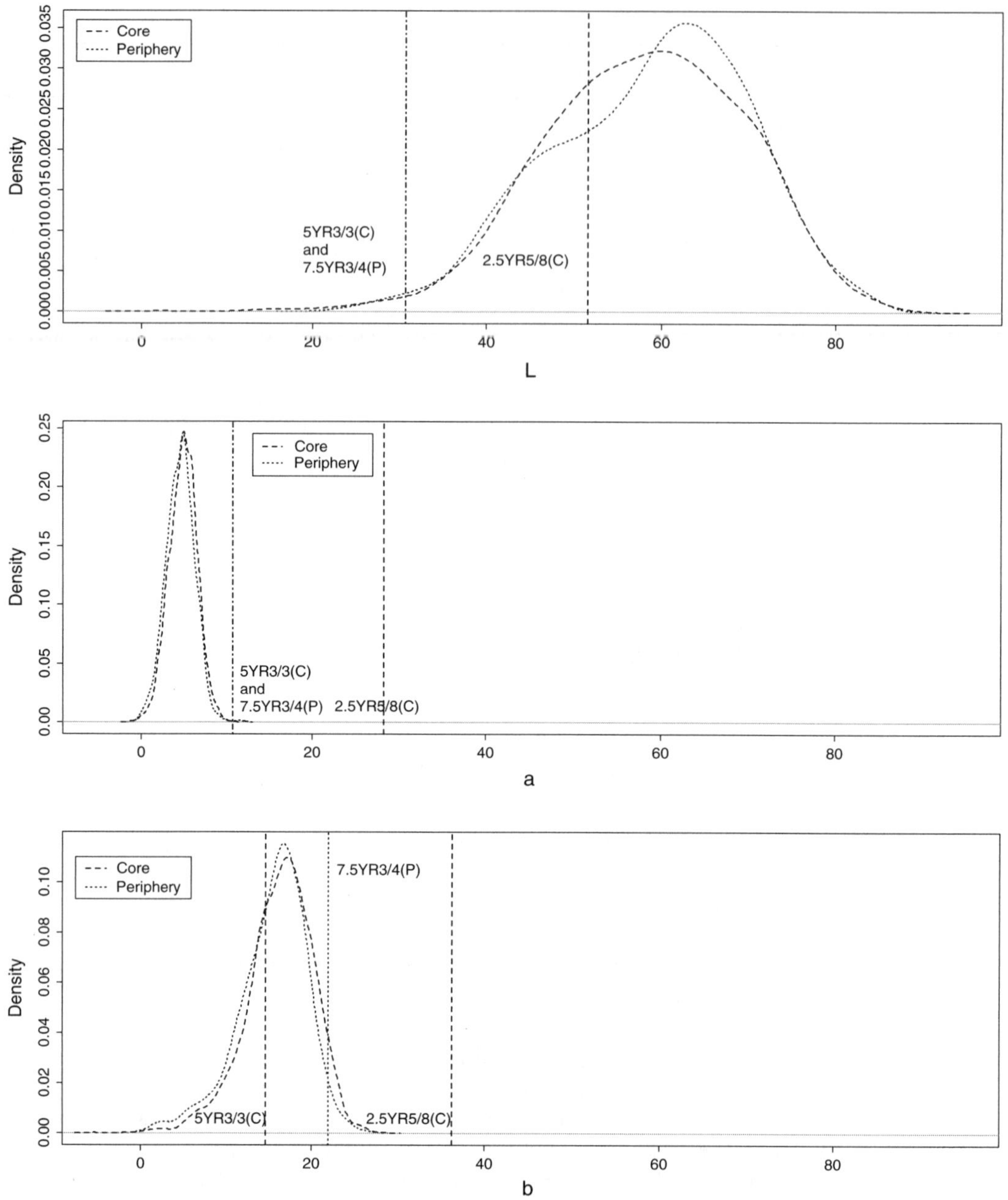

FIGURE 6.4 L*a*b* color density by archaeological context.

a matrix of the direct comparison between divisions for each of the color axes. The greatest difference along the L* axis is between cranial fragments and axial fragments. Mean scores of 64.0 and 46.62, respectively, reflect a significantly whiter color for cranial fragments ($t = 30.30$, $p < 0.05$). Similarly, fragments from the cranium are significantly whiter (higher L* values)

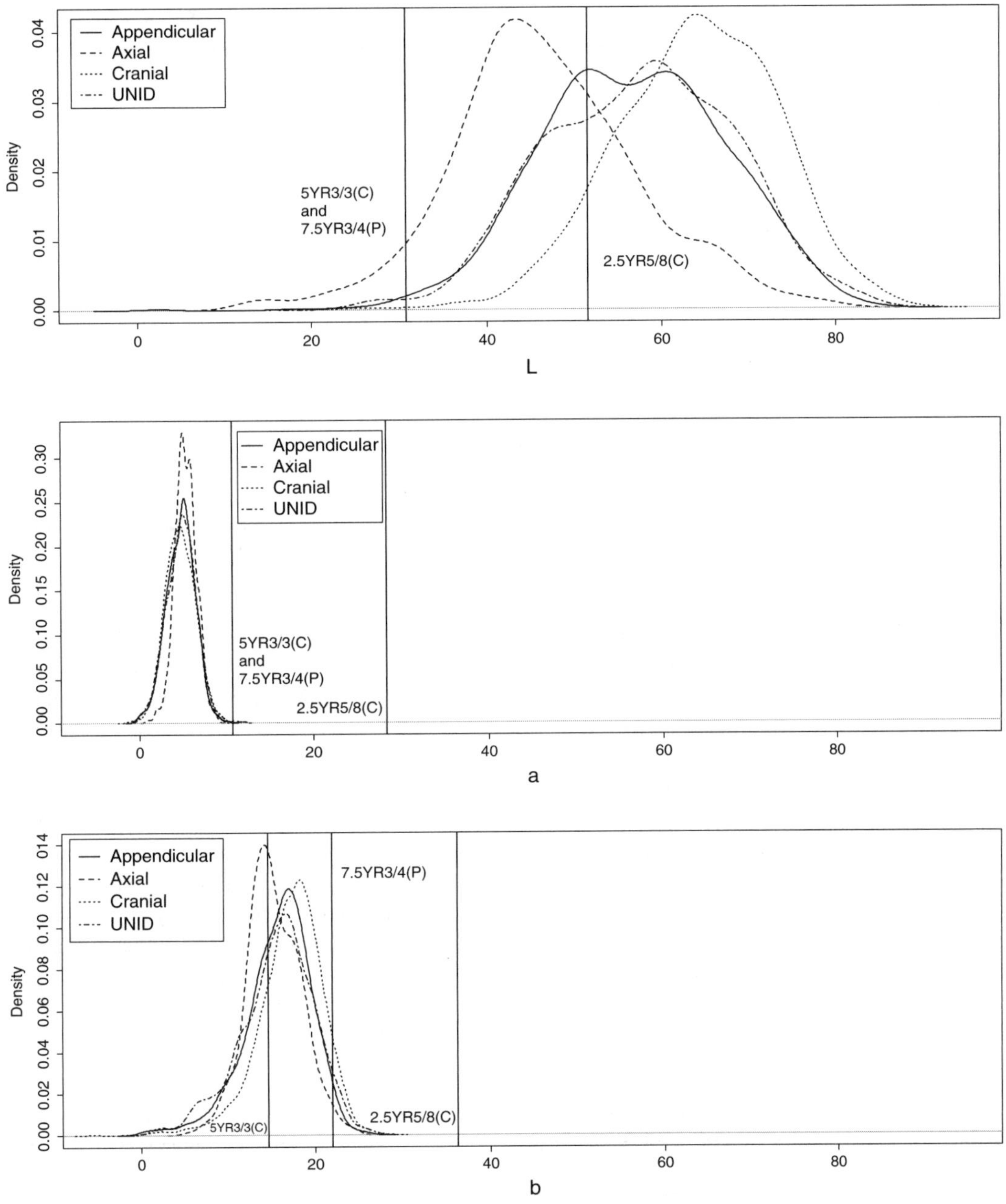

FIGURE 6.5 $L^*a^*b^*$ color densities by skeletal division.

than both appendicular and UNID fragments ($t = 16.81$, $t = 16.47$, $p < 0.05$, respectively). Density values for the a^* axis indicate that axial fragments are significantly different from appendicular, cranial, and UNID fragments ($t = 6.12$, $t = 6.45$, and $t = 5.20$, $p < 0.05$, respectively). Figure 6.5 illustrates the tight distribution of axial fragments that are slightly more red (less green) in color than those from the other regions. A comparable pattern can be seen in density values for the b^* axis. Fragments from the cranium are significantly

TABLE 6.2 T-Test Comparisons of Mean Color Score by Skeletal Division and Color Axis

	Appendicular	Axial	Cranial	UNID
L				
Appendicular	–	1,266	2,087	2,272
Axial	***15.34***	–	1,567	1,752
Cranial	***16.81***	***30.30***	–	2,573
UNID	1.59	***17.12***	***16.47***	–
a				
Appendicular	–	–	–	–
Axial	***6.12***	–	–	–
Cranial	0.54	***6.45***	–	–
UNID	1.10	***5.20***	1.75	–
b				
Appendicular	–	–	–	–
Axial	***2.04***	–	–	–
Cranial	***8.27***	***8.67***	–	–
UNID	0.96	1.23	***9.81***	–

Note: $p < 0.05$ represented as ***2.33***, DFs in upper triangle of L matrix.

different from fragments from the axial, appendicular, and UNID skeletal divisions ($t = 2.04$, $t = 8.27$, and $t = 8.67$, $p < 0.05$, respectively). Axial fragments exhibit minimal significant difference compared to appendicular fragments along the b* axis ($t = 2.04$, $p < 0.05$). Higher values along the b* axis as seen in cranial fragments (mean of 16.50) are indicative of more yellow colors.

Color axis means and standard deviations by skeletal division and archaeological context are provided in Table 6.3. For the L* axis, appendicular, axial, and cranial elements from the periphery are on average whiter than fragments from core areas. Figure 6.6 illustrates this pattern and clearly shows that cranial distributions have the highest average L* values for both the core and the periphery areas. The results of *t*-test comparisons of skeletal divisions by site area are presented in Table 6.4. For the L* axis, only appendicular fragments

TABLE 6.3 L*a*b* Color Values Separated by Site Area and Skeletal Division

	L	(std)	a	(std)	b	(std)	n
Core							
Appendicular	56.03	(10.45)	4.86	(1.67)	15.79	(3.87)	577
Axial	46.25	(11.00)	5.39	(1.26)	15.29	(3.11)	307
Cranial	63.87	(9.51)	4.81	(1.75)	17.34	(3.95)	854
UNID	57.76	(10.24)	4.96	(1.67)	15.98	(4.02)	851
Periphery							
Appendicular	57.88	(10.69)	4.59	(1.58)	15.54	(3.92)	317
Axial	48.32	(10.13)	5.09	(1.23)	14.99	(2.77)	67
Cranial	64.35	(8.70)	4.52	(1.61)	16.50	(3.26)	341
UNID	56.85	(11.62)	4.65	(1.82)	14.82	(4.55)	529

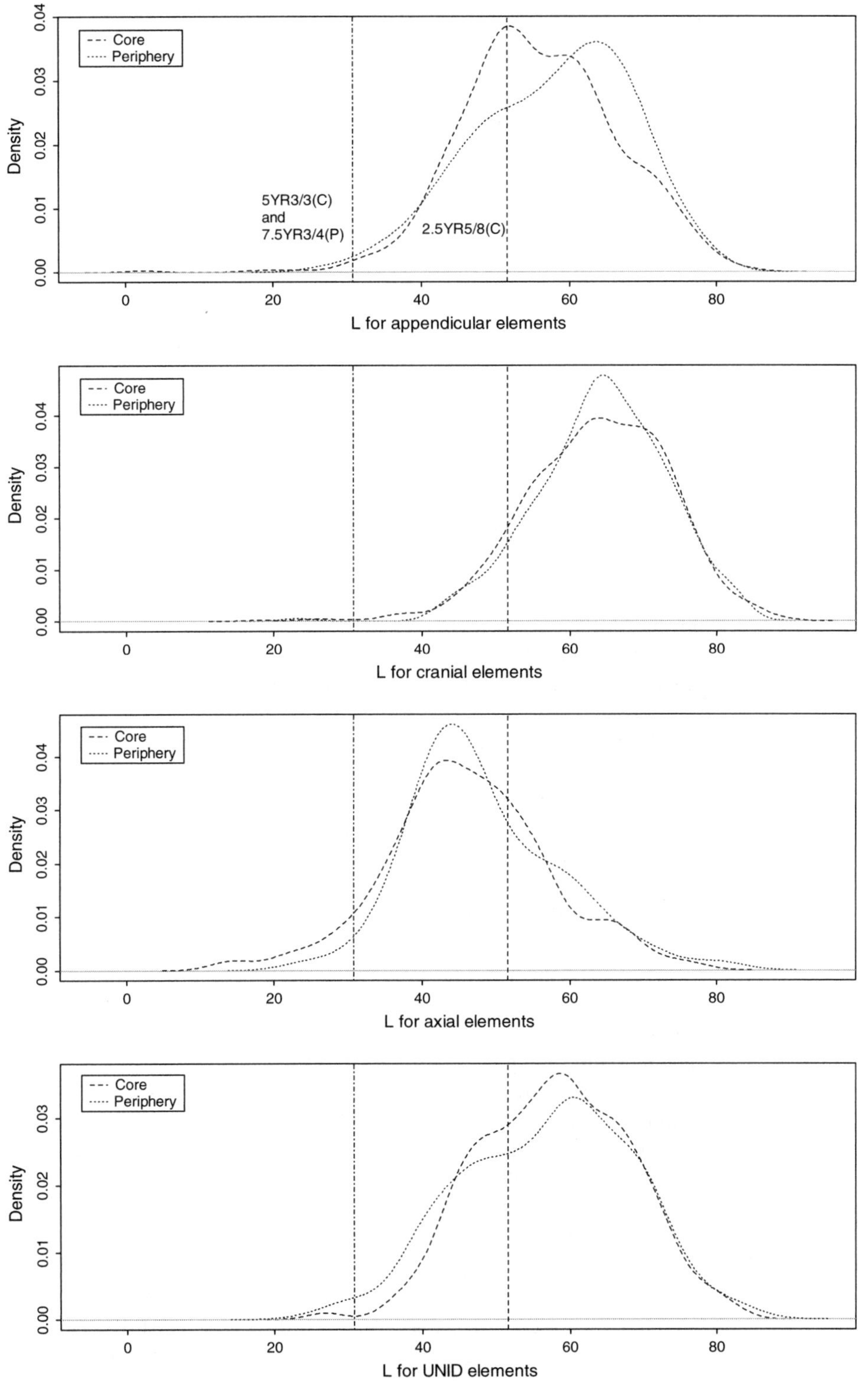

FIGURE 6.6 L* by skeletal division and archaeological context.

TABLE 6.4 T-Test Comparisons of Mean L*a*b* Values by Skeletal Division Within Site Areas

	T-Test Score	DF
L		
Appendicular	***2.51***	892
Axial	1.41	372
Cranial	0.81	1193
UNID	1.52	1378
a		
Appendicular	***2.36***	–
Axial	1.77	–
Cranial	***2.65***	–
UNID	***3.24***	–
b		
Appendicular	0.92	–
Axial	0.73	–
Cranial	***3.48***	–
UNID	***4.95***	–

Note: $p < 0.05$ represented as ***2.33***.

from the periphery are significantly different from the core sample ($t = 2.51$, d.f. = 892, $p < 0.05$).

Mean values for both the a* and b* axis from the core for all skeletal divisions are higher than the means for the site periphery (see Table 6.3 and Figures 6.7 and 6.8). This pattern suggests more red and yellow colors for the core fragments. For the a* axis, appendicular, cranial and UNID samples from the core are significantly different than the periphery ($t = 2.36$, $t = 2.65$, and $t = 3.24$, respectively with $p < 0.05$). For the b* axis, only cranial and UNID fragments from the core are significantly different than the periphery ($t = 3.48$ and $t = 4.95$, respectively with $p < 0.05$). Figure 6.8 clearly shows these differences in these skeletal divisions, with the cranial core density shifted higher on the B* axis and the UNID core density is both more restricted and shifted.

Figure 6.9 illustrates all possible correlations between pairs of color axes for the fragments from Walker-Noe. The plot of a* and b* demonstrates a strong correlation between these axes indicating a trend in surface colors ranging from blue green across the axes to yellow red. Although correlations with the L* axis are not as striking, there are trends in observed surface colors as indicated by the curvilinear plots. The plot of L* and b* illustrates dominant surface colors tend to be blue and less white, more white and yellow, and also more white and less yellow. Of interest, colors identified by researchers (e.g., Shipman *et al.*, 1984) on experimentally burned bone (translated into L* a* b* data) and subsequently compared to the colors observed here are found to be fairly consistent. Although, all comparative bone colors were generated by temperatures in excess of 300°C, it is not the intention to suggest heating temperatures for the archaeological specimens. However, it provides a foundation for considering possible heating conditions in terms of duration of exposure and intensity of heating that potentially generated the Walker-Noe assemblage.

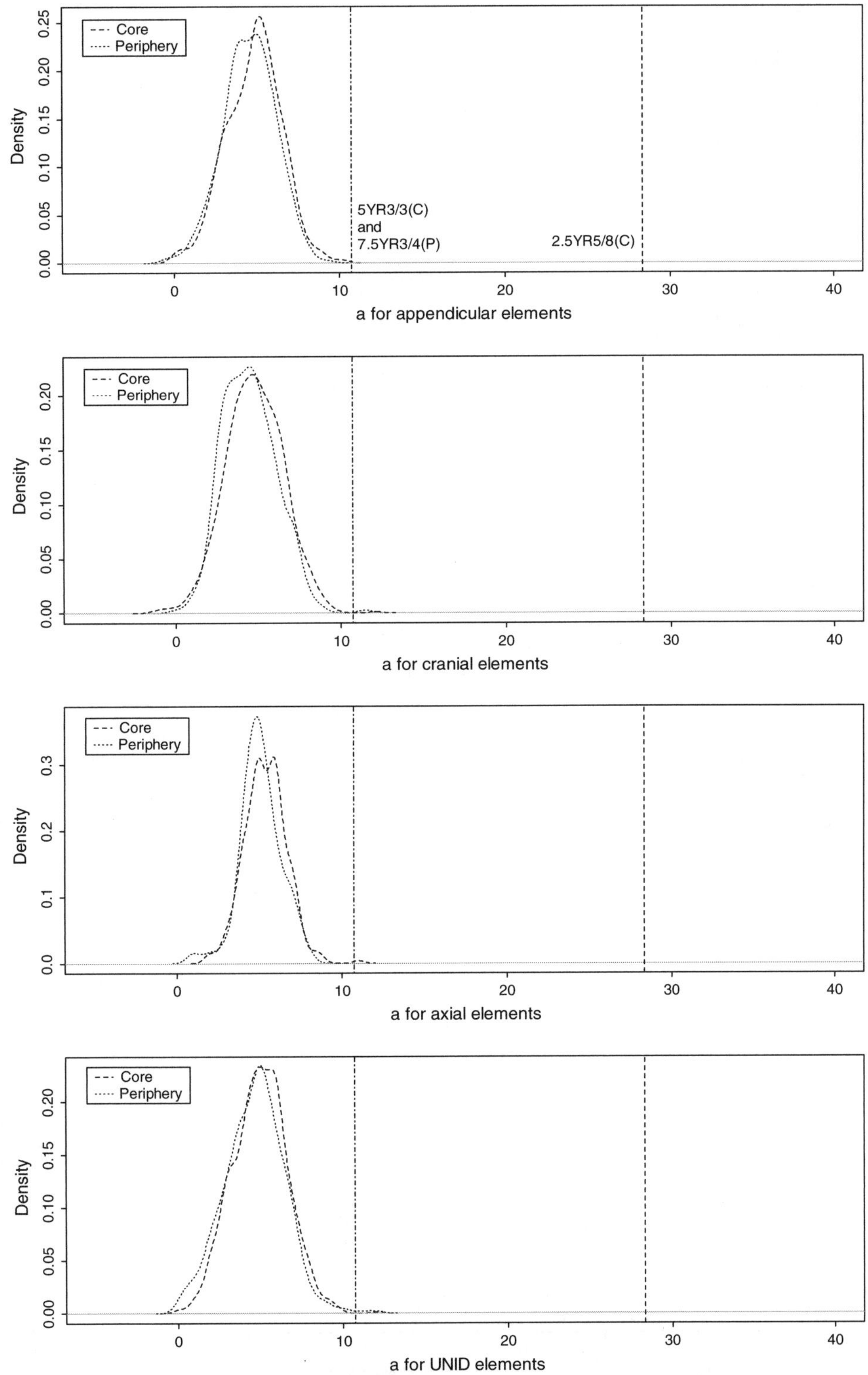

FIGURE 6.7 a* by skeletal division and archaeological context.

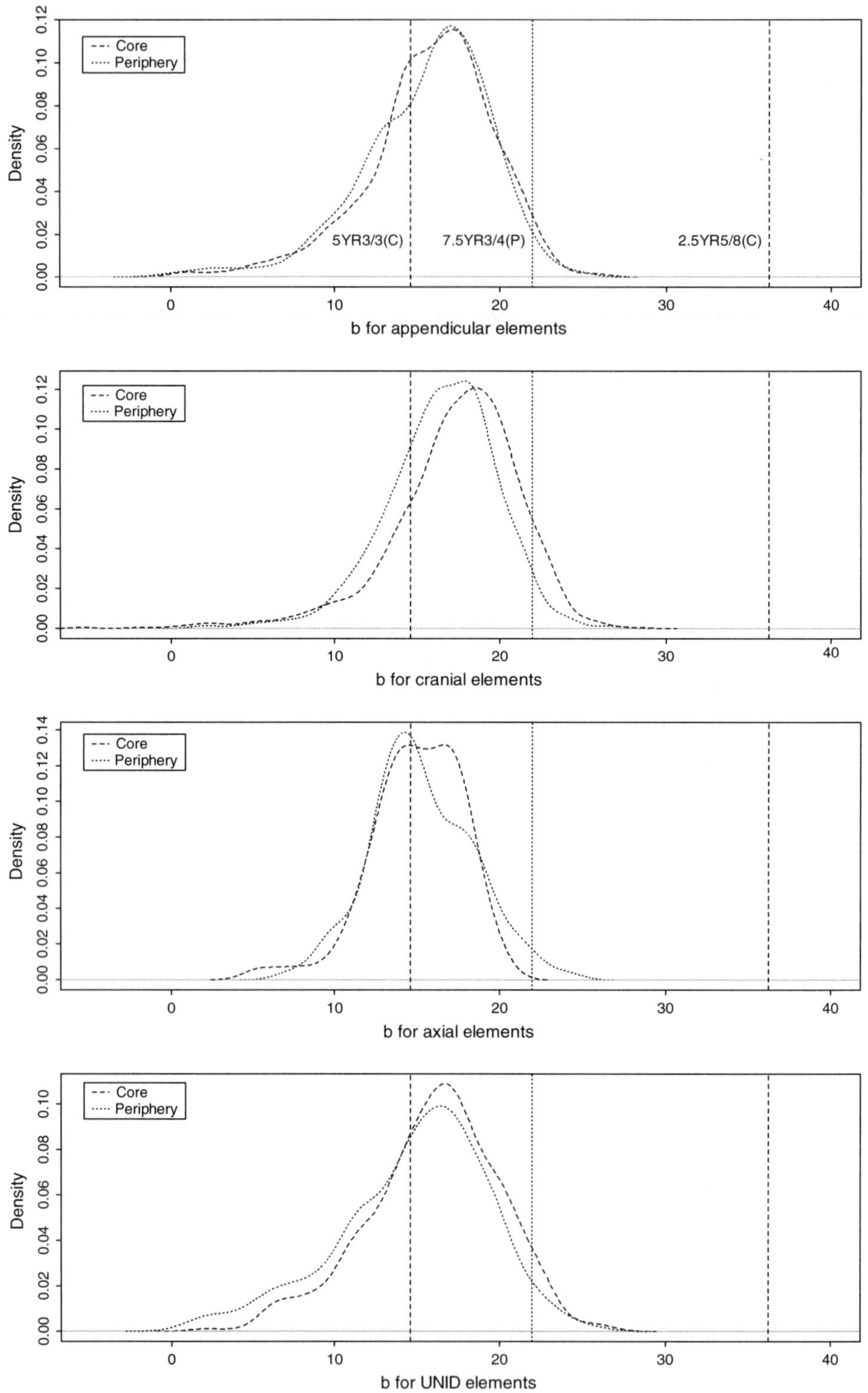

FIGURE 6.8 b* by skeletal division and archaeological context.

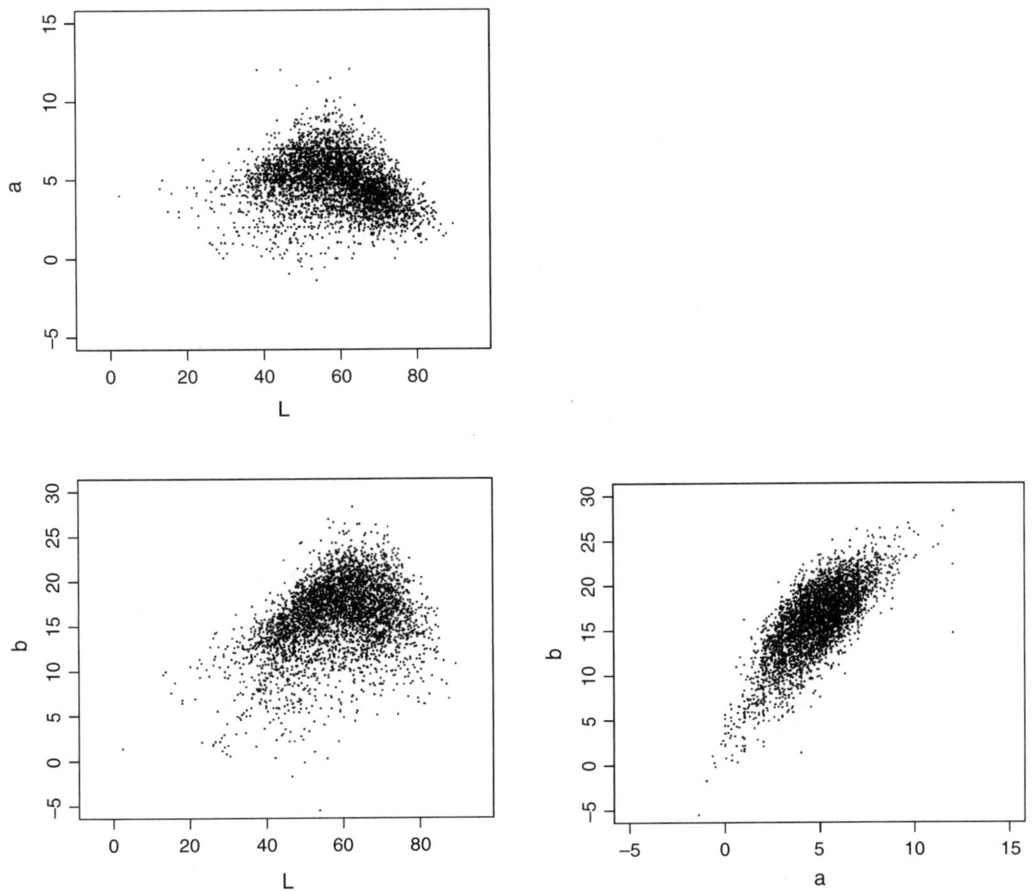

FIGURE 6.9 Bivariate plots of the L*, a*, and b* axes for the entire Walker-Noe sample.

DISCUSSION

Surface color data collected on the archaeologically derived sample from Walker-Noe (15GD56) sheds light on both color patterns of cremated bone and prehistoric crematory activities. Relatively consistent colors for the entire bone deposit and the lack of extremely white fragments are in part the result of the depositional environment at the site and soil staining of the fragments. Nonetheless, there are observable, albeit subtle, patterns that warrant further consideration.

Clearly, the UNID division, by definition, includes fragments from all regions of the body. Interestingly, the color patterns for the UNID fragments tend to overlap with those of all other divisions. As such, this appears to indicate that general burning patterns may be distinct for particular skeletal divisions of material from the Walker-Noe site. There are general tendencies for each skeletal division in terms of each of the three color components, though trends are most apparent on the L* axis. Simply stated cranial fragments exhibit the whitest colors. These fragments are also more yellow in color than others as indicated by values on the b* axis. This pattern suggests a longer duration of exposure to heat or higher temperatures for these elements during the burning process. In general, appendicular elements display less white colors although there is a recognizable difference in surface colors for fragments from the core compared against those recovered from the periphery. The presence of charred wood and heat impacted soils at the site suggests the in situ use of wood as a fire fuel and the maneuvering of burned elements around the site, potentially resulting in repeated heating of particular elements. Variation in surface color related to the location of recovery at the site and skeletal division suggest that it may be possible to draw conclusions about the crematory practices at Walker-Noe.

CONCLUSION

Understanding the color variation in controlled experiments and from forensic and archaeological samples is critical to burned bone studies. The use of CIELAB color coordinate system enables measurement of differences in overall colors and additionally a comparative assessment of the individual axes values. A slight variation is observable in bone surface color between core and peripheral regions of the site. In addition, differences in surface color can be recognized between elements from different skeletal divisions, collectively across the site and also between areas of the site. The interpretation of surface color data for bone fragments recovered from Walker-Noe provides additional insight into site formation processes at this Middle Woodland period crematory. The approach detailed in this study can readily be applied to other archaeological and forensic investigations to address differential burning, site formation issues, and taphonomic factors.

ACKNOWLEDGMENTS

We would like to thank the William M. Bass Endowment at the University of Tennessee Forensic Anthropology Center for funding to purchase the X-Rite CA22 Spectrophotometer. Beatrix Dudzik deserves special thanks for collecting color data from over 3000 bone fragments.

REFERENCES

Baby, R.S. (1954). Hopewell cremation practices. *Ohio Historical Society Papers in Archaeology* **1**, 1–17.

Bennett, B.L. (1999). Thermal alteration of buried bone. *Journal of Archaeological Science* **26**, 1–8.

Bennett-Devlin, J., Herrmann, N., and Pollack, D. (2006). *GIS analysis of cremated and commingled bone*. Poster presented at the 71st Annual Meeting of the Society for American Archaeology, San Juan PR.

Bennett-Devlin, J., Kroman, A., Herrmann, N.P., and Symes, S. (2006). Time temperature and color: heat intensity versus exposure duration. Part 1. Macroscopic influence on burned bone. *Proceedings of the American Academy of Forensic Sciences* **12**, 311–312.

Binford, L.R. (1963). An analysis of cremations from three Michigan sites. *Wisconsin Archaeologist* **44**, 98–110.

Bonucci, E. and Graziani, G. (1975). Comparative thermogravimetric, X-ray diffraction and electron microscope investigations of burnt bones from recent, ancient and prehistoric age. *Atti Memorie Accademia Nazionale die Lincei, Sci Fis Matem. Natur.* **59**, 517–534.

Buikstra, J. and Swegle, M. (1989). Bone modification due to burning: experimental evidence. In *Bone Modification* (R. Bonnichsen and M. Sorg, eds), pp. 247–258. University of Maine Press, Orono, ME.

Corriera, P.M. (1997). Fire modification of bone: a review of the literature. In *Forensic Taphonomy: The Postmortem Fate of Human Remains* (W. Haglund and M. Sorg, eds), pp. 275–293. CRC Press, Boca Raton, FL.

De Graaff, G. (1961). Gross effects of a primitive hearth. *South African Archaeological Bulletin* **16**, 25–26.

DeHann, J. and Nurbakhsh, S. (2001). Sustained combustion of an animal carcass and its implications for the consumption of human bodies in fires. *Journal of Forensic Science* **46**, 1076–1081.

Ebner, M. (2007). *Color Constancy*. John Wiley and Sons Ltd, Chichester, West Sussex, England.

Fairchild, M.D. (2005). *Color Appearance Models*. 2nd ed. John Wiley and Sons Ltd, Chichester, West Sussex, England.

Gilchrist, M. and Mytum, H. (1986). Experimental archaeology and burnt animal bone from archaeological sites. *Circaea* **4**, 29–38.

Herrmann, B. (1977). On histological investigations of cremated human remains. *Journal of Human Evolution* **6**, 101–103.

Herrmann, N., Bennett-Devlin, J., and Pollack, D. (2005). GIS analysis of the cremated skeletal material from the Walker-Noe site, Kentucky. *American Journal of Physical Anthropology* **Supplement 40**, 115.

Herrmann, N., Bennett-Devlin, J., and Pollack, D. (2007). *Walker-Noe: methodological approach and analysis of a commingled Middle Woodland Crematory*. Poster presented at 24th Annual Meeting of the Kentucky Heritage Council, Natural Bridge State Park, Kentucky.

Herrmann, N.P. and Devlin, J.B. (2008). Assessment of commingled human remains using a GIS based approach. In *Recovery, Analysis, and Identification of Commingled Human Remains* (B. Adams and J. Byrd, eds), pp. 257–269. Humana Press Inc., New York.

Marean, C., Abe, Y., Nilssen, P., and Stone, E. (2001). Estimating the minimum number of skeletal elements (MNE) in zooarchaeology: a review and a new image-analysis GIS approach. *American Antiquity* **66**, 333–348.

McCutcheon, P. (1992). Burned archaeological bone. In *Deciphering a Shell Midden* (J.K. Stein, ed.), pp. 347–368. Academic Press, San Diego, CA.

Munsell Color Company (2000). Munsell Soil Color Charts. Munsell Color GretagMacbeth, New Windsor, New York.

Nicholson, R. (1993). A morphological investigation of burnt animal bone and an evaluation of its utility in archaeology. *Journal of Archaeological Science* **20**, 411–428.

Pollack, D., Schlarb, E., Sharp, W., and Tune, T. (2005). Walker-Noe: an Early Middle Woodland Adena Mound in Central Kentucky. In *Woodland Period Systematics in the Middle Ohio Valley* (D. Applegate and R. Mainfort, Jr., eds), pp. 64–74. The University of Alabama Press, Tuscaloosa, AL.

Pope, E. (2007). Burned human remains: myths in forensic sciences. *Proceedings of the American Academy of Forensic Sciences* **13**, 379–380.

Robertson, A.R. (1990). Historical development of CIE recommended color difference equations. *Color Research and Application* **15**, 167–170.

Sharp, W., Pollack, D., Schlarb, E., and Tune, T. (2003). *Walker-Noe: a Middle Woodland Mound in Central Kentucky*. Paper presented at the Sixty-Eighth Annual Meeting of the Society for American Archaeology, Montreal, Canada.

Shipman, P., Foster, G., and Schoeninger, M. (1984). Burnt bone and teeth: an experimental study of color, morphology, crystal structure and shrinkage. *Journal of Archaeological Science* **11**, 307–325.

Spenneman, D. and Colley, S. (1989). Fire in a pit: the effects of burning on faunal remains. *Archaeozoologia* **3**, 51–64.

Thompson, T.J.U. (2005). Heat-induced dimensional changes in bone and their consequences for forensic anthropology. *Journal of Forensic Science* **50**, 1008–1015.

Thurman, M. and Wilmore, L. (1981). A replicative cremation experiment. *North American Archaeologist* **2**, 275–283.

Walker, P.L., Miller, K.W.P., and Richman, R. (this volume). Time, temperature, and oxygen availability: an experimental study of the effects of environmental conditions on the color and organic content of cremated bone. In *The Analysis of Burned Human Remains* (C.W. Schmidt and S. Symes, eds), Elsevier Press, Amsterdam.

Webb, W.S. and Snow, C. (1945). *The Adena People No. 2*. The University of Kentucky Reports in Anthropology and Archaeology #6. Lexington, KY.

Wells, C. (1960). A Study of cremation. *Antiquity* **34**, 29–37.

7

Time, Temperature, and Oxygen Availability: An Experimental Study of the Effects of Environmental Conditions on the Color and Organic Content of Cremated Bone

Phillip L. Walker, Ph.D.[1], Kevin W.P. Miller, Ph.D.[2], and Rebecca Richman, M.S.[1]

[1]*Department of Anthropology, University of California, Santa Barbara, CA*
[2]*Department of Chemistry, California State University, Fresno, CA*

Introduction

The chemical reaction of fire has four requirements: combustible material, adequate ignition temperature, sufficient oxygen, and a suitable environment for maintaining these conditions (Holck, 2005). The extent to which these requirements are met determines how fire affects the human body in cremation. A well-fed, well-ventilated fire can rapidly reduce a body to calcined ashes, while draft and lack of fuel can result in a partially burned body with some areas heavily altered and others virtually unaffected. Likewise, the cremation conditions affect the bones in multiple ways. Heat changes the color, the size, and the weight of the bone and fractures it while modifying the microstructure (Manye Correia, 1997). Like the consumption of the flesh, the extent to which these changes occur depends upon the conditions in the cremation environment.

The color of the cremated bone provides clues to both the condition of the remains at the time of cremation and the environment in which the cremation occurred. As bones go through the four stages of cremation – dehydration, decomposition, inversion, and fusion – their color changes reflect the ongoing chemical processes (Manye Correia, 1997). Cremation experiments have documented the color changes that occur in both 'green' (defleshed or flesh-covered fresh bones) and dried bones as the temperature of the cremation environment increases (Binford, 1963; Shipman *et al.*, 1984; Buikstra and Swegle, 1989). The color of the bone cremated at temperatures as low as 200–300°C begins to change from the ivory or light tan color of unburned bone to dark brown or black as the organic components begin to be carbonized (Manye Correia, 1997). At somewhat higher temperatures, the bone becomes black or dark gray depending on the duration of the heat exposure as carbonation is completed and the carbonates begin to disappear (Manye Correia, 1997). At high temperatures of 800°C or more, the bone becomes 'calcined' and the color changes to blue-gray or white (Van Vark, 1970; Shipman *et al.*, 1984; Buikstra and Swegle, 1989). At such high temperatures, the carbon formed from the organic material bonds with oxygen to form CO_2, calcination occurs, and the bone salts fuse (Manye Correia, 1997). Bonucci and Graziani (1975) found that hydroxyapatite converts to β-tricalcium phosphate at these temperatures, but Shipman *et al.* (1984) dispute this finding.

These color changes vary with the duration of the cremation process and the extent that bones are shielded from direct heat exposure. For example, the exposure of a bone to intense heat for a short period of time may be insufficient to reduce it to a calcined state. Likewise, portions of the skeleton shielded by thick layers of soft tissue or adjacent bones are often less severely affected by burning than bones in less shielded areas (Buikstra and Swegle, 1989; Holck, 2005). Each of the chemical processes requires a specific amount of energy; a short intense exposure to heat may not provide enough energy to break the chemical bonds of the different chemical components of a bone.

The chemical reactions that occur in cremation require both energy and raw material. In cremation, the energy expended depends on the temperature and the duration of the burning. The raw material comes from the body, the fuel, and the air. In particular, oxygen is a vital ingredient of the chemical reactions occurring in cremation (Holck, 2005).

In this paper we present the results of a series of experiments designed to assess the effects that temperature, duration of heat exposure, and oxygen availability in the cremation environment have on the color and collagen content of burned bones. These experiments show that oxygen availability is a key variable influencing the color and collagen and the content of burned bones.

EXPERIMENTAL DESIGN

To determine the effects that different cremation conditions have on the bone color and the preservation of DNA, collagen, and other organic constituents of the bone, we divided a fresh, modern human femur into approximately 1.5-g samples. The samples were obtained by sectioning the diaphysis of the bone transversely and then cutting the resulting rings of bones into sections of appropriate size. These bone samples were then cremated in a muffle furnace under a variety of different conditions. The samples were burned at 100, 200, 300, 400, 500, 600, 700, 800, 900, 1000, 1100, and 1200°C for 1-, 2-, and 3-h intervals.

The environment surrounding the bone also varied. The samples were cremated in open air as well as surrounded by two types of earth that contained different amounts of organic material. One set of samples was surrounded by diatomaceous earth, which is composed largely of silicon dioxide that is practically devoid of organic material. Another was surrounded by topsoil with a high organic content. Before cremation, each sample was weighed, photographed, and scanned on four surfaces (subperiosteal, endosteal, proximal, and distal) using a flatbed scanner to record its color. After cremation, the samples were reweighed, rephotographed, and rescanned.

The color of the bone samples before and after cremation was recorded in two ways. Subjective color assessments were made by matching the predominate color of the surfaces of each bone with one of the pantone (Pantone Inc., 2001) color standards. Quantitative color data were also obtained using the SigmaScan Pro image analysis program (SPSS, 1998) to determine the average red, green, and blue (RGB) pixel values of each bone's four surfaces.

BONE COLOR AND THE CREMATION ENVIRONMENT

Our experiments show that the temperature of cremation, duration of heat exposure, and availability of oxygen and organic compounds in the cremation environment are variables that all have highly significant effects on the bone color as measured by RGB values of postcremation digital images. Analysis of variance results suggests that the temperature bones are exposed to and exposure to air versus different types of soil are especially significant variables (Table 7.1).

The analysis of the color changes in the bones reveals both similarities and differences in the color changes associated with increasing temperatures in different cremation environments (Figures 7.1 and 7.2). At temperatures less than 200°C, all samples showed a gradual darkening of their color to

TABLE 7.1 Analysis of Variance Results Showing the Statistical Relationships Between the Bone Color Measured by the RGB Values of Scanned Images and the Environmental Conditions Varied in the Cremation Experiments

		Average pixel values of bone images					
		Red		Green		Blue	
Independent variables	D.F.	*F*	*P*	*F*	*P*	*F*	*P*
Time	10	5.64	0.005	6.04	0.004	6.61	0.002
Temperature	2	41.10	0.000	49.70	0.000	69.91	0.000
Soil	2	8.82	0.000	9.97	0.000	18.18	0.000
Adjusted R^2			0.810		0.840		0.880

D.F., degrees of freedom, *F*, *f*-value, *P*, probability of test.

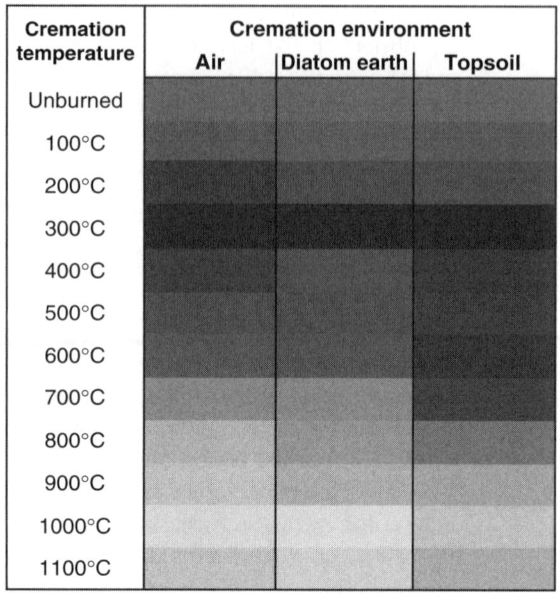

FIGURE 7.1 Colors generated using the average RGB color values of experimental bones samples cremated from 1–3 h in a furnace while surrounded by air, diatomaceous earth, and topsoil. (see Plate 20)

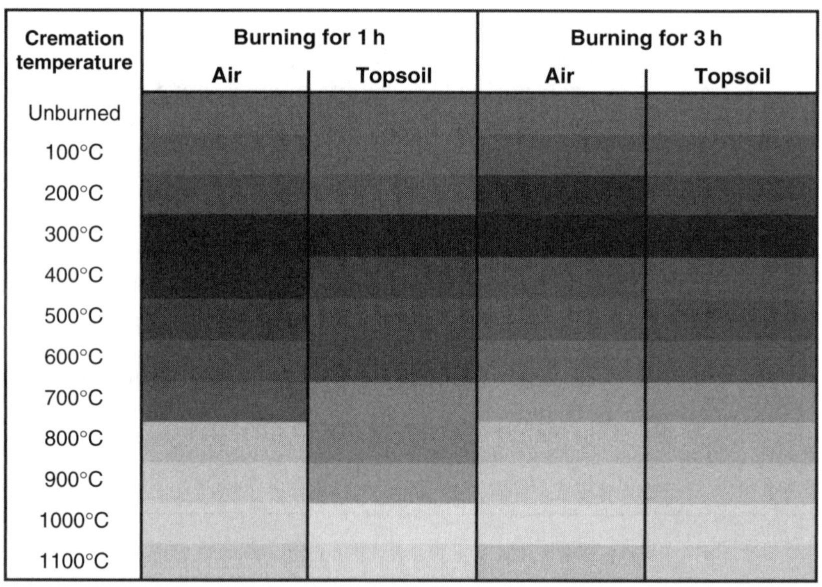

FIGURE 7.2 Colors generated using the average RGB color values of experimental bones samples burned for 1 or 3 h in a furnace while surrounded by air or topsoil with a high organic content. (see Plate 21)

dark brown. At 300°C, there was a striking, transient shift in the color of all specimens from brown to black. This is approximately the pyrolytic (charring) temperature at which much of the noncarbon elements of the organic components of bone disassociate, leaving only carbon.

At temperatures above 300°C, most specimens shift in color from black to tan and then to gray (Figure 7.1). The colors of samples exposed to temperatures above 300°C begin to develop divergent colors depending upon the environmental conditions they were exposed to during burning. The bones burned in air began to become gray at about 600°C and often took on a purple tint at the highest temperature (1100°C) used in the experiment. The bones surrounded by diatomaceous earth showed similar color changes during burning, except for the appearance of purple hues at the highest temperatures. The bones burned in topsoil showed a different pattern. They retained a dark gray color at 800°C, similar to that seen in other specimens, but at a much lower temperature (600°). This may be due to the excess availability of carbon in the topsoil delaying the carbonization of bone until a much higher temperature.

DURATION OF CREMATION

Although highly significant, the duration of the burning process had less influence on the bone color than other experimental variables (Figure 7.2). The bones burned for 1 and 3 h showed similar colors when exposed to the same temperatures. The bones surrounded by topsoil were an exception to this in that after 1 h they retained a dark gray color even at temperatures as high as 900°C, again possibly due to the consumption of oxygen by the carbonized organic material in the topsoil.

Our cremation experiments show that the simplistic view of a one-to-one correlation between the color of cremated bones and the temperature to which they were exposed to is erroneous. The environment surrounding bones during burning has a highly significant effect on the color changes associated with specific cremation temperatures. Oxygen-poor environments inhibit the oxidation process, and this slows the color transformation seen in bones in oxygen-rich environments. We found that cremation environments rich in organic materials are especially effective in inhibiting the color transformation seen in bones burned under other conditions.

CHEMICAL ANALYSIS

Several investigators have recovered amplifiable DNA from bones cremated at temperatures as high as 600°C. Brown *et al.* (1995), for example, analyzed cremated remains from an early Bronze Age cemetery cairn and found more DNA in the cremated remains than in the remains from inhumations at the same site. They suggest that the act of cremation actually increases the chance of DNA survival, because the organic substances that would otherwise provide food for destructive bacteria are unavailable. This unexpected preservation of DNA in remains that have been subjected to such extreme conditions opens the possibility of studying the genetic relationships of ancient people such as those of Bronze Age Europe who generally disposed of their dead through cremation. DNA could also be used to determine the sex of individuals whose remains are too incomplete to clearly assess sexually dimorphic characteristics. The ability to extract DNA from burned bones also has many potential applications in forensic casework.

Since testing for the presence of DNA is expensive and often only a small amount of cremated bone is available for analysis, we conducted a series of experiments to determine the conditions under which cremated remains are likely to contain preserved organic material. Our goal was to develop an inexpensive screening procedure that could be used to judge the potential of a sample for DNA analysis. One way to predict the presence or absence of DNA molecules in the bone is to test the bone for another molecule that also disappears around 600°C and is easier (and less expensive) to analyze. Collagen is perfect for this, as it is pyrolized at 500–600°C resulting in the loss of its birefringence in polarized light causing the bone to become isotropic (Manye Correia, 1997).

COLLAGEN ANALYSIS

A 0.5-g sample was removed from each cremated specimen for collagen analysis. Each sample was ground in a freezer mill cooled with liquid nitrogen. The ground material was then treated in 1 N HCl at 90°C for 20 min. The resulting material was washed to neutrality with 0.125 N NaOH. The residue was mixed with potassium bromide and pressed into pellets, which were analyzed using Fourier Transform Infrared spectrometry (FTIR).

Using FTIR we were able to detect some modified materials produced by the thermal alteration of collagen. Some of the samples retained the birefringence of fresh collagen, seen in the spectrographic profiles as two peaks. Others showed altered spectrographic patterns that lacked one or both of the peaks characteristic of unaltered collagen. We found that neither of the peaks characteristic of intact collagen persisted in samples cremated at temperatures greater than 600°C.

We found that the color of the cremated bone is a good predictor of the bone's collagen content. A discriminant analysis was preformed using red, green, and blue (RGB) values from the scans of the postcremation samples to predict the presence of two, one, or no FTIR collagen peaks. The results of the analysis indicate that RGB values of the cremated bone predict the presence of intact collagen with an accuracy of 85–95%.

We are currently analyzing the cremated samples used in this experiment for the presence of amplifiable DNA. These DNA data will be correlated with our FTIR and bone color observations. Our goal is to develop a screening technique that allows data on collagen content and color to be used to predict the presence of amplifiable DNA. We are also working to resolve the standardization issues that arise when the colors of burned bones are recorded using different types of digital equipment.

CONCLUSION

Depending on the culture, cremation is reserved for the richest, only used by the very poor, never used, or is the main method of body disposal. In areas where fuel is limited, the effort expended to cremate a body is large and may signify high status. In such environments, conditions of cremation may be modified in order to maximize fuel efficiency, while in heavily wooded areas the amount of fuel is much less of a limiting factor. In order to understand a culture's cremation practices from its cremated remains, we must understand how the burning conditions affect the bones that are left behind.

Temperature, duration of burning, and availability of oxygen affect the color of cremated bones as the chemical reactions determining the color require varying amounts of energy and oxygen to occur. Nonetheless, color is a good predictor of the presence of collagen and the possible presence of DNA. The preservation of DNA in cremated remains can be used to show genetic relatedness of cremated individuals, allowing for better understanding of burial practices in the ancient world. It will also allow for the determination of sex in individuals where other assessments are not possible, and increase the ability of forensic caseworkers to identify burned remains. By creating an inexpensive and easy method for determining the presence of DNA in cremated remains, we hope to make future research possible in this area.

REFERENCES

Binford, L.R. (1963). An analysis of cremations from three Michigan sites. *Wisconsin Archeologist* **44**, 98–110.

Bonucci, E. and Graziani, G. (1975). Comparative thermogravimetric, X-ray and electron microscope investigations of burnt bone from recent, ancient and prehistoric age. *Atti Memorie Accademia Nazinale die Lincei Scienze, Fisiche, Matematiche Naturali, Ser. 8 Sec. 2A (Roma)* **59**, 517–534.

Brown, K.A., O'Donoghue, K., and Brown, T.A. (1995). DNA in cremated bone from an Early Bronze Age cemetery cairn. *International Journal of Osteology* **5**, 181–187.

Buikstra, J.E. and Swegle, M. (1989). Bone modification due to burning: experimental evidence. In *Bone Modification* (R. Bonnischen and M.H. Sorg, eds.), pp. 247–258. Peopling of the Americas Publications, Center for the Study of the First Americans, Institute for Quaternary Studies, University of Maine, Orono.

Holck, P. (2005). Cremated bones. Anatomical Institute, University of Oslo, Oslo, pp. 113–119.

Manye Correia, P.M. (1997). Fire modification of bone: a review of the literature. In *Forensic Taphonomy: The Postmortem Fate of Human Remains* (W.D. Haglund and M.H. Sorg, eds.). CRC Press, Boca Raton, FL, pp. 275–293.

Pantone Inc. (2001). *Pantone Tints Uncoated = Pantone Tints Uncoated (non couchâe).* Pantone, Carlstadt, NJ.

Shipman, P., Foster, G., and Schoeninger, M. (1984). Burnt bones and teeth: an experimental study of color, morphology, crystal structure and shrinkage. *Journal of Archaeological Science* **11**, 307–325.

SPSS (1998). *Sigma Scan Pro 5.0.* SPSS, Chicago.

Van Vark, G.N. (1970). *Some Statistical Procedures for the Investigation of Prehistoric Human Skeletal Material.* Unpublished Ph.D. dissertation, Rijksuniversiteit de Groningen, Groningen, The Netherlands.

8

Heat-Related Changes in Tooth Color: Temperature Versus Duration of Exposure

Jeremy J. Beach, M.S.[1], Nicholas V. Passalacqua, M.S.[2], and Erin N. Chapman, B.A.[2]

[1]*Department of Anthropology, University of Indianapolis, Indianapolis, IN*
[2]*Department of Applied Forensic Sciences, Mercyhurst College, Erie, PA*

INTRODUCTION

While a great deal of research has been conducted regarding thermal alterations to bones (e.g., Shipman *et al.*, 1984; Nelson, 1992; Holden *et al.*, 1995; Symes *et al.*, 2005), comparatively little information has been produced specifically regarding the effects of heat on the human dentition (Endris and Berrsche, 1985; Muller *et al.*, 1998; Myers *et al.*, 1999; Schmidt *et al.*, 2005). The study of burned bones has produced standards for documenting the changes in color and texture that bones pass through as it experiences higher temperatures or a greater duration of heat. In general, the bone first darkens, becoming charred. This is eventually followed by calcination of the bone, which renders the bone starkly white as the organic content is lost (Shipman *et al.*, 1984). It is generally assumed that teeth also go through these changes, especially since they do discolor in a manner similar to bones. The following study involved experimentally burning teeth in order to systematically document thermal changes.

Since the make up of the dental hard tissues is notably different than that of bones, reactions to the effects of heating, likewise, should be different. For example, enamel has a much lower organic content than bones, and, therefore, may not go through the same color changes since in bones these changes are related to collagen loss. Moreover, enamel has a different fundamental organization (rods rather than osteons), which may also yield a unique pattern of heat-related alteration. However, Endris and Berrsche (1985) found in their study of burned teeth that the progression in dental color change as a result of thermal exposure was similar to bones. The teeth first turn black and/or brown. These colors then give way to blues and grays, which are eventually replaced with white.

The objectives of the current study are threefold:

(1) To document the gross morphological changes to the dental structures as they are exposed to specific temperatures for specific durations
(2) To understand the relationships of time and temperature in heat-related alterations
(3) To contribute to an updated standard of heat-related alterations specifically developed for teeth

MATERIALS AND METHOD

A sample of 32 adult human cheek teeth (premolars and molars) was obtained from an oral surgeon. The teeth were extracted during routine dental procedures and placed immediately into a saline solution to prevent dehydration. Before heat alterations were conducted, the teeth were individually cleaned of any adhering soft tissue using dental tools and tap water. Once the teeth were processed they were returned to the saline solution until experimentation could begin.

The teeth were weighed on a digital scale accurate to 0.1 g. They were then placed in a BLUE M™ Lab-Heat muffle furnace where they were exposed to a constant temperature for the entire duration of their heat exposure. Each tooth was placed in an individual crucible to prevent commingling of the specimens. The samples were heated in the muffle furnace at temperatures ranging from 204°C to 593°C (400°F–593°F), at 38°C (100°F) intervals (Table 8.1). The samples were heated for either 30 min or 1 h.

TABLE 8.1 Temperature, Duration, and Sample Size for Each Experimental Burn

Temperature, °C (°F)	Duration of heat exposure (min)
204 (400)	30
	60
260 (500)	30
	60
316 (600)	30
	60
371 (700)	30
	60
427 (800)	30
	60
482 (900)	30
	60
538 (1000)	30
	60
593 (1100)	30
	60

Sample size is two (2) for each of the 16 experiments.

After the heating intervals, the teeth were allowed to cool to room temperature. The gross morphological changes and the effects of dehydration (weight loss due to collagen and water loss) were then observed after the teeth cooled to room temperature. One approach to the study of thermally induced dental changes is to use a scanning electron microscope (SEM) (Fairgrieve, 1994; Muller *et al.*, 1998). This device allows for outstanding observations of morphology, but it uses an electron beam rather than a light to produce an image, making it undesirable for the study of color changes in teeth. Therefore, the current study relies upon the direct observation by the unaided eye as well as with a standard stereoscope (6×–40×). A Munsell Soil Color Chart (2000) was used to determine color values of enamel and dentin after incineration.

RESULTS

At 204°C (400°F) no significant morphological differences were noticed after 30 min. After 60 min at 204°C, a slight color change was noted in the dentin. The root structures of both teeth changed from the pale yellow color (5 y 8/3) typically associated with unaltered dentin to a yellow (10 yr 7/8) color. No color change was noted in the enamel. During microscopic examination at low power (between 6× and 40×), slight enamel flaking was noted around the cementoenamel junction (CEJ). No other differences were noted. Average weight losses due to dehydration are presented in Table 8.2.

Major color changes occurred in specimens incinerated at 260°C (500°F) for 30 min. The enamel exhibited a very pale brown (10 yr 8/2) color. The dentin exhibited the greatest amount of change turning a dark reddish brown (2.5 yr 2.5/3) color. This is in stark contrast to the first test at 204°C for 30 min when the dentin showed no signs of color change. Interestingly, the enamel at the CEJ appeared to be translucent. Accordingly, the dark reddish brown color of the dentin could be seen through the enamel. Slight enamel flaking was again noted around the CEJ. Upon microscopic examination it appeared as though the enamel of the crown was being detached from the root structure (Figure 8.1). The only difference between specimens incinerated for 60 min, as opposed to 30 min, was that the enamel appeared to be slightly darker (10 yr 8/2). All other attributes were consistent with the 30-min incineration.

TABLE 8.2 Average Percentage of Weight Loss for Teeth from Each Duration

Temperature, °C (°F)	Average percentage of weight loss at 30 min	Average percentage of weight loss at 60 min
204 (400)	16.3	13.1
260 (500)	17.3	17.3
316 (600)	13.3	23.1
371 (700)	16.9	21.3
427 (800)	37.9	22.5
482 (900)	25.1	30.9
538 (1000)	27.4	24.1
593 (1100)	36.0	33.3

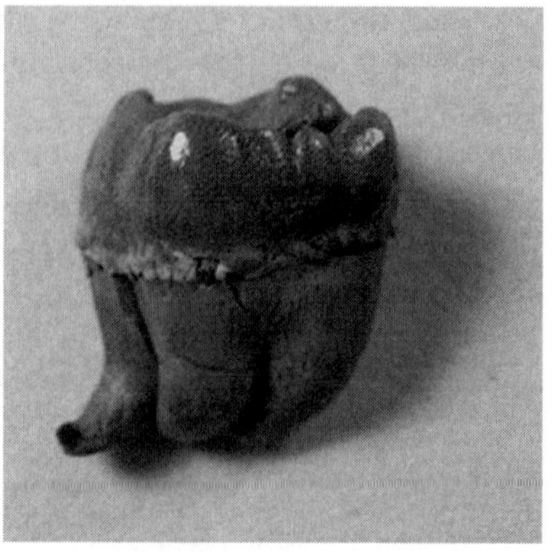

FIGURE 8.1 Crown patina and flaking/separation of crown at cementoenamel junction (CEJ). (see Plate 22)

No notable changes occurred between samples incinerated at 206°C and 316°C (600°F) for a period of 30 min. The enamel and dentin discoloration remained consistent and flaking around the CEJ was still apparent.

Several major changes occurred during the 371°C (700°F) interval. The most significant change was to the color of the dental hard tissues. After the incineration the enamel exhibited a very dark grayish brown (10 yr 3/2) color with a glossy appearance. The gloss of the enamel gave it a metallic appearance. Upon magnification, many small nonuniform cracks occurred over the majority of the enamel. This patina-like appearance occurred on nearly every aspect of the crown (Figure 8.1), with the exception of the CEJ. Major changes continued to occur around the root structure. The color of the root dentin turned from very pale brown to black (GLEY-1 2.5/N) color. No major morphological differences were noted between the time intervals. The only difference between time intervals at this temperature was the percentage of weight lost.

Several major degradational changes affected the dental hard tissues of both the crown and the root structures at 427°C (800°F). The crowns in three of the four specimens separated entirely from their roots. The CEJ around the fourth specimen showed extreme enamel flaking and appeared to be almost completely separated from the root. The color of all the crowns remained constant from the tests conducted during the 371°C interval (10 yr 3/2). Likewise, the patina-like appearance continued to be present. It should also be noted that the crowns were extremely friable and were easily subject to breakage during the postincineration handling and analysis. The root structures of the 30 minute incineration remained unchanged. However, the roots from the 60-min incinerations exhibited an olive brown (2.5 y 4/3) color. A significant change was the light gray color (5 y 7/1) of the mantle dentin from the olive brown color exhibited by the remaining dentin of the root structure (Figure 8.2). The morphology of the root structures remained fairly unaltered with the exception of a few cracks.

For the most part, at 482°C (900°F) the color of the specimens remained unchanged from the previous incineration cycle. The only difference worth noting is the color of the crown between the two time intervals. While specimens that were incinerated for 30 min remained unchanged, the specimens

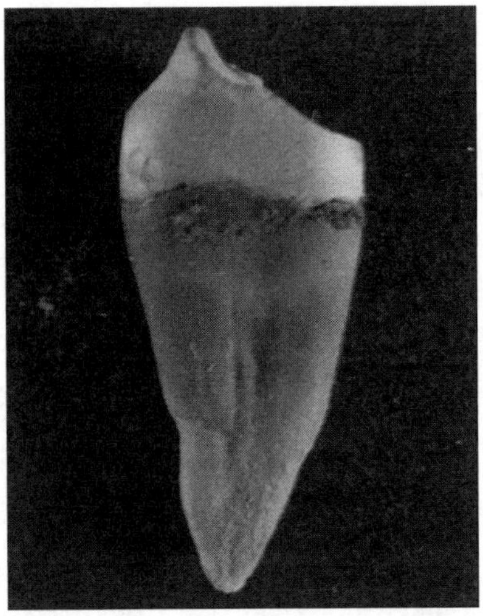

FIGURE 8.2 Differences in dentin color above and below the cementoenamel junction (CEJ). (see Plate 23)

that were incinerated for 60 min exhibited a light gray (5 y 7/1) color. In addition, the crowns of the specimens incinerated at 482°C had separated from their roots and were extremely friable. The glossiness of the crown noted at lower temperatures was absent from these crowns. The structural integrity of the roots remained similar to that described for the 427°C group.

Four changes were noted during the 538°C (1000°F) incinerations all of which applied to all specimens regardless of the time interval. Two of these changes were apparent in the crown. First, the crown color changed to a light gray (5 y 7/1) color with patches of grayish brown (2.5 y 5/2) and gray (2.5 y 6/1). Second, the integrity of the crowns was severely compromised. In all specimens the enamel was completely disintegrated into many small fragments. Third, the color of the root structures was in the process of changing from olive brown to patches of light gray (5 y 7/1) and gray (2.5 y 6/1) colors. Finally, the root apices were highly friable and broken off from the rest of the root.

Only minor changes were observed when comparing the 538°C and 593°C (1100°F) incinerations. The crowns from all specimens, regardless of the time interval, were highly fragmented and still exhibited the light gray color previously described. The color of the root structures for all specimens was still in transition from an olive brown to a light gray (5 y 7/1) to the white color typically associated with calcined bones (the Munsell edition that was utilized during the experiment did not have a color value that accurately described the calcined while color). The root apices on all specimens were completely disintegrated.

DISCUSSION

The general progression of the dentin color change from lower temperature to higher was (1) typical light yellow color of unaltered dentin; (2) black; (3) brown/olive; (4) gray; and (5) white. This finding is consistent with what was reported by Endris and Berrsche (1985). A steady change was observed in the color of the dentin, although the mottling of the dentin (Figure 8.1) may cause confusion if the root has become fragmented. Therefore, it is recommended that a careful examination be conducted upon the root structure if it has become fragmented.

It is worth noting that when working with dental hard tissues that have been heat-altered, the root structure seems to be the most reliable source of information for two reasons. First, dentin seems to be much less susceptible to the effects of heat alteration than enamel. The examination of specimens altered in the upper heat ranges revealed that the root structure remained relatively intact. The only exception to this observation is the apical portion of the root structures that tended to disintegrate at higher temperatures. Second, the integrity of the enamel tissue begins to minimally flake away from the cementoenamel junction as early as half an hour at 204°C. At 427°C–482°C the crown enamel is extremely friable and susceptible to breakage. Similarly, Endris and Berrsche (1985) noted that the structural integrity of the root was much more reliable than that of the crown enamel. In addition, the extreme friability and small enamel fragments may make color assessment very difficult when compared to an entire specimen of dentin. Finally, the progression of colors exhibited on the roots is much more static and tend to exhibit only one color at a time. If multiple colors are exhibited on the root, which from our experience is not more than two colors, a majority of the root is dominated by one color. Conversely, some enamel fragments exhibited as many as three colors during some of the upper-limit heat tests. A complete listing of observed colors and morphologies is given in Tables 8.3 and 8.4.

TABLE 8.3 Enamel Dentin Color and Features at 30 min

	Enamel		Dentin	
Temperature, °C (°F)	Munsell value	Feature	Munsell value	Feature
204 (400)	No change	CEJ flaking	5 y 8/3	No change
260 (500)	10 yr 8/2	CEJ flaking	2.5 yr 2.5/3	No change
316 (600)	10 yr 8/2	CEJ flaking	2.5 yr 2.5/3	No change
371 (700)	10 yr 3/2	Enamel gloss Patina	GLEY-1 2.5/N	No change
427 (800)	10 yr 3/2	Separation of crown Extreme CEJ flaking Patina	GLEY-1 2.5/N	No change
482 (900)	10 yr 3/2	Crown separation Extreme CEJ flaking Patina	GLEY-1 2.5/N	No change
538 (1000)	Multiple colors	Enamel Disintegration	5 y 7/1 and 2.5 y 6/1	Apical deterioration
593 (1100)	Multiple colors	Enamel disintegration	5 y 7/1 and 2.5 y 6/1	Extreme apical deteriorations

TABLE 8.4 Enamel and Dentin Color and Features at 60 min

	Enamel		Dentin	
Temperature, °C (°F)	Munsell value	Feature	Munsell value	Feature
204 (400)	No change	CEJ flaking	10 yr 7/8	No change
260 (500)	10 yr 8/2	CEJ flaking	2.5 yr 2.5/3	No change
316 (600)	10 yr 8/2	CEJ flaking	2.5 yr 2.5/3	No change
371 (700)	10 yr 3/2	Enamel gloss Patina	GLEY–1 2.5/N	No change
427 (800)	10 yr 3/2	Separation of crown Patina	2.5 y 4/3 (mantle = 5 y 7/1)	No change
482 (900)	5 y 7/1	Extreme enamel degradation	2.5 y 4/3 (mantle = 5 y 7/1)	No change
538 (1000)	Multiple colors	Enamel disintegration	5 y 7/1 and 2.5 y 6/1	Apical deterioration
593 (1100)	Multiple colors	Enamel disintegration	5 y 7/1 and 2.5 y 6/1	Extreme apical deteriorations

One of the most notable findings during experimentation was the continuity exhibited between specimens incinerated for 30 min and specimens incinerated for 60 min. A majority of the morphological changes that were observed during the half-hour interval were also observed after an hour of incineration. This seems to suggest that a longer incineration period does not necessarily mean that there will be more observable changes. The greatest amount of change was observed within the first 30 min of incineration. No significant changes occurred during the last 30 min of incineration suggests that a longer incineration period does not necessarily cause more alterations to dental hard tissues.

However, a difference does exist between 30- and 60-min incineration intervals between the average losses of weight due to dehydration. Overall, the teeth incinerated for 60 min lost more weight than the teeth incinerated for 30 min. For example, teeth incinerated for 1 h lost an average of 20% of their weight at 316°C and above. This 20% loss did not occur in the 30-min specimens until temperatures reached 427°C–483°C. Finally, one shared characteristic between the two incinerations is the steady, progressive weight loss from the lowest temperature tested to the highest temperature. On average, both incinerations lost between 13–16% of their weight at 204°C and 33–36% of their weight at 593°C.

CONCLUSION

The application of heat to dental hard tissues produces changes in color that are similar, but not identical, to those reported for bones. A steady change in the color of both dentin and enamel was observed as temperatures were increased. In addition, the loss of weight due to dehydration is directly proportional to a rise in temperature and increased duration of heat exposure. In general, the dentin provides a more sensitive indicator of heat alteration than the enamel, which eventually shatters and becomes difficult to analyze.

When comparing temperature and duration, it appears that temperature is the more important variable when it comes to interpreting tooth color. Teeth that were burned for a longer period of time, but at a lower temperature, did not mimic teeth burned at higher temperatures, despite the fact that the dehydration increased.

Additional research is needed to determine if the observed trends presented herein remain consistent after very long exposures to lower temperatures (i.e., over 1 h). This study should use time intervals of less than half an hour in order to determine more precisely when the degradation of dental hard tissues commences.

LITERATURE CITED

Endris, R. and Berrsche, R. (1985). Color change in dental tissue as a sign of thermal damage. *Zeitschrift Für Rechtsmedizin* **94**, 109–120.

Fairgrieve, S.I. (1994). SEM analysis of incinerated teeth as an aid to positive identification. *Journal of Forensic Sciences* **39**, 557–565.

Holden, J.L., Phakey, P.P., and Clement, J.G. (1995). Scanning electron microscope observations of incinerated human femoral bone: a case study. *Forensic Science International* **74**, 17–28.

Muller, M., Berytrand, M.F., Quatrehomme, G., *et al.* (1998). Macroscopic and microscopic aspects of incinerated teeth. *Journal of Forensic Odonto-stomatology* **16**, 1–7.

Myers, S.L., Williams, J.M., and Hodges, J.S. (1999). Effects of extreme heat on teeth with implications for histologic processing. *Journal of Forensic Sciences* **44**, 805–809.

Nelson, R. (1992). A microscopic comparison of fresh and burned bone. *Journal of Forensic Sciences* **37**, 1005–1060.

Schmidt, C.W., Reinhardt, G., Nawrocki, S., and Hill, M. (2005). Analysis of burned dental remains. *American Journal of Physical Anthropology* **Suppl. 40**, 183.

Shipman, P., Foster, G., and Schoeninger, M. (1984). Burnt bones and teeth: an experimental study of color, morphology, crystal structure and shrinkage. *Journal of Arachaeological Science* **11**, 791–795.

Symes, S.A., Kroman A.M., Rainwater C.W., *et al.* (2005). Bone biomechanical considerations in perimortem vs. postmortem thermal bone fractures: fracture analysis on victims of suspicious fire scenes. *American Journal of Physical Anthropology* Suppl. 40, 208.

9

Investigations on Pre-Roman and Roman Cremation Remains from Southwestern Germany: Results, Potentialities and Limits

Joachim Wahl, P.D.Dr.

Regierungspräsidium Stuttgart, Landesamt für Denkmalpflege, Arbeitsstelle Konstanz, Osteologie, Stromeyersdorfstraße 3, D-78467 Konstanz, Germany

INTRODUCTION

The investigation of cremation remains has a long tradition in Germany. As far back as the sixteenth century, urns and burned bones were interpreted correctly as remnants of cremation burials. In 1679 J. von Mellen deduced the presence of a double burial from the presence of bigger and smaller skull bones found together in one urn. In 1719 A.A. Rhode recognized pieces of a young child. Even though the church, reacting against the ideas of enlightenment, instrumentalized the combustion of the corpse as a foretaste of the inextinguishable hellfire, by the end of the nineteenth century, the scientific view of cremations gradually began to gain ground. The main goal was to achieve information on the sociological and ethnical relationships of buried persons. During and after the middle of the 1930s, the first competent works were published in which the percentual occurrence of distinct bone fragments as well as age at death, sex and number of individuals were recorded (Krumbein, 1934, 1935; Thieme, 1937, 1939).

In Europe in the 1950s, the study of prehistoric cremations boomed, and many fundamentals for the understanding and handling of calcined bones were developed. Today, these investigations remain the basis for the treatment and interpretation of cremations. Well-known authors in this context include Breitinger (1954), Dokládal (1969, 1970), Gejvall (1955, 1963), Grimm (1953, 1961), Herrmann (1972, 1973a,b, 1977a,b), Kloiber (1953, 1963),

Kühl (1966, 1979), Lisowski (1968), Malinowski and Porawski (1969), Müller (1958, 1964), Rösing (1976, 1977), Schaefer (1961, 1964), van Vark (1974, 1975) and Wells (1960). The emphasis was, amongst other things, on research into basic phenomena of combustion, metrics, reconstruction of body height and involved the remains of teeth as well as a consideration of demographic parameters. Later especially detailed methodological questions and population studies based upon larger series took place (e.g. Kunter, 1980; Caselitz, 1981; Wahl, 1981, 1982; Grupe and Herrmann, 1983; Schutkowski and Hummel, 1987; Wahl, 1988a–c; Schutkowski and Hummel, 1991; Caselitz, 1998; Wahl, 1998; Kunter, 2000; Kokabi and Wahl, 2001; Großkopf, 2004; Kunter, in press; Kunter and Malek, in prep.); moreover, Pusch *et al.* (2000) have pursued detailed studies of DNA analyses in cremated remains.

Baden-Württemberg in the southwestern part of Germany is an archaeological rich region. Cremation burials are present from the early Neolithic times (Trautmann and Wahl, 2005), from the Bronze Age, particularly from the Pre-RomanIron Age (Hallstatt age and LaTene period) and from the Imperial Roman Period. Several human bones showing traces of charring or calcination come from the finds of the late Neolithic "Michelsberger" culture. They do not document burial–rituals but show catastrophies caused by fire or warlike attacks (Wahl, 1999). A fragment of an adult femur from a settlement pit dating to the early LaTene period in Bondorf exhibits signs of charring as well as violence carried out with sharp blades (Kokabi *et al.*, 1994). Perhaps it indicates a religious background.

SKELETAL REMAINS

In a previous compilation of individual epochs of the Pre-Roman Iron Age in southwestern Germany, nearly 800 graves from approximately 100 locations were registered (Wahl, 1988d). Today, as a result of recent inventory, the number has more than doubled. Up till now, approximately 750 individuals from about 75 places of burial have been investigated anthropologically. The whole complex can be divided into inhumation graves and cremation graves, showing a ratio of nearly 2:1. In the late Hallstatt age (Ha C and D; *c.* 750–450 BC) as well as in the LaTene period (*c.* 450–15 BC), both burial types appear contemporary. This fact has to be further examined with regard to more precise dating, associated grave goods, regional distribution, etc. Present research indicates that the decision on whether a decedent should be cremated or buried unburned depended neither on age at death nor on sex and in contrast to previous assumptions, nor on the social status of the individual. Some scientists assume that we are dealing exclusively with graves from upper class people in general.

Several dozen inhumation graves have been found from the preceding early Hallstatt period (Ha A and B; *c.* 1200–750 BC), also known as the Urn Field culture because of the predominance of cremation burials. A gradual transition from inhumation rites to cremation burials has already been registered in the late Bronze Age. Accordingly, we can identify the commingled use of both burial types as a tradition more than 1000 years old by the time the Romans arrived.

At the moment, the stock of graves from the Imperial Roman Period in Baden-Württemberg might be far more than 4000. The biggest contingents that have so far been investigated anthropologically are the graveyards from Stettfeld (Wahl, 1988c), Schwäbisch Gmünd-Schirenhof (Volk *et al.*, 1988), Altlußheim-Hubwald (Parsche *et al.*, 1994) and Welzheim (Plhak, 1987) with a total of more than 900 graves. They date mostly to mid-second to mid-third century. Presently work is taking place on the necropoles from Heidelberg-Neuenheim with nearly 1300 graves (Berszin and Wahl, in prep.) and from Rottweil with more than 570 graves (Burger-Heinrich, in prep.). Other larger cemeteries are known from Sontheim/Brenz, Weil am Rhein and Osterburken, but these have not yet been examined anthropologically. An older investigation exists for the Roman graveyard from Stuttgart-Bad Cannstatt (Nierhaus, 1959), but unfortunately the publication does not contain much data comparable to other locations. Additionally, we know about numerous smaller burial groups and single burials (e.g., Asskamp *et al.*, 1987; Gaubatz *et al.*, 1988; Hensen *et al.*, 2004), so that our knowledge on the Imperial Roman Period rests on solid evidence. Most of the larger cemeteries from this epoch are mixed in use, i.e., they include inhumation graves as well as cremation burials, the relation being strongly in favour of cremations. On average, the proportion of unburned funerals is about 10% and is in contrast to the Pre-Roman Iron Age; we see them underlying a special selection (see below).

METHODOLOGICAL ASPECTS

The typical appearance of cremated bones is characterized by shrinking, deformation and fragmentation. These nuisances must be taken into consideration in regard to the methodological spectrum as well as to the interpretation of the results. Depending on the intensity of burning, they affect nearly all aspects of anthropological investigation.

To determine age at death in prehistoric cremations, one can usually use the same phenomena of maturation used for unburned skeletal remains: the obliteration of cranial sutures, the fusion of epi- and apophyses and the status of teeth. But the influence of heat may force sutures, which are in an early stage of fusion, to separate and may misleadingly indicate a younger age at death. Isolated parts of joints from subadults have very thin cortices. Therefore, in many cases they do not survive the combustion and/or the physical weight of the deposit. The same is true for the metaphyses. Sometimes the change of teeth cannot be determined in detail because high temperatures cause the dental enamel to shatter into microscopic fragments.

Crowns usually survive intact if they are still within their crypts, i.e., if not yet erupted. Additionally, in many cases, teeth fall out of their sockets during cremation. Their rarity in most of the examined series shows that they were overlooked or neglected by the buriers when selecting skeletal remains from the remnants of the funeral pyre (see Chapter 3). Further clues are offered by histology, the manufacture of thin sections for counting the density of osteons in the compacta of long bones or the tooth cementum annulation (TCA). In practice these methods are highly time-consuming if hundreds of cremations have to be examined. As far as TCA is concerned, the first results for burned teeth are now available. It is, however, apparent that the burning intensity plays an important role in this method and the physiological background of these rings is not yet sufficiently clarified. The divergence between chronological and biological age may be as much as 10 years or more for late mature or older individuals (Jankauskas *et al.*, 2001; Pilloud, 2004). An alternative possibility for the aging of cremated children and juveniles is provided by metrics. Corresponding correlations have been found for the thickness of the vault wall as well as for the thickness of the compacta from the femur and other long bones. They even allow the recognition of the known growth spurts, which today occur at the ages of around 6 and 12–13 years and in former times can be measured at the ages of 7–8 and 13–15 years (Wahl, 1983, 1988a: 75).

To determine sex in cremation remains, the more voluminous characteristics of cranium and pelvis are not available. It is seldom the case that, for example, *inclinatio frontale*, *forma orbitae, incisura ischiadica major* or *angulus pubis* can be judged. The focus is more on *glabella, margo supraorbitale*, *processus mastoideus*, *protuberantia occipitalis externa* or *sulcus praeauricularis*. Empirical values show that even these anatomical regions are not generally preserved. Once again measurements are of greater value (e.g., thickness of the vault wall, transverse and sagittal diameter of *dens axis*, *pars petrosa ossis temporalis*, thickness of diaphyseal compacta of humerus and radius). Besides the diagnostic features, their use makes it possible to recognize and detect a remarkable sex dimorphism (Gejvall, 1963; van Vark, 1975; Wahl, 1996; Graw *et al.*, 2005; Norén *et al*, 2005). For example, the cremation remains from Stettfeld yielded significant metrical distances between men and women for 25 of 29 measurements (Wahl, 1988c: 105).

In comparison with unburned skeletal remains, the calculation of body height of cremated individuals is of much greater uncertainty because two correlations are involved. Firstly, between the diameter of the proximal epiphyses of humerus, radius and femur and the corresponding bone length and secondly between the bone length and the body height. Both are dependent on separate tolerances. In many cases the above-mentioned skeletal parts are only fragmentarily preserved. Most famous in this context is the so-called "Nomogramm" (Rösing, 1977: 71) into which an overall shrinking factor of 12% is incorporated.

With regard to morbidity, skeletal remains principally document only diseases that leave traces of bones or teeth. Indeed this is merely a small portion of the whole spectrum of illnesses our ancestors had to endure. Concerning cremated material, the situation is even more unfavourable, because usually only small pieces of the skeleton are available. Teeth crowns that could show carious defects cannot be judged because they split into microscopic pieces above temperatures of 700°C. In these cases, carious defects can be detected only if they have reached the dentine portion. Vertebrae and joints, which perhaps carry degenerative alterations, are often missing. Hence, the record of pathological conditions is barely more than casuistic. Only large series with plenty of cremation material may occasionally yield tendencies enabling us to determine, for example, different frequencies of degenerative symptoms between males and females or between distinct age groups (see below; Wahl, 1988c: 120).

As already mentioned, the possibilities of determination are highly dependant on the preserved quantity of cremated material as well as on the combustion intensity of the cremations. The complete cremation remains of an adult may weigh up to 2000 g or more. It was the exception rather than the rule that prehistoric or early historic cremations reached this weight. For instance, eight Roman series from Germany and Switzerland exhibit average weights between 199.3 and 814.1 g for males and between 18.8 and 555.0 g for females (Wahl, 1988c: 89). These losses may depend on the burial rites, conditions within the deposit soil (bioturbation, microorganisms, chemical and physical influences), excavation, or the subsequent preparation (e.g., purifying of the calcined bones from adherent mud). Cremations that are exceptionally voluminous in relation to the existing averages are principally suspected to belong to individuals of higher social rank, to include a large portion of animal bones, or to document a double or multiple burial. For subadults a correlation between age at death and weight of cremation remains has been repeatedly documented (Wahl, 1988a: 52; Caselitz, 1995).

To estimate the intensity or the grade of combustion, the scheme suggested by Chochol (1961) is frequently used, but this pattern is not totally congruent with the graduation reported in more recent investigations by Malinowski and Porawski (1969), Herrmann (1977a) and Wahl (1981) (see Table 9.1). The distinction between human and animal bones in an uncremated and relatively completely preserved material is not difficult for specialists, but in burned

TABLE 9.1 Combustion Grades After Observations in a Modern Crematorium

Combustion grade	Colour of bone fragments	Observations	Temperatures	Graduation after Chochol '61	Graduation after Malinowski/ Porawski '69
I	Yellowish white	Like fresh unburned bone	To 200°C		
	Ivory-coloured	First shrinking (about 1–2%)	Around 250°C		
	Glassy light grey				
I–II		Roots of teeth brown till dark brown			
		Crowns of teeth without injuries			
II	Brown	–	Around 300°C	(d)	2
	Dark brown	–			
	Black	Incomplete combustion resp.	Around 400°C		
		Charring of organic substances		e	3
III	Grey	Compacta inside sometimes still black	Around 550°C		
	Milky light grey				
	Bluish grey				
III–IV		Roots of teeth milky-grey till grey			
		Crowns of teeth black with microscopical cracks			

(Continues)

TABLE 9.1 (*Continued*)

Combustion grade	Colour of bone fragments	Observations	Temperatures	Graduation after: Chochol '61	Graduation after: Malinowski/ Porawski '69
IV	Milky white	Chalky surface	On and after 650–700°C	a	4
	Mat cretaceous	Bone calcined, light and less resistant On and after 750°C continuous shrinking			5
V	Old white	Hard and brittle	On and after 800°C	c	
	Cream-coloured	Appearance of parabolic heat cracks and deformations			
	Brownish, greyish, ochre	Outside colour corresponding with soil conditions Compacta: surface of fracture always white Spongiosa sometimes yellowish-ochre Crowns of teeth survive only if not yet erupted Average shrinking 10–12% Maximum shrinking up to 25%			

Graduation from Wahl (1983) in comparison to other authors (Chochol, 1961; Malinowski and Porawski, 1969), modified and completed.

skeletal remains, the fragmentation and the potential deformation sometimes make them difficult to address. For example, specific anatomical regions (e.g., metapodials, phalanges, the proximal epiphysis of femur from subadults) from human, pig and bear are markedly similar, particularly in fragments, and much experience is necessary to distinguish them morphologically (Wahl, 2001: 161). Pigs in particular, or parts of them, are commonly mixed in Roman cremations. So far results have shown that at least 80% of all cremation burials contain animal bones that were burned together with the human corpse. Identifying them correctly is of exceptional importance, e.g. avoiding the misinterpretation of an adult human cremation containing bone fragments of a young pig as a double burial containing an adult and a child. Meticulous examination in cooperation with an archaeozoologist shows, in several cases, which parts from which animals have been cocumbusted on the funeral pyre (Hensen *et al.*, 2004). Sometimes the quantity of animal bones represents a larger portion of the cremation than the human remains (Asskamp *et al.*, 1987).

Another method of distinguishing human bone fragments and animals of various species from each other is histological investigation. Remarkable differences can be determined by considering the size, morphological details and the frequency of osteons (Rämsch and Zerndt, 1963; Dittmann, 2003). The technical, instrumental (and financial) effort of doing this is unquestionable in forensic cases, but may be difficult to justify in cases where perhaps hundreds of possible animal bones are present within one prehistoric cremation that, in turn, belongs to a graveyard of hundreds of cremation burials.

The identification of double burials, especially distinguishing double burials from the so-called "cremation displacements", and the consideration of the combustion grade are more problematical. Larger Roman cemeteries regularly contain one or more ustrinas – combustion places upon which the funeral pyre for public use was built and used over many years or even decades by various families to cremate their deceased relatives. After burning, the cremation remains were collected and buried elsewhere in the graveyard. If skeletal residues from the preceding combustion had been overlooked by the previous funeralists and gathered by their successors, the corresponding cremation burial would indeed contain parts from a second individual. But this situation must not be determined as a double burial *sensu stricto*. Admixtures like this are called cremation displacements. As criteria of exclusion of double burials, either the quantitative relation (portions of the weight of the according bone remains) or the representation (delivery of all major parts of the skeleton) from both individuals has to be estimated (Wahl, 2001: 160). Cremation displacements have been found among the burials on the Roman cemetery from Stettfeld in more than 14% of the cremation graves.

In the cremation displacement context, a distinction based on differing combustion grades is usually unsuitable, because cremations with a funeral pyre often do not produce remains of homogenous appearance. At the edges of such a pile, lower temperatures may prevail so that head, arms, hands or feet perhaps may suffer lower combustion grades. Consequently, they exhibit a smaller shrinking rate than the bones from the rest of the corpse. Less shrunken remains look more robust in comparison with completely burned body parts and therefore can be easily misinterpreted as belonging to a different individual. If the funeral pyre collapses during the combustion or the combustion is affected by wind and weather, all regions of the skeleton can in principle be affected in this way.

Real double burials are indeed relatively rare on graveyards from the Imperial Roman Period. Their percentage mostly lies below 5%. Thereby one can find not only the classical combination of an adult together with a child, but also two adults or two children of possibly different ages (Wahl, 1988c: 117; Caselitz, 2000). Another normal variant is the combination of a cremated adult together with an unburned infant. The comprehensive problem of double burials is of particular relevance, because in this point a (possibly systematical) false diagnosis can lead to considerable distortions under demographical aspects.

Equally, determination and definition of special burials are difficult (Wahl, 1994). The parameters that lead to a special treatment of the deceased (e.g., age at death, circumstances of death, religious background or ethnical descent) are markedly varied and are only seldom detected from the archaeological context alone. As far as inhumation graves are concerned, such anomalies as skeletons lying on the belly or dislocated body parts within supine burials were regularly thus explained. Constructional and/or equipmental divergences in the grave are mostly due to social differences. And in examining cemeteries that contain cremation burials combined with inhumation graves, the question principally arises as to which population group was cremated and which was left unburned.

Finally, human bones bearing traces of violence provide interesting insights. Finds such as those from the settlements or wells from Lomersheim and Pforzheim (Wahl, 1997) exhibit remarkable and spectacular parallels in Regensburg (Schröter, 1985), Bonn (Wahl *et al.*, 2002/2003) and Augst in Switzerland (Kaufmann and Furger, 1988). The vault fragment from Walheim that had been used as profane digging tool for preparing sand holes into which tip-bottomed amphoras were put (Wahl and Planck, 1989) gives an insight to the attitude towards unburned human skeletal remains, if in fact their human origin were really known.

In addition, DNA analysis has been successfully performed on cremated bones (Pusch *et al.*, 2000; see Walker *et al.*, this volume). As anticipated, it seems that replicable fragments of DNA survive only under milder combustion conditions. Experimental combustions and investigation of cremation remains have shown that prehistoric cremation temperatures reached 1000°C and more (Swillens *et al.*, 2003; Wahl *et al.*, in prep; see also Becker *et al.*, 2005).

ANTHROPOLOGICAL INVESTIGATION

WEIGHT OF CREMATIONS AND GRADE OF COMBUSTION

The average weight of 65 cremations from the Urn field culture is 393 g, from the Hallstatt age/LaTene period is 347 g ($n = 102$) and from the Imperial Roman Period is 317 g ($n = 850$; compare Table 9.2), displaying an apparent continual decrease of about 20%. But looking at the averages for males, females and nonadults, we can recognize an increase in the cremation quantity from the Urn field culture to the Roman times for all three groups. The drop in mean values therefore obviously depends on the decrease in weights of the cremations from sexually undetermined adults. Within the Urn field culture and the Imperial Roman Period, male cremation remains are on average around 30% heavier than the females and during the Hallstatt/LaTene times more than 40%. Thereby the quantities within the single series vary enormously, e.g., male cremations in Stettfeld from 50 to 1955 g, the female ones from 15 to 1960 g, the undetermined adults from 1 to 1160 g and the subadults from 1 to 680 g. The largest quantity in Welzheim indeed is only 580 g. Such

TABLE 9.2 Average Weights of Cremation Remains from Males, Females, Subadults (Children and Juveniles) and Totals (Undetermined Adults Inclusive) in Different Cultures Resp. Periods

	Ufc	Ha/LT	IRP
Males	562 (19)	572 (24)	638 (192)
Females	438 (20)	401 (23)	479 (177)
Subadults	87 (4)	94 (20)	106 (39)
Total	393 (65)	347 (102)	317 (850)

Data in grams. Number of individuals in brackets. Ufc, Urn field culture; Ha, Hallstatt age; LT, LaTene period; IRP, Imperial Roman Period).

differences obviously show that the Roman burial rites were not accomplished with the same accuracy everywhere. In addition, therein may be a concealed social ranking. In contrast, looking at the single series from the Hallstatt age, it becomes clear that the male cremations are by no means always heavier than the female ones and some of these conflicting results could depend on relatively small numbers of individuals from certain cemeteries.

Concerning the grade of combustion, no consistent difference can be established. In several Roman series, the cremation remains of females are often found more homogenous in grade V (around 800 degrees or more) than cremations of men, and incompletely burned parts are typical of cremation burials from the Pre-Roman Iron Age, although this distinction cannot yet be demonstrated as mathematically significant. The same is true for the tendency within a single grave. Animal bones usually exhibit a slightly lower combustion grade than the human remains. It is possible that they were thrown into the fire later or laid at the edge of the funeral pyre. For the Pre-Roman Iron Age, such divergences between sexes are more subtle, with male cremations perhaps indicating increased burning.

PORTION OF SUBADULTS

The percentage of children and juveniles is one of the most important parameters concerning the demographic structure of prehistoric series. In many cases, neonates and children up to an age of 2–3 years are clearly underrepresented in comparison to the expected values. One possible explanation could be special burials for this age group. Another possibility is that they are less resilient and therefore less likely to survive. The portion of subadults in 71 cremations from the Urn field culture is 8.5%, in 133 cremations from the Hallstatt age is 21.1%, and in 673 cremations from the Imperial Roman Period is 30.4% (see Table 9.3), showing a marked increase. Until now only two cremation burials from adults have been reliably dated as belonging to the LaTene period.

Within the inhumation graves, the proportions of children or juveniles are still clearer: Urn field culture 8.7% ($n=30$), Hallstatt C/D 31.5% ($n=575$), LaTene period 38.1% ($n=21$) and Imperial Roman Period 52.5% ($n=59$). Thereby, it has to be taken in consideration that the Roman inhumations do not reflect a representative cross-section of the population; most are neonates or young infants, while many of the rest are older adults. These are often buried in unusual body postures and declared as special burials.

TABLE 9.3 Percentage of Subadults in Different Cultures Resp. Periods and Burial Types

	Ufc	Ha C/D	LT	IRP
Subadults inhumations	8.7% (30)	31.5% (575)	38.1% (21)	52.5% (59)
Subadults cremations	8.5% (71)	21.1% (133)	0% (2)	30.4% (673)
Subadults total	8.7% (101)	29.5% (708)	34.8% (23)	32.6% (732)

Ufc, Urn field culture; LT, LaTene period; IRP, Imperial Roman Period; Ha C/D, late Pre-Roman Iron Age. Total number of individuals in brackets.

Contemporary documentary evidence reports that infants were not cremated but interred uncremated if they died before their first teeth erupted. Inhumated adults reveal fewer grave goods than does cremated adults, perhaps indicating that they belonged to less privileged groups, whose circumstances of death excluded a cremation burial. Metric investigations on the material from Stettfeld have shown a strong correlation that inhumated and cremated individuals at least belonged to the same population substrate. Therefore, taking the percentage of children and juveniles from cremation burials combined, according to their unequal proportion of quantities between both burial types, their presence decreases to 32.6% ($n = 732$). The overall picture shows an increase in the proportion of subadults from the Urn field culture to the LaTene period. Only a relatively small number of individuals of the LaTene period could qualify this statement. Especially between the Urn field culture and the Hallstatt C/D we can observe a significant increase. While the percentages of subadults within inhumation graves and cremation burials from the Urn field culture are nearly identical, differences do exist between both burial types within the Hallstatt age.

AGE AT DEATH

The average age at death is another important indicator in estimating the living conditions of a (pre)historic population sample, including factors influencing health, ecologic conditions, and additionally, a comparison between genders. The mean age at death of all individuals from Baden-Württemberg of the Urn field culture is 35.1 years ($n = 99$), Hallstatt age is 28.6 years ($n = 650$), LaTene period is 32.2 years ($n = 23$) and the age for the Imperial Roman Period is 34.7 years ($n = 792$; Table 9.4). Because of the large sample size of the Hallstatt period, the low life expectancy value would be assumed to be accurate – at least in relation to the Imperial Roman Period and reservedly in comparison with the Urn field culture. Even if the small sample sizes from the LaTene period are carefully interpreted, they fit the general trend of males and the total population. A decrease of mean age at death in the Hallstatt C/D and an increase in the Imperial Roman Period can be determined both for men and women, as well as for the overall population. Hence, from about 800 to 500 BC the life expectancy increased until the birth of Christ and continued to do so during the first century AD. The proportions indicate that all three groups are similar: 6.3 years for males, 7.3 years for females and 6.1 years in total. As expected, women reached a lower mean age than men throughout all

TABLE 9.4 Average Age at Death (Years) of Males, Females and Totals (Children, Juveniles and Undetermined Adults Inclusive) in Different Cultures Resp. Periods

	Ufc	Ha C/D	LT	IRP
Males	38.2 (46)	37.2 (178)	41.8 (7)	43.5 (221)
Females	37.0 (19)	31.9 (150)	48.8 (6)	39.2 (247)
Total	35.1 (99)	28.6 (650)	32.2 (23)	34.7 (792)

Ufc, Urn field culture; LT, LaTene period; IRP, Imperial Roman Period; Ha C/D, late Pre-Roman Iron Age. Total numbers of individuals in brackets.

observed epochs. The data for women from the LaTene period are outliers. They are solely based on a small series of females from Nebringen aged far above average.

A comparison of the contemporary dated unburned and cremated individuals, either of the Urn field culture or of the Hallstatt period, shows no remarkable distinctions between the sexes. The age of cremated men from Hallstatt reached mean ages that were only insignificantly lower than those of their inhumated contemporaries. There is a difference of about 2 years among the females. The cremated men during the time of Urn field culture were a little older than the unburned ones. An adequate differentiation for the Imperial Roman Period is not particularly meaningful because of the unrepresentative composition of the group of inhumation graves already mentioned. Nevertheless, more men than woman are present amongst the predominantly older unburned adults.

BODY HEIGHT

Many of the factors influencing body height are fixed genetically. The rest is determined primarily by nutrition and physical burden during the growth phase. Additionally body height is considered as one of the factors influencing the choice of partner. Another phenomenon traced within the complexes from the Roman times in South West Germany is the dependence of height on the social status of the buried individuals, where the social status and rank are reflected by grave construction and/or furnishing. Whilst, in general, taller individuals are found in socially higher ranking groups, such correlation has been seldom found within the investigated material. In some cases even a reversed correlation is present. It seems that in some series, smaller bodied individuals – possibly from the Mediterranean region – commingled with resident individuals. The cremated women from Stettfeld may be an indication of foreign elements or populations of differing origin. An examination of 39 women revealed their mean body height to be about 1.60 m, but their distribution showed two peaks (an absolute maximum at 1.56–1.57 m and a second, relative maximum at 1.62–1.63 m), as well as a wide variation. The accompanying 37 men reached a mean value of 1.71 m, but exhibited a smaller variation and an average corresponding with the maximum (Wahl, 1988c: 100).

A diachronic comparison of men results in the following mean values of body height: Urn field culture 171.4 cm ($n=28$), Hallstatt C/D 171.0 cm ($n=61$), LaTene period 168.3 cm ($n = 6$) and Imperial Roman Period 170.6 cm ($n=64$; see Table 9.5; calculated for cremation remains after Rösing (1977) and for unburned funerals after Bach (1965), Breitinger (1938) and/or Olivier *et al.* (1978)). Hence, no noteworthy tendency can be stated, but the LaTene period is represented sparsely with only six individuals. The average body height of women varies only between 159.5 and 161.7 cm within the same time span.

In separating inhumation graves and cremation burials we found an inclination to slightly lower body heights in cremated males and females of the Urn field culture and the Hallstatt age. But this difference may possibly depend on methodological aspects. For the Imperial Roman Period, the data are almost identical in both burial types. Independent of methodological problems, the

TABLE 9.5 Average Body Height (cm) of Males and Females (Only Adults) in Different Cultures Resp. Periods and Burial Types

	Ufc	Ha C/D	LT	IRP
Males inhumations	171.8 (21)	170.9 (55)	168.3 (6)	170.0 (15)
Males cremations	171.0 (7)	167.7 (6)	–	170.8 (49)
Males total	171.4 (28)	171.0 (61)	168.3 (6)	170.6 (64)
Females inhumations	162.0 (1)	161.3 (53)	161.7 (3)	160.0 (4)
Females cremations	159.4 (12)	159.6 (9)	–	160.2 (46)
Females total	159.5 (13)	161.1 (62)	161.7 (3)	160.2 (50)

Ufc, Urn field culture; LT, LaTene period; IRP, Imperial Roman Period; Ha C/D, late Pre-Roman Iron Age. Total numbers of individuals in brackets.

lower social status of the inhumated individuals combined with the "infiltration" of cremations by small-bodied persons could have led to an alignment of the values.

PATHOLOGICAL CONDITIONS

Within cremation remains, pathological conditions can be documented only in a casuistic manner. Only larger series with comparatively voluminous cremation quantities allow a statistical evaluation of diagnoses. In this context, the cremation graves from the Roman cemetery from Stettfeld offer some trends (compare Wahl, 1988c: 120). Corresponding to other authors' experiences, mastication apparatus alterations were most common. Pathological conditions of the parodontium appear in roughly 64% of all adults. Contrasting to expectations, the rate of incidence does not increase with advancing age. With reference to individual sets of teeth, the molars, in particular, in the upper jaw and the incisors and canines in the lower jaw are often similarly affected. In Stettfeld, almost 50% of all individuals with dental remains exhibit signs of calculus, although certainly some calculus was lost during combustion. The difference between men and women is negligible. Cervical caries is found in 5% of all adults, but carious defects in crown regions are likely to have been much more frequent.

Apical suppurations at the root increase with advancing age at death: in adult individuals *c.* 22%, in mature individuals about 32%, and in senile individuals nearly 43%. Altogether 33% of the male and around 25% of the female jaws, more often the front teeth rather than the premolars and molars, are affected. Advanced stages (abscesses, fistulas or cysts) can be observed in nearly 22% of all adults. Antemortem tooth loss, in most of the cases, is probably caused by caries. Tooth loss almost always involves premolars and molars, most frequently the first molars. This phenomenon occurs in males more often than in females, but nevertheless in *c.* 41% of all adults.

As with other cremation series, *spondylosis deformans* is the most frequent pathological condition of the postcranium. It appears in about 36% of male burials and 21% of female ones. In most of the cases, the initial stadium spondylosis stage 1 was diagnosed. In only one case within the whole graveyard was a block of cervical vertebrae preserved (spondylosis stage

4). The distribution clearly shows that women suffered predominantly in the cervical and thoracic vertebrae and men in the cervical and lumbar vertebrae. This seems to correlate with gender-specific burdens. The dependence of spondylosis on age at death is also clear. Whilst the rate of incidence in females increases from 11% in adult to more than 32% in mature and 53% in senile individuals, the corresponding values in males reach 25, 49 and 66%, respectively. Similar tendencies can be shown for *spondylarthrosis deformans*. An association with specific social groups still remains to be examined in this context.

CONCLUDING REMARKS

On the basis of these partial results, it is not yet possible to create a complete picture of the peoples of the Pre-Roman Iron Age and the Imperial Roman Period in southwestern Germany, but shadowy contours are already perceptible. At the completion of the present and planned anthropological research, more detailed knowledge will be gained. We can also expect the development of more advanced methodological tools for the handling and interpretation of cremation remains.

Finally, several other sources that yield essential clues useful for reconstructing the appearance and living conditions over the centuries in question should be mentioned, but have yet to be exploited: the bog bodies from Northern Germany, Scandinavia and Great Britain as well as numerous preserved paintings, mosaics and relief images. Last but not least, the grave monuments found all over the *Imperium Romanum* give evidence of high mobility and the manifold mixture of populations produced.

ACKNOWLEDGMENTS

Many thanks go to the editors Christopher W. Schmidt and Steven A. Symes for asking me to participate in this volume and for giving me the opportunity to present results of cremation remains' research from "Old Europe".

I would like to thank Felix Wahl for helping me approach the first step of translation from German to English and I am particularly grateful to my very good friend and colleague David Bibby, who has invested much time and effort in polishing up the original rough version.

Furthermore I thank all who directly or indirectly have helped with detailed information, criticism and support.

REFERENCES

Asskamp, R., Kokabi, M. and Wahl, J. (1987) Das römische Brandgrab von Altdorf, Gde. Ettenheim, Ortenaukreis. *Fundber. aus Baden-Württemberg* **12**, 325–363.

Bach, H. (1965). Zur Berechnung der Körperhöhe aus den langen Gliedmaßenknochen weiblicher Skelette. *Anthrop. Anz.* **29**, 12–21.

Becker, M., Döhle, H.-J., Hellmund, M., Leineweber, R. and Schafberg, R. (2005). Nach demgroßen Brand. Verbrennung auf dem Scheiterhaufen - ein interdisziplinärer. *Ansatz. Ber. Röm.-Germ. Kommission* **86**, 61–195.

Berszin, C. and Wahl, J. (in prep.). Anthropologische Untersuchung der Skelettreste aus dem römischen Gräberfeld von Heidelberg-Neuenheim. Working title.

Breitinger, E. (1938). Zur Berechnung der Körperhöhe aus den langen Gliedmaßenknochen. *Anthrop. Anz.* **14**, 249–274.

Breitinger, E. (1954). Das Brandknochengemenge von Dauborn. *Nass. Ann.* **65**, 48–61.

Burger-Heinrich, E. (in prep.). Das römische Gräberfeld von Rottweil „Kapellenösch" – Anthropologische Befunde. Forsch. u. Ber. z. Vor- u. Frühgesch. in Baden-Württemberg.

Caselitz, P. (1981). Die Ergebnisse der anthropologischen Untersuchung der Leichenbrände eines Gräberfeldes der vorrömischen Eisenzeit auf Gemarkung Krummesse, Hansestadt Lübeck. Lübecker Schriften z. *Archäologie u. Kulturgeschichte* **5**, 61–80.

Caselitz, P. (1995). Quantitative Unterschiede im Leichenbrandgewicht von Kindern der römischen Kaiserzeit. Hinweis auf eine abweichende Bestattungssitte? *Arch. Korr.bl.* **25**, 321–329.

Caselitz, P. (1998). Die menschlichen Leichenbrände des ältereisenzeitlichen Gräberfeldes von Godshorn. In *Neue bronze- und eisenzeitliche Gräberfelder aus dem Regierungsbezirk Hannover. Materialh. Ur- u. Frühgesch.* (E. Cosack, ed.), pp. 177–216, Niedersachsen A 26.

Caselitz, P. (2000). Die jüngerkaiser- bis völkerwanderungszeitliche Bevölkerungsstichprobe von Satels, Gde. Wingst, Ldkr. Cuxhaven. *Die Kunde N.F.* **51**, 63–94.

Chochol, J. (1961). Anthropologische Analyse menschlicher Brandreste aus den Lausitzer Gräberfeldern in Ústí Nad Labem-Strekov II und in Zírovice, Bezirk Cheb. In. *Die Lausitzer Kultur in Nordwestböhmen. Monumenta Arch.* 8 (E. Plesl, ed.), pp. 273–290.

Dittmann, K. (2003). Histomorphometrische Untersuchung der Knochenmikrostruktur von Primaten und Haustieren mit dem Ziel der Speziesidentifikation unter Berücksichtigung von Domestikationseffekten. *Anthrop. Anz.* **61**, 175–188.

Doklàdal, M. (1969). *Über die heutigen Möglichkeiten der Personenidentifikation auf Grund von verbrannten Knochen. Aktuelle Kriminol.*, pp. 223–246.

Doklàdal, M. (1970). Ergebnisse experimenteller Verbrennungen zur Feststellung von Form- und Größenveränderungen von Menschenknochen unter Einfluß von hohen Temperaturen. *Anthropologie* **8**, 3–17.

Gaubatz, A, Wahl, J. and Kokabi, M. (1988). Römische Brandgräber aus Lützelsachsen und Sulzbach, Stadt Weinheim, Rhein-Neckar-Kreis. *Fundber. aus Baden-Württemberg* **13**, 249–266.

Gejvall, N.G. (1955). The cremations at Vallhagar. In *Vallhagar, a Migration Period settlement on Gotland, Sweden* (M. Stenberger and O. Klindt-Jensen, eds), pp. 700–723. Munkgaard, Kopenhagen.

Gejvall, N.G. (1963). Cremations. In *Science in Archaeology* (D. Brothwell and E. Higgs, eds), pp. 468–479. Thames and Hudson, London.

Graw, M., Wahl, J. and Ahlbrecht, M. (2005). Course of the meatus acusticus internus as criterion for sex differentiation. *Forensic Science International* **147**, 113–117.

Grimm, H. (1953). Feststellungen an Leichenbränden aus den Urnen von Berlin-Britz. *Berliner Bl. Vor- u. Frühgesch.* **2**, 59–61.

Grimm, H. (1961). Der gegenwärtige Stand der Leichenbranduntersuchungen. *Ausgr. u. Funde* **6**, 299–308.

Großkopf, B. (2004). Leichenbrand – Biologisches und kulturhistorisches Quellenmaterial zur Rekonstruktion vor- und frühgeschichtlicher Populationen und Funeralpraktiken. Diss. Leipzig.

Grupe, G. and Herrmann, B. (1983). Über das Schrumpfungsverhalten experimentell verbrannter spongiöser Knochen am Beispiel des Caput femoris. *Z. Morph. Anthrop.* **74**, 121–127.

Hensen, A., Wahl, J., Stephan, E. and Berszin, C. (2004). Eine Ärztin aus dem römischen Heidelberg. *Arch. Korr.bl.* **34**, 81–100.

Herrmann, B. (1972). Zur Beurteilung von Kohlenstoff verfärbungen bei Leichenbränden. *Ausgr. u. Funde* **17**, 275–277.

Herrmann, B. (1973a). Das Combe Capelle-Skelet. *Ausgr. in Berlin* **3**, 1972, 7–69.

Herrmann, B. (1973b). Möglichkeiten histologischer Untersuchungen an Leichenbränden. *Mitt. Berliner Ges. Ethnol. u. Urgesch.* **2**, 164–167.

Herrmann, B. (1977a). Anthropologische Bearbeitung der Leichenbrände vom Gräberfeld Lengerich-Wechte, Kreis Steinfurt. *Bodenaltertümer Wesfalen* **15**, 97–118.

Herrmann, B. (1977b). Über die Abhängigkeit der Schrumpfung vom Mineralgehalt bei experimentell verbrannten Knochen. *Anthrop. Anz.* **36**, 7–12.

Jankauskas, R., Barakauskas, S. and Bojarun, R. (2001). Incremental lines of dental cementum in biological age estimation. *Homo* **52**, 59–71.

Kaufmann, B. and Furger, A.R. (1988). Menschenknochen. *Forschungen in Augst* **9**, 178.

Kloiber, Ä. (1953). Die Skelettreste aus dem römerzeitlichen Brandgräberfeld in der Wurmstraße (Kreuzschwestern). *Jahrb. d. Stadt Linz* 1952, 491–538.

Kloiber, Ä. (1963). Anthropologische Untersuchungen an Leichenbränden des hallstattzeitlichen Gräberfeldes von Leoben-Hinterberg. *Schild von Steier* **11**, 17–27.

Kokabi, M. and Wahl, J. (2001). Über die Aussagemöglichkeiten verbrannter Tier- und Menschenknochen anhand ausgewählter Befunde aus der vorrömischen Eisenzeit und der römischen Kaiserzeit. In *Archäologisches Zellwerk. Festschrift für Helmut Roth* (E. Pohl, U. Recker and C. Theune, eds.), pp. 61–70. Verlag Marie Leidorf GmbH, Rahden/Westf.

Kokabi, M., Amberger, G. and Wahl, J. (1994). Die Knochenfunde aus der Villa rustica von Bondorf. In: A. Gaubatz-Sattler, Die Villa rustica von Bondorf (Lkr. Böblingen). *Forsch. u. Ber. z. Vor- u. Frühgesch in Baden-Württemberg* 51. Stuttgart, pp. 285–335.

Krumbein, C.N. (1934). Anthropologische Untersuchungen an urgeschichtlichen Leichenbränden. *Forsch. u. Fortschr.* **10**, 411–412.

Krumbein, C.N. (1935). Die anthropologische Bestimmung der Leichenbrände des Urnenfeldes von Loccum. *Die Kunde* **3**, 188–191.

Kühl, I. (1966). Beobachtungen an vorgeschichtlichen Leichenbränden aus Schleswig-Holstein. *Offa* **23**, 130–140.

Kühl, I. (1979). Die Leichenbrände vom Brandgräberfeld auf der Düne Wissing, Gemeinde Haldern, Kreis Wesel (früher Kreis Rees). Schleswig-Kiel.

Kunter, M. (1980). Analyse von Leichenbränden aus dem keltisch-römischen Gräberfeld von Wederath-Belginum. *Funde u. Ausgr. im Bezirk Trier* **12**, 41–45.

Kunter, M. (2000). Zeitgruppenvergleich bei der keltisch-römischen Leichenbrandserie von Wederath-*Belginum*. In *Kelten, Germanen, Römer im Mittelgebirgsraum zwischen Luxemburg und Thüringen* (A. Haffner and S. von Schnurbein, eds), pp. 345–348. Kolloquien zur Vor- und Frühgeschichte Bd. 5. Dr. Rudolf Habelt GmbH, Bonn.

Kunter, M. (in press). Leichenbrandreste aus einem hallstattzeitlichen Brandgräberfeld bei Bopfingen.

Kunter, M. and Malek, C. (in prep.). Die Leichenbrände des keltisch-römischen Gräberfeldes von Wederath-Belginum im Hunsrück.

Lisowski, F.P. (1968). The investigation of human cremations. In *Festschrift für Karl Saller, Anthropologie und Humangenetik*, pp. 76–83. Gustav Fischer Verlag, Stuttgart.

Malinowski, A. and Porawski, R. (1969). Identifikationsmöglichkeiten menschlicher Brandknochen mit besonderer Berücksichtigung ihres Gewichts. *Zacchia* **44**, 1–19.

von Mellen, J. (1679). Historia Urnae sepulchralis Sarmaticae anno 1674 repertae. Jena.

Müller, C. (1958). Schätzung der Körperhöhe bei Funden von Leichenbränden. *Ausgr. u. Funde* **2**, 52–58.

Müller, C. (1964). Methodisch-kritische Betrachtungen zur anthropologischen Untersuchung von Leichenbränden. *Prähist. Zeitschr.* **42**, 1–29.

Nierhaus, R. (1959). Das römische Brand- und Körpergräberfeld „Auf der Steig" in Stuttgart-Bad Cannstatt. Veröff. Staatl. Amt Denkmalpfl. Stuttgart A 5, Stuttgart.

Norén, A, Lynnerup, N, Czarnetzki, A and Graw, M. (2005). Lateral angle: a method for sexing using the petrous bone. *American Journal of Physical Anthropology* **128**, 318–323.

Olivier, G., Aaron, C., Fully, G. and Tissier, G. (1978). New estimations of stature and cranial capacity in modern man. *Journal of Human Evolution* **7**, 513–518.

Parsche, F., Maurer, U. and Ziegelmayer, G. (1994). Die anthropologischen Ergebnisse der Brandbestattungen von Altlußheim-Hubwald. In *Das römische Gräberfeld von Altlußheim-Hubwald (Rhein-Neckar-Kreis)* (G. Dreisbusch, ed.), pp. 116–125. Materialh. z. Archäol. in Baden-Württemberg 24.

Pilloud, S. (2004). Läßt sich mittels der Altersbestimmung anhand des Zahnzementes auch bei älteren Individuen ein signifikanter Zusammenhang zwischen histologischem und reellem Alter finden? *Anthrop. Anz.* **62**, 231–239.

Plhak, M. (1987). Anthropologische Untersuchungen des Leichenbrandes aus dem römischen Brandgräberfeld Flur „Burgfeld" in Welzheim (Rems-Murr-Kreis). Manuscript 1987.

Pusch, C., Broghammer, M. and Scholz, M. (2000). Cremation practices and the survival of ancient DNA: burnt bone analyses via RAPD-mediated PCR. *Anthrop. Anz.* **58**, 237–251.

Rämsch, R. and Zerndt, E. (1963). Vergleichende Untersuchungen der Havers'schen Kanäle zwischen Menschen und Haustieren. *Archiv Kriminol.* **131**, 74–87.

Rhode, C.D. (1719). *Cimbrisch-Hollsteinische Antiquitäten-Remarques*. Liebezeit und Felginer, Hamburg.

Rösing, F.W. (1976). Die Leichenbrände der eisenzeitlichen Gräberfelder von Bargstedt I, Harsefeld und Issendorf III (Kreis Stade). Materialh. *Ur- u. Frühgesch. Niedersachsen* **11**, 131–143.

Rösing, F.W. (1977). Methoden und Aussagemöglichkeiten der anthropologischen Leichenbrandbearbeitung. *Arch. u. Naturwiss.* **1**, 53–80.

Schaefer, U. (1961). Grenzen und Möglichkeiten der anthropologischen Untersuchung von Leichenbränden. *Ber. V. Kongr. Vor- u. Frühgesch. Hamburg 1958*, 717–724.

Schaefer, U. (1964). Beiträge zum Problem der Leichenbranduntersuchung. *Z. Morph. Anthrop.* **55**, 277–282.

Schröter, P. (1985). Skelettreste aus zwei römischen Brunnen von Regensburg-Harting als archäologische Belege für Menschenopfer bei den Germanen der Kaiserzeit. *Archäol. Jahr in Bayern 1984, Stuttgart*, pp. 118f–119.

Schutkowski, H. and Hummel, S. (1987). Variabilitätsvergleich von Wandstärken für die Geschlechtszuweisung an Leichenbränden. *Anthrop. Anz.* **45**, 43–47.

Schutkowski, H. and Hummel, S. (1991). Vorgeschichtliche Bevölkerungen in Schleswig-Holstein. *Offa-Bücher 48*, Neumünster, 133–262.

Swillens, E., Pollandt, P. and Wahl, J. (2003). Zur Quantifizierung von Knochenverbrennungstemperaturen durch Röntgenbeugungs-Intensitätsmessungen. In *Beiträge zur Archäozoologie und Prähistorischen Anthropologie IV* (N. Benecke, ed.), pp. 204–210. Gesellschaft für Archäozoologie und Prähistorische Anthropologie, Konstanz.

Thieme, U. (1937). Leichenbranduntersuchungen, in *R. Dehnke, Burgundische Gräber von Lichtenberg, Kreis Torgau. Prähist. Zeitschr.* **28/29**, 265–284.

Thieme, U. (1939). Leichenbrände. In Ein Urnenfriedhof der älteren Eisenzeit vom Knickbrink, Gem. Krankenhagen (J. Erdniß, ed.). *Die Kunde* **7**, 30–36.

Trautmann, I. and Wahl, J. (2005). Leichenbrände aus linearbandkeramischen Gräberfeldern Südwestdeutschlands – Zum Bestattungsbrauch in Schwetzingen und Fellbach-Oeffingen. *Fundberichte aus Baden-Württemberg 28, Stuttgart*, 1–13.

van Vark, G.N. (1974). The investigation of human cremated skeletal material by multivariate statistical methods. I. Methodology. *Ossa* **1**, 63–95.

van Vark, G.N. (1975). The investigation of human cremated skeletal material by multivariate statistical methods. II. Measures. *Ossa* **2**, 47–53.

Volk, P., Niedermann, H. and Schneider, B. (1988). Frauen am römischen Limes. Anthropologische Rekonstruktion der Bevölkerung von Kastell und Vicus Schirenhof. In *Festschrift für H.-G. Hillemanns* (M. Hilgarth and M. Mönig-Schuth, eds), pp. 549-587. Konstanz.

Wahl, J. (1981). Beobachtungen zur Verbrennung menschlicher Leichname. *Arch. Korrbl.* **11**, 271–279.

Wahl, J. (1982). Leichenbranduntersuchungen. *Prähist. Zeitschr.* **57**, 1–125.

Wahl, J. (1983). A contribution to metrical age determination of cremated subadults. *Homo* **34**, 48–54.

Wahl, J. (1988a). Süderbrarup. Ein Gräberfeld der römischen Kaiserzeit und Völkerwanderungszeit in Angeln. II. Anthropologische Untersuchungen. Urnenfriedhöfe Schleswig-Holsteins 11/2, Neumünster.

Wahl, J. (1988b). Die Leichenbrände des römischen Gräberfeldes von Schankweiler, Kreis Bitburg-Prüm. *Trierer Zeitschr.* **51**, 367–422.

Wahl, J. (1988c). Menschenknochen. In Das römische Gräberfeld von Stettfeld I. *Forsch. u. Ber. z. Vor- u. Frühgesch. in Baden-Württemberg 29, Stuttgart* (J. Wahl and M. Kokabi, eds), pp. 46–223.

Wahl, J. (1988d). Prähistorische Anthropologie. In *Archäologie in Württemberg* (D. Planck, ed.) pp. 439–464. Konrad Theiss Verlag, Stuttgart

Wahl, J. (1994). Zur Ansprache und Definition von Sonderbestattungen, in *Forschungen u. Ber. z. Vor- u. Frühgesch in Baden-Württemberg 53, Stuttgart*, 85–106.

Wahl, J. (1996). Erfahrungen zur metrischen Geschlechtsdiagnose bei Leichenbränden. *Homo* **47**, 339–359.

Wahl, J. (1997). Römerzeitliche Menschenknochen mit Spuren von Gewalteinwirkung und Manipulation. In *Beiträge zur Archäozoologie und Prähistorischen Anthropologie I* (M. Kokabi, ed.), pp. 77–85. Gesellschaft für Archäozoologie und Prahistorische Anthropologie, Konstanz.

Wahl, J. (1998). Anthropologische Untersuchung der spätbronzezeitlichen Leichenbrände aus Singen, Kreis Konstanz. In *Die mittel- und spätbronzezeitlichen Grabfunde auf der Nordstadtterrasse von Singen am Hohentwiel. Forsch. u. Ber. z. Vor- u. Frühgesch. Baden-Württemberg 67* (W. Brestrich, ed.), pp. 395–419. Konrad Theiss Verlag, Stuttgart.

Wahl, J. (1999). Menschliche Skelettreste aus Erdwerken der Michelsberger Kultur. In *Beiträge zur Archäozoologie und Prähistorischen Anthropologie II* (M. Kokabi and E. May, eds), pp. 91–100. Gesellschaft für Archäozoologie und Prahistorische Anthropologie, Konstanz.

Wahl, J. (2001). Bemerkungen zur kritischen Beurteilung von Brandknochen. In *Beiträge zur Archäozoologie und Prähistorischen Anthropologie III* (E. May and N. Benecke, eds.), pp. 157–167. Gesellschaft für Archäozoologie und Prahistorische Anthropologie, Konstanz.

Wahl, J. and Planck, D. (1989). Ein menschliches Kalottenbruchstück als Schöpf- oder Grabgerät. *Fundberichte aus Baden-Württemberg* **14**, 373–385.

Wahl, J., König, H.G. and Wahl, S. (2002/2003). Die menschlichen Skelettreste aus einem Brunnen des Legionslagers in Bonn "An der Esche 4". *Bonner Jahrbücher* **202/203**, 199–226.

Wahl, J., Keefer, E., Baumeister, R. and Schmidt-Lawrenz, S. (in prep.). Experimentelle Verbrennungen zur Rekonstruktion prähistorischer Scheiterhaufen.

Wells, C. (1960). A study of cremation. *Antiquity* **34**, 29–37.

10

In the Heat of the Pyre: Efficiency of Oxidation in Romano-British Cremations – Did it Really Matter?

Jacqueline I. McKinley, B.Tech.

Wessex Archaeology, Portway House,
Old Sarum Park, Salisbury, Wiltshire

Introduction

Full oxidation of the body's soft tissues and, occasionally with some minor but acceptable variations, the organic components of the bone is a requisite of modern Western cremation, borne of the desire to sanitize and render inert the decomposing corpse that lay at the forefront of the aims of those involved in the early struggle to reintroduce cremation in Europe at the end of the nineteenth century (Thompson, 1889: 1–2; Parsons, 2005). Most cremations in Western societies are currently undertaken in energy-efficient, temperature- and oxygen-controlled cremators fuelled by gas (predominantly), electricity or oil (Chamberlain, 2005; Davies, 2005; Mates, 2005). In many countries, certainly in Europe and North America, the cremation in itself does not form an intrinsic part of the ritual surrounding the committal of the body; there is no 'belief' upholding its undertaking and it essentially comprises a means to an end.

This 'deritualized' view of cremation and its attendant requirement for full oxidation are not followed by some contemporary cultures (Barber, 1990: 381–387; Perrin, 1998; Downes, 1999: 23 and 28), nor would it necessarily have been required by other cremating cultures within Europe's past. Where it was the 'magic' of transformation from one state (the corpse, recognizable as the individual) to another (burnt, clearly altered and 'purified' remains) that was required, the degree of oxidation attained may have been of little or no consequence (McKinley, 2006). Minor, and occasional major, variations in oxidation of the bone observed in the remains of archaeological cremation burials across the temporal range of the rite in Britain (Neolithic to Norse;

c. 4000 BC to tenth century AD) suggest a complacent or possibly simply pragmatic attitude to the level of oxidation.

The Romans, possibly as a result of their belief that the soul left the body with the final breath (Toynbee, 1971: 43), in part seem to have shared the modern Western 'ritual-free' attitude to cremation, seeing it as a means of rendering inert a corpse that would otherwise putrefy and could potentially be desecrated, providing a more convenient 'package' for final disposal and as an excuse for 'a good show' (Noy, 2005); some familiar concepts to Western sensibilities. The common inclusion of pyre goods other than those simply forming part of the deceased's burial attire, items to accompany the dead to the other world, does, however, suggest that some ritual significance was attached to the transformation affected in the cremation process (Toynbee, 1971: 63, 291; Alcock, 1980: 62; Noy, 2005; McKinley, 2006). Being almost a millennia ahead in their use of cremation as a sanitary option for disposal of the dead, did the Romans share the modern obsession with a need for full oxidation and how far did the contemporary written theory (Toynbee, 1971: 43–61; Noy, 2005) hold with practice in the provinces?

The aim of this paper is to focus on the archaeological evidence from the Roman–British period (43–410 AD) to assess the levels of oxidation, the potential factors affecting variation and ultimately to try and ascertain what importance was laid on the level of oxidation achieved. The process of cremation – both in modern crematoria and on an open pyre – will first be considered, highlighting the factors affecting its efficiency and the appearance of the final product. A brief outline of the mortuary rite of cremation in Roman Britain will be followed by an assessment of the observed variations in the oxidation of bones in the light of a variety of intrinsic and extrinsic factors.

THE CREMATION PROCESS

Cremation is a process of oxidation and dehydration and is effected by the interaction between three basic requirements: sufficient temperature to ensure that the body will burn; sufficient oxygen supply; and sufficient time for the organic components of the body to be oxidized (Holck, 1989: 42; McKinley, 1994a: 72–78, 2000a, 2004a: 293). In modern cremators, these requirements are carefully monitored either by the operator or in the most up-to-date facilities by computer-linked sensors. Should the temperature drop at the wrong stage within the process, an adjustment to the heat can be made; if the atmosphere within the cremator becomes oxygen-depleted and reducing conditions start to prevail, the air flows can be adjusted to ensure reoxygenation. Although generally undertaken within a predictable timescale, in all cases cremation will continue until full oxidation is achieved (McKinley, 1994a: 72–75).

Cremators are constructed of heat-retentive brickwork and the air flows, in addition to refreshing the oxygen, are designed to help create a uniform temperature throughout the main chamber. British cremators are fuelled by gas jets, generally two focused on the skull and the axial area of the body, which are applied until the chamber achieves the required operational temperature (*c*. 700–850°C), after which it will be sustained and retained by the heat produced by the burning corpse and the efficient structure of the cremator (McKinley, 1994a, Figures 16–17). The bodies of young (infants and children) and emaciated individuals produce little heat during cremation and the external heat source may need to be maintained or reapplied to ensure the required operational temperature and efficient cremation (Holck, 1989: 39; McKinley, 1994a: 72). Most cremators have operating temperatures of between 800 and 1000°C, with temporal fluctuations largely related to the amount of latent heat produced and a significant drop occurring when the majority of the soft tissues have been burnt away after *c*. 45 min (Figure 10.1(a); Holck, 1989, Figure 3; McKinley, 1994a: 72–75; Fengming, 2005, Figure 1). Depletion in the oxygen supply could create reducing conditions, obstructing or curtailing the cremation process and resulting in charring rather than burning of the body irrespective of the temperature within the chamber. The body is not a good conductor of heat. Soft tissues form a dense, damp mass of material (Fengming, 2005, Table 2), variable thicknesses of which overlay the bone within different parts of the body. The soft tissues not only restrict heat transfer but also effectively cut off the oxygen supply to the underlying bone. The variability in the thickness of soft tissues within different parts of the body means that some bones will be exposed to higher temperatures and oxygen before others, e.g., the cranium, the forearm and legs as compared with the axial skeleton and the thigh (in the later part of the cremation process, some thick layers of soft tissues may fall away from the bone and continue to burn independently, e.g., the mass of the muscle around the thigh and buttocks). Consequently, the temperature experienced by any individual bone – as indicated microscopically by the crystal structure (Shipman *et al*., 1984; Holden *et al*., 1995 a,b; Hiller *et al*., 2003) – may not be indicative of the maximum temperature attained within the cremator (or pyre – see below) or bones from other parts of the body. Unless broken open, bones will oxidize from the outside, hence the two-tone or the 'sandwich' effect commonly observed in the cremated bone (Figure 10.2; Holden *et al*., 1995a; McKinley, 2000a; see below). Trabecular bone tends to have a greater infiltration of additional organic materials within its structure (marrow, cartilage and blood vessels), which may result in such bones taking longer to oxidize than the compact bone.

While the factors necessary for full oxidation may be controlled within a modern cremator, a variety of intrinsic and extrinsic variables may affect the efficiency of oxidation in a pyre cremation (McKinley, 1994a,b, 2000a). Experimental pyres have demonstrated that temperatures similar to those seen in modern crematoria are commonly achieved (Figure 10.1(a) and (b)). From a cold start, the pyre reaches a temperature similar to the cremator in the first half-hour and higher temperatures within an hour. Slightly different mechanisms are at work, however. Unlike in the cremator, most of the heat is lost to the atmosphere and a constant external heat source is needed to maintain the temperature. Since heat rises, the body is normally placed at or close to the top of the pyre, which is also the position of the greatest oxygen supply; the oxygen at lower levels within the pyre is rapidly consumed in burning the fuel (McKinley, 1994a: 79–81). The heat supply is directed to only one side of the body – from below – unlike in the cremator where various

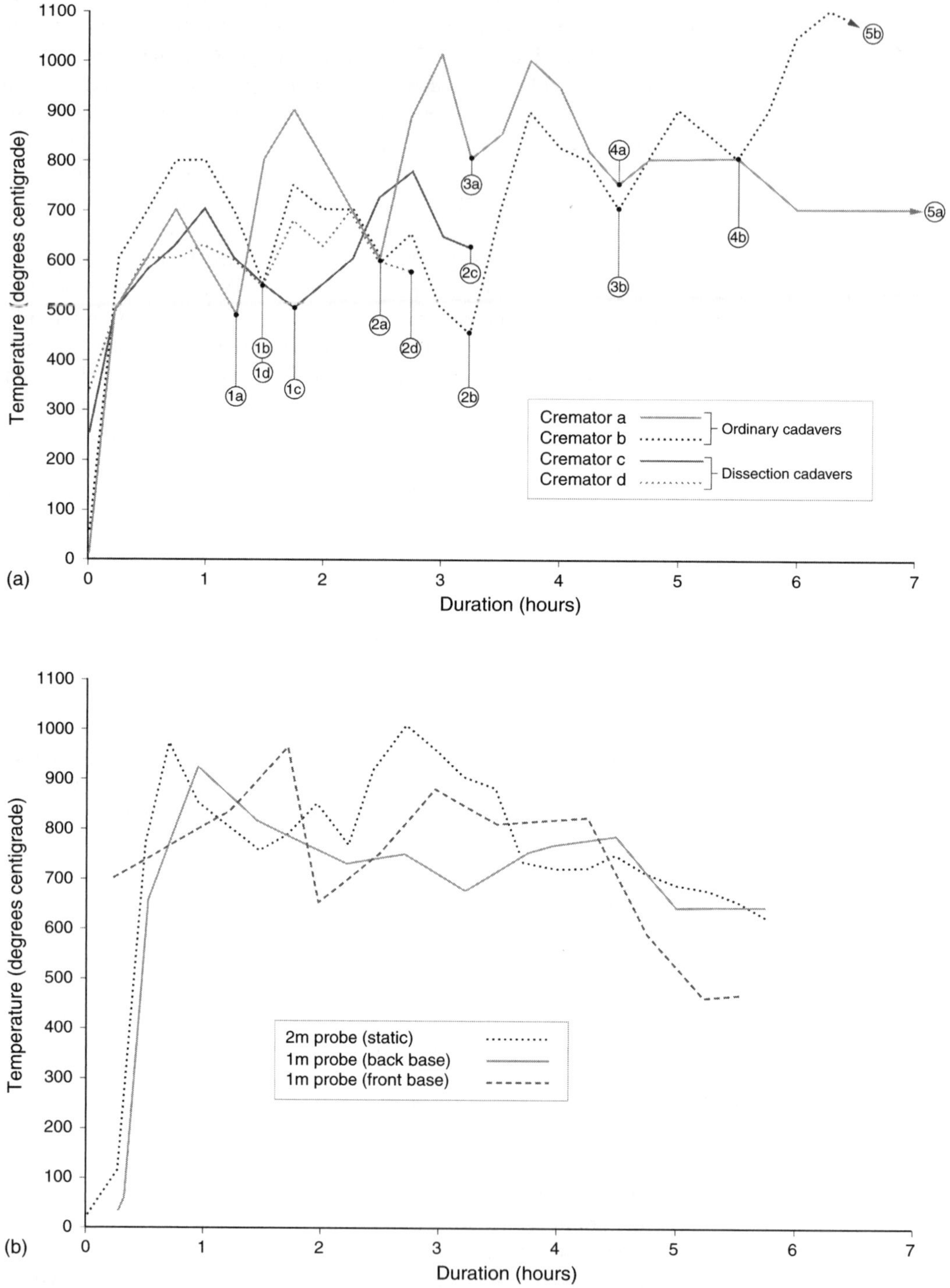

FIGURE 10.1 (a) Temperatures recorded at 15 minute intervals in four modern cremators (a–d); numbers 1–5 indicating the final readings taken for each successive cremation. (Reproduced from McKinley, 1994a Figure 18). (b) Temperatures recorded at 15 minute intervals in a single experimental pyre cremation; readings collected from three separately located probes.

FIGURE 10.2 Long bone shaft fragment from a Romano-British burial showing the 'sandwich' effect of differential levels of oxidation through the bone thickness. (see Plate 24)

air flows ('recuperator air' from above, 'cross hearth air' across the main hearth, 'reverse flow air' and 'under hearth air' from below) help to ensure a multi-directional action of heat on the body, including from below up through the air bricks forming the main hearth (McKinley, 1994a, Figures 16–17). A uniform temperature cannot be maintained across the pyre or through time (Figure 10.1(b)). The peripheries will be cooler than the central areas; wind strength and directional changes will affect how evenly distributed and how high temperatures will be within any given part of the pyre. The body itself, as already discussed, is not a good conductor of heat, so if the heat source itself is not directed at any one part of the body, there may be insufficient temperature in that area to effect efficient cremation.

In the open air, an oxygen supply is guaranteed but access to facilitate the cremation of the corpse may be inhibited by extrinsic mechanisms as well as the intrinsic one of the soft tissue coverage. The pyre structure is generally of an open, layered log construction that enhances oxygen availability and heat transfer (Figure 10.3). If the body was laid on a solid-based couch/bed or bier, this would act as a screen defecting the flame around the body until sufficient time had elapsed for the board to burn. Wrapping the body in or laying it on thick furs or leather would have a similar insulating effect; such materials might have affected only specific parts of the body, e.g., a pillow or cap/hood muffling the head. Crossing, and by implication binding, the hands and the forearm across the chest would shield them from the heat source for longer time than other parts of the body. As the pyre collapses, the body retains its position in relation to the fuel. After about 2 h, the main structure of the pyre will have burnt down and the remains – including much charred soft tissue – will rest on the hot ash bed, cremation potentially continuing for many hours at this level; in the experiments conducted by the author, temperature readings were generally taken over a 6–7-h period and the pyre was then left overnight (a similar time period and practice as apparently followed in many Roman cremations; Noy, 2005). If parts of the incompletely oxidized remains

FIGURE 10.3 Experimental pyre cremation undertaken in Shetland, northern Scotland, showing pyre structure and effects of veering strong winds. (NB: The pyre structure is more open than normal due to half the fuel in this case comprised a central peak stack.) (see Plate 25)

were buried in the dense fuel ash at this stage and not lifted out by tending, the oxygen supply would be curtailed.

The length of time required for the pyre to burn depends largely on the quantity of wood used to build it (the author has generally used about 700–900 kg; Holck, 1989: 43; McKinley, 1994a: 78–79) and, to some extent – particularly in temperate climates at certain times of year – on the weather (McKinley, 2006). Strong winds would result in the pyre burning faster and more fiercely but not necessarily more efficiently since burning would be uneven and result in the collapse of the structure (Figure 10.3); windbreaks can assist in cutting down the risk and attendants on hand could help redress the problem. Heavy rain would inevitably result in the cessation of cremation, whilst moderate and/or persistent rain would at the very least reduce the temperature.

MEASURE OF LEVEL OF OXIDATION

The degree of oxidation of the bone is reflected macroscopically in its colour, ranging from the black colour of charred bone, through various hues of blue and grey, to the white colouration of the oxidized bone. Laboratory experiments, undertaken using defleshed green bone, have monitored colour changes at different temperatures within given time spans (Shipman *et al.*, 1984; Holden *et al.*, 1995a,b; Hiller *et al.*, 2003). It should be remembered, however, that even within a modern crematorium – dealing with the cremation of full, fleshed corpses – a variety of colours may be observed in the bone from a single cremation undertaken at 700–900°C over the standard *c.* 1.5 h period, for the various reasons outlined above (Figure 10.4). The changes

FIGURE 10.4 Cremated bone from a modern cremation showing colour variations indicative of different levels of oxidation.

resulting from dehydration may also be observed macroscopically – shrinkage, fissuring and warping (Krogman, 1939; Webb and Snow, 1945; Baby, 1954; Binford, 1963; Thurman and Wilmore, 1981; Buikstra and Swegle, 1989; McKinley, 1994a: 75–76, 2000a) – temperature-related changes to the crystal structure requiring microscopic examination (Shipman *et al.*, 1984; Grupe and Hummel, 1991; Holden *et al.*, 1995a,b; Hiller *et al.*, 2003).

CREMATION IN ROMAN BRITAIN

The mortuary rite of cremation was already well-established in parts of Britain at the beginning of the Romano-British period in AD 43 (Whimster, 1981; Philpott, 1991: 8). Many cemeteries – particularly in south-east England – show an unbroken continuum in practice across the Late Iron Age/Early Romano-British range, some burials possessing insufficient distinguishing features to enable their date to be confidently set in either the decades preceding or those subsequent to commencement of the Roman occupation.

Cremation formed the predominant rite for the disposal of the dead in the Early Romano-British period (AD 43–150), with a gradual shift to inhumation burial in the mid-second century (Philpott, 1991: 8). By the late third century AD and throughout the Late Romano-British period (AD 250–410), disposal by inhumation was the norm, though cremation did persist in some places (Philpott, 1991: 50–52). In the Northern Frontier zone, adjacent to Hadrian's Wall (Figure 10.5), cremation commonly remained the rite of choice (Philpott, 1991), e.g., the mid-second- to third-century cemetery at Brougham, Cumbria, with a minimum of 132 excavated cremation burials (Cool, 2004).

FIGURE 10.5 Map of England showing distribution of sites/number of deposits included in the sample by county.

Philpott (1991: 50–52) recorded a scatter of late cremation burials in south-east and central-southern England, and recently further burials of this date have been found in London (Barber and Bowsher, 2000) and Winchester (Birbeck and Moore, 2004). The persistence of the rite was, however, clearly more widespread than was once thought, since recent finds from rural sites in England – often culturally indistinct burials dated by radiocarbon analysis – have proved to be of third- or fourth-century date (e.g., Bellamy, 2000).

The majority of Romano-British cremation-related deposits excavated in Britain comprise the remains of burials. Those made within a ceramic container, the variety of vessel forms being covered by the generic term 'urn' to describe their function within the funerary rite, probably represent the most numerous burial type (Philpott, 1991: 22–25, 35–43; see sample below); on rare occasions, glass vessels also acted as containers for the bones (Philpott, 1991: 26–27). Unurned burials were also common, the majority (see *busta* below) being originally made within an organic container for which little extant evidence generally survives. In most cases, the container would have comprised a bag of textile, skin or possibly basketry, the surviving evidence for which is the discrete concentration of bones within the grave fill. This confined distribution is often not visible during excavation due to the common practice of incorporating pyre debris in the grave fill, masking the distribution of the bone; whole-earth recovery of the grave fill in quadrants enables the distribution of the bone to be deduced and the deposit type to be confirmed (McKinley, 1998, 2000b). Sometimes wooden caskets were also used as containers for the bones in burial (Philpott, 1991: 12–21).

Most Romano-British cemeteries also appear to have functioned as crematoria, though the position of the *ustrina* – the area in which cremations were undertaken – is not always evident. The ephemeral nature of pyre sites constructed on a flat ground surface, where the heat of the pyre will have penetrated only *c.* 50–100 mm (monitored in experiments) and, therefore, have been easily erased by plough damage or other disturbances, means that relativity few conclusive Romano-British pyre sites have been found (McKinley, 2000c). Some of the latter include *busta*-type pyre sites, though the author believes that most British examples functioned only as the pyre site and not also – as the true *busta* should – as the place of burial. Since the bone remaining at the end of cremation was never collected in its entirety for inclusion in the burial – a standard feature of the rite across the temporal range – pyre debris remaining at uncleared or incompletely cleared pyre sites, or amongst debris redeposited elsewhere, would invariably include some bone fragments (McKinley, 2000c). In the common absence of the pyre sites themselves, their original presence in the vicinity of the burials is often demonstrated by the presence of redeposited pyre debris in a variety of locations including cremation grave fills, pre-existent features, spreads and apparently deliberately excavated features (McKinley, 2000c).

DATA

The following observations have been gathered over the past 20 years during the analyses of cremated remains undertaken by the writer from 60 Romano-British sites in England (Table 10.1; Figure 10.5); to the writer's knowledge, *c.* 70% of these sites still await publication. The majority of the materials derived from contexts of first- to mid-second-century date, with a small proportion of Late Iron Age/Early Romano-British (3.1%) deposits, and *c.* 16.6% later Romano-British contexts predominantly from the Northern Frontier Fort (NFF) cemetery at Brougham, Cumbria (Cool, 2004; McKinley, 2004a).

The analysis of the cremated bone followed the writer's standard procedure (McKinley, 1994a: 5–21, 2004b), variations in colour from the white of full

TABLE 10.1 Types of sites and deposits from which the cremated bone in the sample derived

	Towns	Rural	NFF	Total
Number of sites	15	42	3	60
Urned burials	566 (47.6%)	124 (53%)	140 (47.3%)	830 (48.2%)
Unurned burials	338 (28.5%)	64 (27.3%)	18 (6.1%)	420 (24.4%)
Casket burials	10 (0.8%)	1 (0.4%)		11 (0.6%)
Combined urned and unurned burials	5 (0.4%)	5 (2.1%)		10 (0.6%)
Pyre sites	11 (0.9%)	4 (1.7%)		15 (0.9%)
Other[a]	260 (21.8%)	36 (15.4%)	138 (46.6%)	434 (25.2%)
Total	1190 (69.2%)	234	296	1720

[a]Includes deposits of pyre debris and deposits that formed either redeposited pyre debris (rpd) or unurned burials with rpd but that were not excavated in sufficient detail to allow the deposit type to be conclusively deduced (see above).

oxidation being noted for the deposit in general and for specific identifiable skeletal elements, including siding where possible. The level of attainable detail in the identification of skeletal elements and the age and sex of individuals is highly variable, largely dependent on the quantity of bone available for examination (which may range from a few grams to over 1000 g) and the surviving size of the bone fragments (McKinley, 1993, 1994b, 2000a). Consequently, adult age ranges may be very broad and overlapping, and it is commonly not possible to attribute a sex to *c*. 50% or more of the adult population. These factors have limited the meaningful detailed breakdown of groups of data for consideration in this overview.

Sample Bias

All the sites included in this study were excavated as a result of commercial development – rather than reflecting a targeted research design – the scope of the excavation being limited by the margins of the development that did not necessarily correspond with the boundaries of the cemetery. Pipelines or service routes may cut a swathe through a variety of archaeological remains, providing a random sample of features and deposits. Consequently, in some instance, the undertaking of a cremation within the vicinity of an area of archaeological investigation may be illustrated only by the presence of redeposited pyre debris or a pyre site, the burials related to which lay outside the area of excavation. The number of features excavated from any one site ranges from 1 – singletons being frequently found on rural sites – to 480 from one of the several cemeteries serving the Roman small town of Baldock, Hertfordshire (Figure 10.5).

The remains within the sample are predominantly from the south-east of England (Figure 10.5), 61.6% of the sites containing 70.4% of the deposits having been excavated from the area contained within the arc extending from Hampshire, via Oxfordshire to Norfolk. This, at least in part, reflects the pattern of property development in the United Kingdom over the past approximately 30 years, most cemeteries having been excavated as a result of housing or road building. There is, therefore, a potentially in-built bias within the data since, although providing an arbitrary sample, it is neither equally distributed nationwide nor necessarily a representative of the population in the Roman period in Britain.

The size of assemblages from individual cemeteries/sites within the town sample range from 3 to 480, 93% being recovered from two settlements in Hertfordshire, Baldock (three cemeteries), and St. Albans, and from the cemeteries of East London. The rural sites include those associated with small settlements, individual farmsteads and villas, where single burials were often made on field boundaries; individual assemblage sizes ranged from 1 to 31. The three NFF sites were all situated in Cumbria at the west end of Hadrian's Wall (Figure 10.5); assemblage sizes ranged from 3 to 270.

VARIATIONS IN OXIDATION

In general, the majority of bones in all assemblages were white, indicative of full oxidation of the bone (Holden *et al.*, 1995a,b). A variable proportion of bones within individual deposits and deposits within most individual assemblages, however, showed different colouring, indicative of incomplete oxidation. A number of potential variables have been considered in an attempt to ascertain patterns in the distribution of different degrees of oxidation, both within and between deposits, including location, deposit type, date, age and sex of the individual, and skeletal element.

One possible intrinsic bias, which should be noted, is that of variable bone survival. Cremated bones commonly survive in soil types where most or all unburnt bones will be destroyed, i.e., one that is either too acidic or too alkaline. Cremated bones may have the benefit of additional protection from the detrimental effects of an aggressive burial environment by being placed inside a ceramic vessel (these were usually lidded, though organic lids degraded with time) or with large quantities of pyre debris altering the burial microenvironment (McKinley, 1994b). Trabecular bones are subject to preferential destruction where compact bones may appear unaltered, though in highly acidic conditions the latter will attain a worn and chalky appearance (Nielsen-Marsh *et al.*, 2004; McKinley, 1997: 245). It is possible, if not likely, that in these burial conditions, the less well-oxidized bone may also be prone to preferential destruction.

One other factor that may skew our view of the levels of oxidation across the skeleton is the quantity of bone included in the deposit. As discussed above, this is highly variable in the Romano-British period as much as any other (e.g., McKinley, 2004a: 295–298); though generally, even where the overall quantity of bone is small, the burial will include some fragments from each of the four skeletal areas. It is possible, however, that in some instances, less of the poorly oxidized bone was included in the burial (see section *Deposit Type*). Loss of bone from the context due to disturbance, whilst a potentially common phenomenon, is unlikely to distinguish between the well and poorly oxidized bone.

LOCATION

Although there could be considerable variation within and between sites, on average, 53.8% of deposits from the towns showed some level of variable oxidation compared with 44.8% of those from rural locations and 34.8% of those from the NFFs. The great range in individual assemblage sizes limits the validity of comparing those in which all the recorded bones were white but they include 13.3% of the town sites (representing *c.* 0.9% of town deposits), 35.7% of rural sites (*c.* 19% of rural deposits) and one of the three NFF sites (*c.* 7.8% NFF deposits); one other NFF site in Cumbria currently under assessment also shows no colour variations. In purely geographic terms, there was no consistent difference in the frequency of lower levels of oxidation other than in the relatively few (16) burials from the northern counties of Lancashire, Yorkshire and Cheshire (Figure 10.5). Only one of these burials (6.2%) contained bones showing any variation in colour from the white of full oxidation.

DEPOSIT TYPE

The frequency of variable levels of oxidation observed in bones from the different deposit types in different locations does not suggest any great consistency (Table 10.1). The urned burials from town sites most commonly show a higher proportion with less well-oxidized bones in comparison with the unurned burials, e.g., 35%, 66% and 44% compared with 24%, 50% and nil from one of the Baldock cemeteries (Hertfordshire; unpublished), East London (2000d) and Wall, Staffordshire (in prep.), respectively. Data from the rural sites show no consistent pattern, while that from the NFF sites suggest a closer similarity between the two burial types; e.g., 9.7% of urned burials compared with 10.1% of unurned burials from Brougham, Cumbria (2004a).

In some cases, the bone from the pyre sites/redeposited pyre debris is less well oxidized than that from the corresponding burial, possibly suggesting that the more easily visible white bone is preferentially collected from the pyre site for burial. Conversely, in other cases, the bone from the debris is more fully oxidized, perhaps indicating that it remained burning slowly on the pyre following the extraction of the bone for burial. Both observations have potential implications with respect to the mode of collection of bones for burial; was the pyre deliberately cooled prior to the collection of bones? Were individual bone fragments collected by hand or using tongs? Was the bone raked-off and 'winnowed'? (McKinley, 2000a, 2004a; Noy, 2005).

DATE

Deposits of Late Iron Age/Early Romano-British date generally show frequencies of variation in colour similar to that seen in the Romano-British deposits from the same sites. Where a site contained deposits of both Early and Late Romano-British date, differences were commonly apparent; the bone from one phase was all white and the other showed variations, but the division was not consistent to one phase across all sites. No general temporal change in practice is indicated; there may have been local variations or, the numbers involved being relatively small (Table 10.1), it could simply reflect the vagaries of a change in the personnel undertaking cremation.

AGE/SEX

Individuals of both sexes and across the age range – from neonate to older adult – have been identified from the remains of Romano-British cremation burials in England. The inclusion of neonates (0–6 months) in formal Roman cemeteries/burial locations is rare, and their deposition in various non-cemetery locations is viewed as more reflective of the 'cultural norm' (Philpott, 1991: 97–102). Clearly, there were exceptions, however, in addition to the six sites in this sample (town sites in Hertfordshire and Staffordshire; rural sites in Norfolk, Berkshire, Wiltshire and Essex) from which small numbers of neonates were recovered, either as singletons or from burials containing the remains from a dual cremation. Stirland (1989) identified three neonates within the Romano-British phases at King Harry Lane, St. Albans, Hertfordshire.

Some very broad age-related variations are apparent. No variation was observed in the neonatal (0–6 months) remains. Variations in infant (0.5–5 years) and juvenile (5–12 years) remains were relatively uncommon and, where present, were of limited extent and severity (slight blue-grey); of the 15 sites with individuals of 0–12 years, the bone was all white in 68.2%. The frequency of colour variation in the subadult (13–18 years) remains was similar to that seen in the adults (>18 years), though at some sites it was lower. Figures from one of the Baldock sites (Hertfordshire; town) provide an example, with variations in 19% of infant/juvenile remains, 4% of subadult and 37% of adult remains. Overall, there was great variability amongst the adult remains, particularly on sites with small numbers of deposits, but amongst the larger assemblages, a range of 26–87% of adults with some variation in the bone colour was recorded. No attempt was made to break down the adult group further due to the commonly applied broad age ranges and frequent overlapping (see above).

Variable levels of oxidation were observed in the remains of both sexes but there is some indication that the sex of the individual may have been a factor in the levels of oxidation in some cases. Several sites show a higher percentage of males with variable levels of oxidation compared with the females, e.g., 53%, 67% and 25% compared with 40%, 50% and 18% from Baldock Area 15 (unpublished), Wall (in prep.) and Brougham (2004a), respectively. At one other rural site (outside Baldock, Hertfordshire), although the proportions of males and females were similar, the variations observed in the male remains were more extensive and extreme.

Taken together, these figures suggest that the body mass, perhaps unsurprisingly, was a factor affecting the levels of oxidation achieved, though further implication is that sufficient adjustments were not necessarily made to accommodate these differences during cremation.

SKELETAL ELEMENTS

Observed variations ranged from brown/unburnt (very rare), black (charred), through hues of blue and grey, most commonly the latter two. A substantial minority of fragments demonstrated the 'sandwich' effect, the exposed surface/s of the bone being white whilst the central/internal part was

TABLE 10.2 Frequency and Distribution of Incompletely Oxidized Bone from Town and Rural Cemeteries

	Skull	Axial skeleton	Upper limb	Lower limb
Town sites	Range: 50–60%	Range: 31–32%	*c.* 41%	Range: 44–49%
Rural sites	Range: 24–42%	Range: 8–16%	Range: 24–42%	Range: 40–64%

black/blue/grey (Figure 10.2); this was commonly seen in both long bone shaft fragments and the skull vault where the endo- and exocranial surfaces were white and the diplöe coloured.

In general, only one or two skeletal areas were affected, with elements from all four areas (skull, axial skeleton, upper and lower limb) showing variations in a minority of cases; e.g., *c.* 2–8% of town sites, *c.* 8–13% of rural sites, and an even smaller proportion of the NFF sites. In *c.* 48% of cases from all sites, only one or two skeletal elements from any one skeletal area were affected, with >10 elements involved in only 13% of cases. There appears to be some slight difference between the town and rural sites with respect to the skeletal area most commonly affected (Table 10.2; from sample of the larger assemblages in each group); the skull followed by the lower limb most frequently show variations in the town sites, the frequencies being reversed in the rural sites. In the majority of assemblages, the axial skeleton was least affected, followed by elements of the upper limb. Although variations were observed on occasions in almost all skeletal elements, in all assemblage types, those elements that most frequently showed colour variations include the femur and the tarsal bones from the lower limb area, the cranial vault of the skull, and the carpals in the upper limb.

Whatever be the distribution, it is common for only one or two skeletal elements to be affected in any one skeletal area (48%), with more than 10 elements being involved in only *c.* 13% of cases. Generally, only fragments of any one skeletal element were affected, very rarely the entire bone (Figure 10.6; see section *Partial Cremation*).

Most of these variations suggest general factors that affect the overall efficiency of oxidation such as a short fall in the quantity of fuel used to build the pyre, leading to insufficient temperature over a sufficient length of time to fully oxidize all the bones (particularly that with extensive soft tissue coverage) and or/the insufficient temperature applied to skeletal elements on the peripheries of the pyre, e.g., the feet, the hands (placed to the sides of the body) and the head. It is also possible that some small, individual bones/fragments may have fallen outside the confines of the pyre and stopped burning due to the drop in temperature, or have fallen through the pyre structure to be buried in the fuel ash effectively cutting off the oxygen supply.

SPECIFIC VARIATIONS

Several burials from six sites show variations in the levels of oxidation to discrete areas of the skeleton suggestive of some specific problem during cremation. Four adults, including a minimum of one female, from the town site

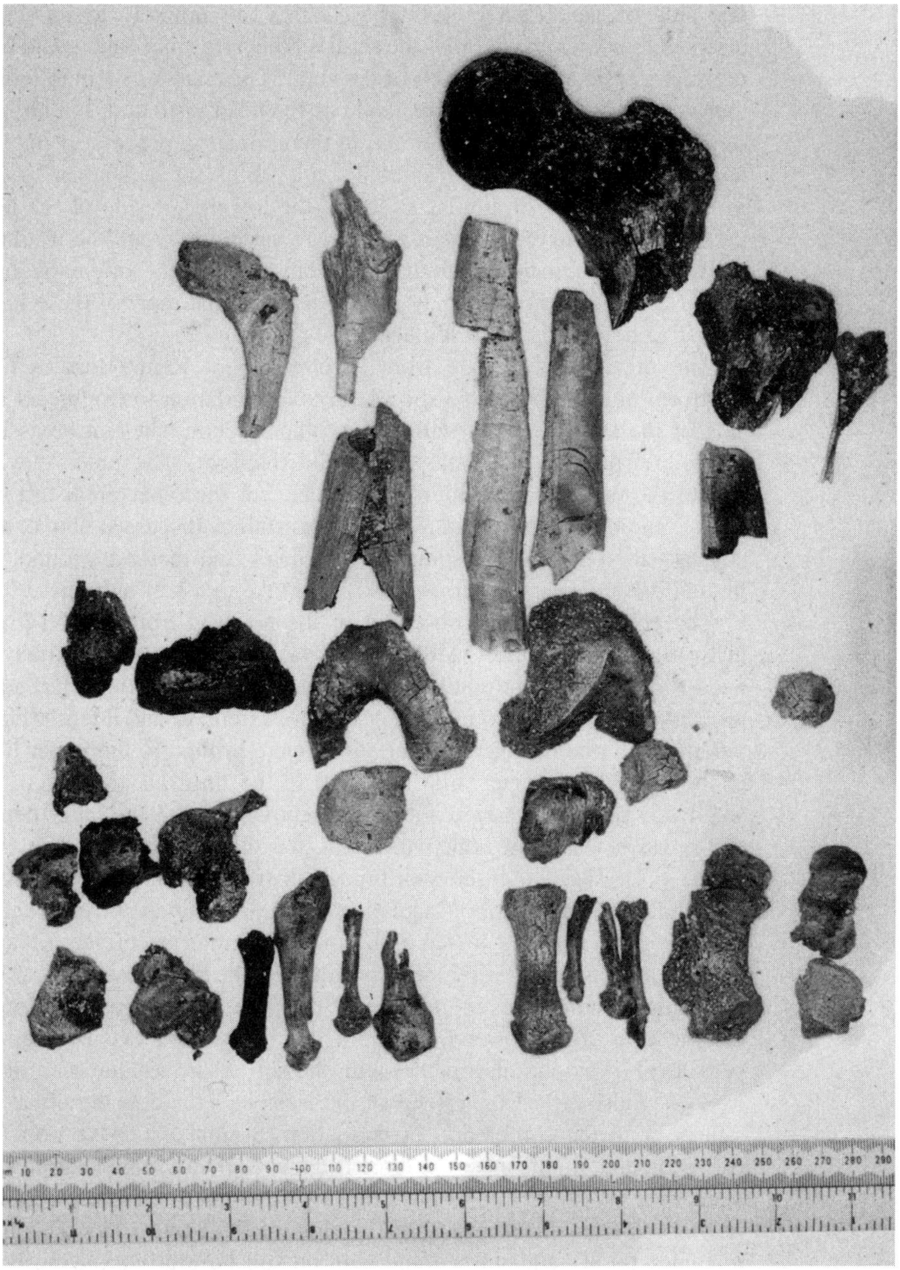

FIGURE 10.6 Indentified elements of lower limb from the Romano-British burial from Purton, Wiltshire, showing varying degrees of oxidation to the bone. (see Plate 26)

of St. Stephens', St. Albans, Hertfordshire (unpublished) show reduced levels of oxidation to only one side of the skull. Similar variations were observed in the remains of two other adults, each from a rural site and including at least one female (Fenny Lock and Aylesbury-Chalgrove, both Buckinghamshire; unpublished). One other adult female from a rural site in Gloucestershire (Totterdown Lane) shows variation to most of the cranial vault and to only one side of the facial bones. These cases are unlikely to reflect just the peripheral (cooler) position of the skull on the pyre since this would have been unlikely to affect only one side of the skull. The changes seem to indicate that there was something around one side of the head insulating it, cutting off the oxygen supply and thereby inhibiting the cremation process. If the head was laid on a cushion of dense material – e.g., thick fur or leather – or encased in a hood or a hat of similar material, the lowermost side of the head/face, particularly if laid against part of the pyre structure, would be insulated from the flame; this situation could have been maintained for long enough to effect the observed variations. It may be of relevance that most of these individuals appear to have been adult females.

One other adult female from a rural site in Kent (Beadles Car Park, Dartford, unpublished) also shows variable oxidation to fragments from one side of the skull together with a few other specific skeletal areas including the elbow region, the distal femur and the foot. The latter may well be due to the peripheral position of the feet, but the other areas are not those most commonly involved in a general shortfall as discussed above, and again suggests the potential for some form of insulating material around the body, though it is unclear what form this would take to affect only these few areas.

The colour variations observed in the remains from the NFF cemetery at Brougham, Cumbria (McKinley 2004a), were generally limited in extent and of low intensity (slightly blue/grey), but three individuals (all adults, one probable female and two males) show more extensive and intense (black/blue) variation. In one instance the variations were limited to the upper limb only, in the second the lower limb only and in the third all areas were affected; small fragments of charred soft tissue recovered with the latter demonstrate that it was not only the bone which was not fully oxidized (McKinley, 2004a, Figure 6.4). The two cases with limited distribution of poorly oxidized bone may indicate some form of insulating effect as outlined above. Alternatively, the case involving the lower limb could illustrate the effects of too strong a wind, resulting in preferential burning of the pyre towards the proximal end with eventual uneven collapse of the pyre, and a consequent deficiency in oxidation of the lower limb elements. The overall extensive and intense poor level of oxidation, together with the soft tissue residue seen in the third case, may also reflect the effects of the weather – the dousing effect of heavy rain prior to the completion of cremation. Similar extensive poor levels of oxidation and, in this instance, large fragments of surviving charred soft tissue residue were observed in an elaborate and high-status cremation burial from a walled cemetery at Purton, Wiltshire (Figures 10.6 and 10.7). This burial had remained totally undisturbed and with no soil infiltration, having been made in a large globular glass vessel deposited in a lead canister, all held within a stone sarcophagus. Given the wealth of the deposit, it is unlikely that there was insufficient fuel provided for cremation and it is most likely that rain curtailed the process. The fragments of charred soft tissue were extremely brittle and unlikely to survive in most burials, even urned ones, though very small fragments of this material have been found in a few other Romano-British burials (one from Worcestershire and one from Essex). Similar residues have been observed in several experimental pyre cremations, surviving either as separate fragments or still attached to the bone (Figure 10.8).

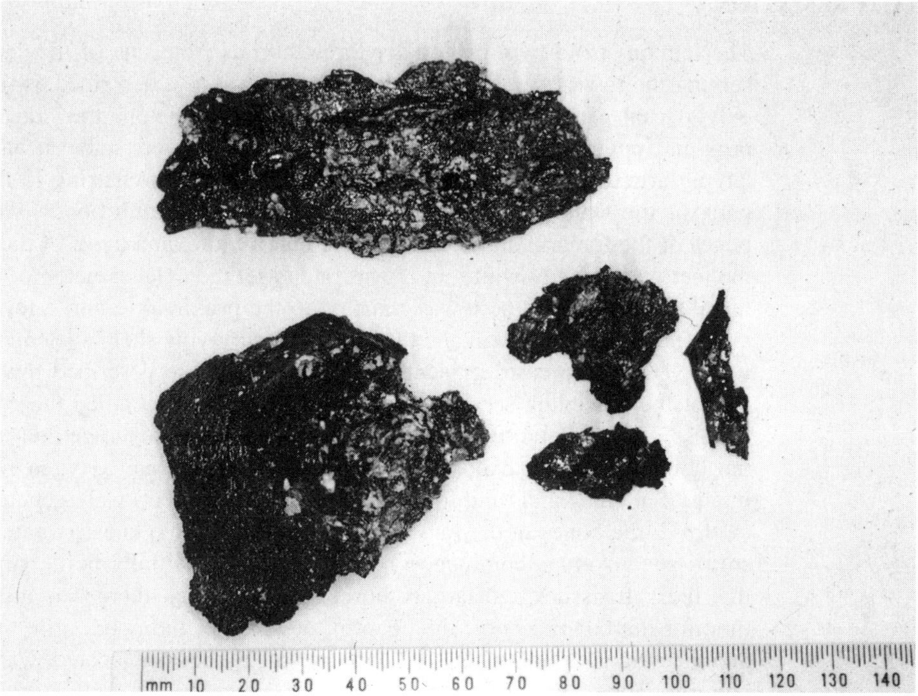

FIGURE 10.7 Fragments of charred soft tissue (soft tissue residue) from the Romano-British burials from Purton, Wiltshire.

FIGURE 10.8 Section of cremated sheep spine from an experimental pyre cremation; cremated vertebral held articulated by charred longitudinal ligament. (see Plate 27)

PARTIAL CREMATION

The remains from two, potentially three, burials from one of the cemeteries serving the town of Baldock in Hertfordshire are best described as reflecting only partial cremation. In one case, the axial skeleton, the cranium, the proximal upper and lower limbs of an adult male were unburnt and found largely articulated within a small grave cut; there was charring to the basal parts of the skull, the elbow and the knee areas of limb bones, whilst the bones of the forearm and the hand were burnt white, and those of the leg and the feet were mostly white. In the second case, the axial skeleton of an adult female was unburnt, there was charring to the proximal femur, a few cranial vault fragments and tarsal, and the rest of the surviving skeletal elements were white. These two cases present the extreme of those described towards the end of the preceding section reflecting an event that curtailed the cremation process. The skeletal elements with little soft tissue coverage and in which the bones would be exposed to the effects of heat and oxygen relatively rapidly (i.e., the hands and forearm, the legs and the feet) were generally well oxidized; the bones in the region of the elbows and the knees, where heavier soft tissue coverage commences, were charred or partially burnt, suggesting that the soft tissues had largely burnt away, but that there was insufficient time for the bone to burn; the bone in the areas of the upper arm, the thigh, and the axial skeletal, where the mass of soft tissues are located, was largely unburnt. There may, of course, have been a variety of factors involved in these cases, but a major event had curtailed cremation long before completion, and the most obvious cause would be the weather. Cremation in towns was generally undertaken by professional *ustores*, and if, contrary to the dictates of Roman tradition, no relatives were on hand to collect the remains for burial (Toynbee, 1971: 50) and to object to their condition, burial of these partially cremated remains might have followed a pragmatic course. Rare but similar examples of partial cremation have also been observed from at least one other town cemetary (Wheeler, 1985, 231–232).

CONCLUSIONS

Contemporary written sources indicate that for the Romans, incomplete cremation was '...to be deplored, being regarded as in insult to the deceased and...not enabling the soul to reach the afterlife...' (Noy, 2005). What is unclear is what, exactly, they would have regarded as 'incomplete'. As there are contemporaneous records of cremations lasting for up to 8 h (Noy, 2005), similar to the timing of the writer's experimental cremations, one would anticipate a similar level of oxidation had the process run its full course; but even after 8 h, some bones and soft tissues can remain incompletely oxidized. Judgement of how complete cremation was would have been based on the visual appearance. The use of the word *ossilegium* to refer (at least in late Latin; Noy, 2005) to the remains suggests that what was chiefly required was to be able to distinguish the remains as bones; variations in their colour may have been incidental, as may have been the occasional presence of fragments of charred soft tissues unidentifiable as specific organs.

The main cause of minor variability in the degree of oxidation seen in Romano-British cremated remains within the sample appears to relate to body mass; adults, and amongst them a slightly higher proportion of males (generally larger body mass), show the greatest frequency, range and extent of variability. This suggests that the quantity of wood and/or size of the pyre was not always adjusted to accommodate variations in the size of the deceased. This may result in an overall deficiency in the amount of time and related sustained temperature necessary to facilitate oxidation of all the body's organic components and in the corpse's extremities lying too close to the cooler peripheries of an under-sized pyre. There are indications that incomplete oxidation of the bone occurred more frequently and to a greater extent in the towns compared with the rural areas, overall both appearing less effective than the cremations undertaken in the Northern Frontier zones. In the towns, cremations would have been undertaken by professional *ustores* and payment would have been made for the quantity of wood to be used (Toynbee, 1971: 45; Noy, 2005). Inevitably, the poor, unable to afford sufficient fuel, appear to have been less well cremated than the better-off (Morris, 1992: 43), a problem still experienced in contemporary cultures (Barber, 1990: 380). Cremations in rural areas are more likely to have been undertaken by the family or their retainers, wood may have been – but not necessarily – more directly easily obtainable, and there may have been more consistent care in tending the pyre. The apparent overall greater efficiency in the oxidation of remains from the NFF cemeteries may have a link with military efficiency and freer access to sufficient fuel supplies for military personnel and their dependants/associates.

Roman cremations were probably mostly undertaken within two to seven days of death (Toynbee, 1971: 45; Noy, 2005). In Britain, particularly at certain times of year, this must have presented difficulties with regard to suitable weather conditions (McKinley, 2006). Fuel could be well seasoned and kept dry up until the point the pyre was lit, but a sudden unexpected downpour or the need to catch a break in the weather in prolonged periods of rainfall would, undoubtedly, have lead to problems at times, particularly in towns where the holding of a corpse for any extensive length of time would have potentially caused health risks as well as concern on the part of the relatives/associates of the deceased. These things did not always go according to plan and these pragmatic decisions that sometimes had to be taken are indicated by those less well-cremated remains recovered from several cemeteries, particularly, but not exclusively, those in the towns.

ACKNOWLEDGMENTS

Figures 10.1 and 10.5 were prepared by Rob Goller, Wessex Archaeology.

REFERENCES

Alcock, J.P. (1980). Classical religious beliefs and burial practice in Roman Britain. *Archaeological Journal* **137**, 50–85.

Baby, R.S. (1954). Hopewell cremation practices *Papers in Archaeology*, Vol. 1–7. Ohio Historical Society, Columbus, OH.

Barber, P.T. (1990). Cremation. *The Journal of Indo-European Studies* **18** (3–4), 379–388.

Barber, B. and Bowsher, D. (2000). *The Eastern Cemetery of Roman London*. Museum of London Archaeological Services Monograph 5.

Bellamy, P. (2000). Excavations at Penny's Farm, Cranbourne. *Proceedings of the Dorset Natural History and Archaeological Society* **122**, 83–97.

Binford, L.R. (1963). An analysis of cremations from three Michigan sites. *Wisconsin Archaeologist* **44**, 98–110.

Birbeck, V. and Moore, C. (2004). Preservation and investigation of Roman and Medieval remains at Hyde Street, Winchester. *Proceedings of the Hampshire Field Club and Archaeological Society* **59**, 77–110.

Buikstra, J.E. and Swegle, M. (1989). Bone modification due to burning: experimental evidence. In *Bone Modification* (R. Bonnichsen and M.H. Sorg, eds.). Centre for the study of the First Americans, Institute for Quaternary Studies, University of Maine, Orono, ME.

Chamberlain, C. (2005). European cremator development. In *Encyclopaedia of Cremation* (D.J. Davies and L.H. Mates, eds.), pp. 146–148. Ashgate, Aldershot, England and Burlington, USA.

Cool, H. (2004). *The Roman Cemetery at Brougham, Cumbria.* Britannia Monograph Series 21.

Davies, D.J. (2005). Fuel. In *Encyclopaedia of Cremation* (D.J. Davies and L.H. Mates, eds.), p. 212. Ashgate, Aldershot, England and Burlington, USA.

Downes, J. (1999). Cremation: a spectacle and a journey. In *The Loved Body's Corruption* (J. Downes and T. Pollard, eds.), pp. 19–29. Cruithne Press, Glasgow.

Drool, J. and Wheeler, H. *et al.* (1985). Roman Derby: Excavations 1968–1983. *Derbyshire Archaeological Journal* **105**, 222–280.

Fengming, L. (2005). Cremation process: China. In *Encyclopaedia of Cremation* (D.J. Davies and L.H. Mates, eds.), pp. 132–135. Ashgate, Aldershot, England and Burlington, USA.

Grupe, G. and Hummel, S. (1991). Trace element studies on experimentally cremated bone. I. Alteration of the chemical composition at high temperatures. *Journal of Archaeological Sciences* **18**, 177–186.

Hiller, J.C., Thompson, T.J.U., Evison, M.P., *et al.* (2003). Bone mineral changes during experimental heating: an X-ray scattering investigation. *Biomaterials* **24**, 5091–5097.

Holck, P. (1989). *Cremated Bones. A Medical–Anthropological Study of an Archaeological Material on Cremation Burials*. Antropologiske skrifter nr. 1. University of Oslo.

Holden, J.L., Phakley, P.P. and Clement, J.G. (1995a). Scanning electron microscope observations of incinerated human femoral bone: a case study. *Forensic Science International* **74**, 17–28.

Holden, J.L., Phakley, P.P. and Clement, J.G. (1995b). Scanning electron microscope observations of heat-treated human bone. *Forensic Science International* **74**, 29–45.

Krogman, W.M. (1939). *A Guide to the Identification of Human Skeletal Material.* Federal Bureau of Investigation Law Enforcement Bulletin 8.

Mates, L.H. (2005). Siemens, Schneider and Kilgenstierna Cremators. In *Encyclopaedia of Cremation* (D.J. Davies and L.H. Mates, eds.), pp. 148–151. Ashgate, Aldershot, England and Burlington, USA.

McKinley, J.I. (1993). Bone fragment size and weights of bone from modern British cremations and its implications for the interpretation of archaeological cremations. *International Journal of Osteoarchaeology* **3**, 283–287.

McKinley, J.I. (1994a). *Spong Hill Part VIII: The Cremations*. East Anglian Archaeology 69, East Dereham, Norfolk.

McKinley, J.I. (1994b). Bone fragment size in British cremation burials and its implications for pyre technology and ritual. *Journal of Archaeological Sciences* **21**, 339–342.

McKinley, J.I. (1997). The cremated human bone from burial and cremation-related contexts. In *Archaeological Excavations on the Route of the A27 Westhampnett Bypass, West Sussex, 1992 Volume 2*. (A.P. Fitzpatrick, ed.), pp. 55–73. Wessex Archaeology Report No. 12.

McKinley, J.I. (1998). Archaeological manifestations of cremation. *The Archaeologist* **33**, 18–20.

McKinley, J.I. (2000a). The analysis of cremated bone. In *Human Osteology* (M. Cox and S. Mays, eds.), pp. 403–421. Greenwich Medical Media, London.

McKinley, J.I. (2000b). Putting cremated human remains in context. In *Interpreting Stratigraphy: Site Evaluation, Recording Procedures and Stratigraphic Analysis* (S. Roskams, ed.), pp. 135–140. BAR International Series 910.

McKinley, J.I. (2000c). Phoenix rising: aspects of cremation in Roman Britain. In *Burial, Society and Context in the Roman World* (M. Millett, J. Pearce and M. Struck, eds.), pp 38–44. Oxbow Books, Oxford.

McKinley, J.I. (2000d). Cremation burials. In *The Eastern Cemetery of Roman London* (B. Barber and D. Bowsher, eds.), pp. 264–277. Museum of London Archaeological Services Monograph 5.

McKinley, J.I. (2004a). The human remains and aspects of pyre technology and cremation rituals. In *The Roman Cemetery at Brougham, Cumbria* (H. Cool, ed.), pp. 283–309. Britannia Monograph Series 21.

McKinley, J.I. (2004b). Guidelines for recording cremated bone. In *Guidelines to the Standards for the Recording of Human Remains* (M. Brickley and J.I. McKinley, eds.), pp. 9–13. British Association for Biological Anthropology and Osteoarchaeology and the Institute of Field Archaeologists joint on-line publication.

McKinley, J.I. (2006). Cremation…the cheap option? In *The Social Archaeology of Funerary Remains* (C. Knusel and R. Gowland, eds.). pp. 81–88. Oxbow Books, Oxford.

Morris, I. (1992). *Death-Ritual and Social Structure in Classical Antiquity*. Cambridge University Press, Cambridge.

Nielsen-Marsh, C., Gernaey, A., Turner-Walker, G., *et al.* (2004). The chemical degradation of bone. In *Human Osteology* (M. Cox and S. Mays, eds.), pp. 439–454. Greenwich Medical Media, London.

Noy, D. (2005). Romans. In *Encyclopaedia of Cremation* (D.J. Davies and L.H. Mates, eds.), pp. 366–369. Ashgate, Aldershot, England and Burlington, USA.

Parsons, B. (2005). *Committed to the Cleansing Flame. The Development of Cremation in Nineteenth-Century England.* Spire Books Ltd., Reading, MA.

Perrin, J. (1998). Great goddess Ganges. *The Daily Telegraph; Travel*, 31.1.1998, pp. 1–2.

Philpott, R. (1991). *Burial Practices in Roman Britain*. British Archaeological Reports British Series 219.

Shipman, P. Forster, G. and Schoeninger, M. (1984). Burnt bones and teeth: an experimental study of colour, morphology, crystal structure and shrinkage. *Journal of Archaeological Sciences* **11**, 307–325.

Stirland, A. (1989). The cremations from the Iron Age cemetery. In *Verulamium: The King Harry Lane site* (I.M. Stead and V. Rigby, eds.), pp. 240–249. English Heritage Archaeological Report 12, London.

Thompson, H. (1889). *Modern Cremation: Its History and Practice*. Kegan Paul, Trench & Co., London.

Thurman, M.D. and Wilmore, L.J. (1981). A replicative cremation experiment. *North American Archaeologist* **2**, 275–283.

Toynbee, J.M.C. (1971). *Death and Burial in the Roman World.* Johns Hopkins University Press, London.

Webb, W.S. and Snow, C.E. (1945). The Adena People. *Reports in Archaeology and Anthropology VI.* University of Kentucky, Lexington.

Whimster, R. (1981). *Burial Practices in Iron Age Britain*. British Archaeological Reports British Series 90 (i and ii).

11

Fire as a Cultural Taphonomic Agent: Understanding Mortuary Behavior at Khuzhir-Nuge XIV, Siberia

Misty A. Weitzel, Ph.D.[1] and Hugh G. McKenzie, Ph.D.[2]

[1]*Department of Anthropology, Oregon State University, Corvallis, OR*
[2]*Department of Archaeology, University of Sheffield, Sheffield, UK*

Introduction

Located within the Lake Baikal region of Siberia, Russia, are hundreds of hunter-gatherer cemeteries dating between the Late Mesolithic and the Bronze Age (~7000–1000 BC). Sites such as these are unusual for most of Siberia and, indeed, for hunter-gatherers across the northern hemisphere and provide a unique opportunity to investigate the culture, particularly the mortuary behavior, among prehistoric foragers. One of these sites, Khuzhir-Nuge XIV (KN XIV), is an early Bronze Age cemetery located on a south-facing slope 15–30 m above Lake Baikal, in a shallow cove on the northwest coast of the Little Sea (Figure 11.1).

The data for this investigation were generated from original excavations at KN XIV collected under the auspices of the Baikal Archaeology Project (BAP) – a multidisciplinary and international research team investigating processes of prehistoric culture change and continuity in the region. Six seasons of excavation (1996–2001) produced archaeological data on 79 graves (G), including the remains of 89 individuals or burials (B). These graves seem to have been constructed by digging a pit approximately 40 cm deep, placing the body in the pit, and building a large stone cairn directly on top of the body. The burials were usually single with the exception of seven double and two triple burials. Individuals were generally placed in an extended, supine position although some individuals were flexed and lying on their sides. In terms of

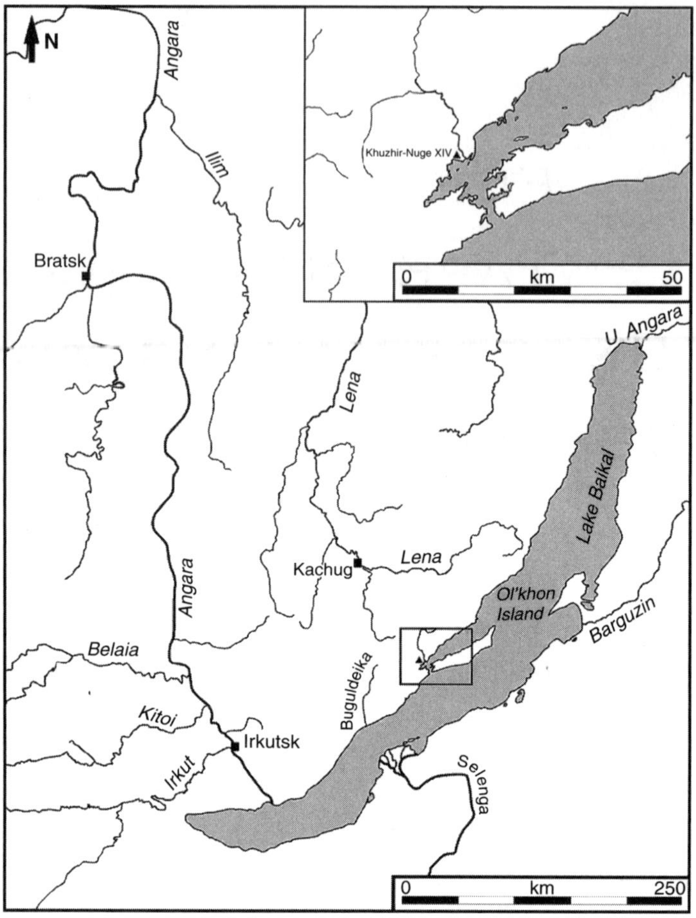

FIGURE 11.1 Map of Lake Baikal region showing location of Khuzhir-Nuge XIV (McKenzie, 2006).

grave and burial orientation, one grave (G 7) exhibited characteristics of the late Neolithic Serovo mortuary tradition, while 78 graves (containing 88 individuals) were identified as belonging to the early Bronze Age Glazkovo mortuary tradition. The Neolithic and the Bronze Age people discussed in this paper were of hunting–gathering–fishing cultures and, so far as we know, did not engage in any animal husbandry or horticulture. It is important to recognize that for Siberian archaeologists, the term Neolithic refers to the period between the introduction of pottery and the introduction of metallurgy, rather than to the adoption of animal and plant domestication as it does in other parts of the world.

At least 54 of the 79 graves show some evidence of fire, but this evidence was extremely variable, affecting both the features (stones and sediment) and 20 of the burials, but not necessarily both simultaneously (Weitzel, 2005). While many of the burned bones are calcined, most are more aptly described as charred, therefore charred is used throughout this paper to describe these 20 burials. In this study, we examine the mortuary use of fire at KN XIV as a cultural taphonomic agent to help interpret its use as a funerary practice, i.e., to help understand the practice within the larger context of mortuary archaeology.

The preliminary analysis of the spatial, temporal, and demographic context of fire use within the cemetery suggests that cremation was one means among many by which social distinctions were expressed. McKenzie (2005) examined mortuary data from KN XIV using a holistic approach that acknowledged both the multiscalar and multidimensional nature of mortuary variability (e.g., Brown, 1995; Cannon, 2002; Charles and Buikstra, 2002). Taking the spatial representation of death as his focus (e.g., Goldstein, 1980, 1981; Silverman and Small, 2002), he argued that the spatial representation offers a context within which to link individual beliefs and actions with larger patterns of social organization and ideology (Cannon, 2002). In particular, he noted that graves containing charred skeletons were spatially distinct from graves that did not contain charred individuals.

Graves at KN XIV were classified within the main concentration according to their spatial relationship with neighboring graves. The distribution of graves in rows, graves in groups, and scattered graves formed the primary basis for dividing the main concentration of graves into three spatial clusters: West, Center, and East (Figure 11.2). Graves in the West Cluster were generally scattered, graves in the Center Cluster were primarily arranged in north–south rows, and graves in the East Cluster were grouped. The charred burials were found predominantly in the Center Cluster, with only two burials from the West Cluster (B 9, 24) and one from the East Cluster (B 82) affected. When the cemetery as a whole was considered, statistically significant relationships

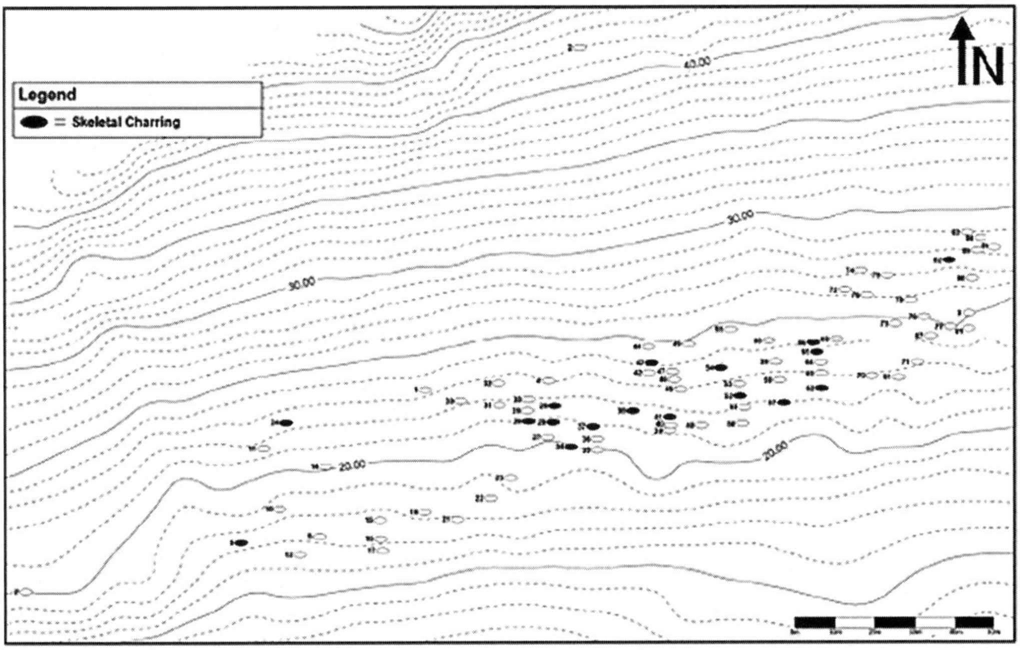

FIGURE 11.2 Spatial distribution of skeletal charring at Khuzhir-Nuge XIV (modified from McKenzie, 2005).

existed between skeletal charring, spatial cluster, and interment in rows. Two-sided Fisher's Exact Tests and Cramer's V were used to evaluate the significance and the strength of associations. However, when the Center Cluster was examined on its own, the association between rows and charring disappeared, since charred individuals from this cluster were just as likely to be found in rows as not. Similarly, despite the concentration of multiple graves in the Center Cluster, the proportion of charred individuals found in single and multiple graves was virtually identical. Despite the abundance of subadult burials in the Center Cluster, 11 charred individuals from this area were adult, while only 2 were children and only 3 were adolescent. The concentration of skeletal charring in the Center Cluster, then, seems to be unrelated to the concentration of subadults, rows, and multiple graves in the same cluster. This provides evidence that the spatial clusters at KN XIV reflect an intersection of multiple social distinctions and not just the use of fire alone.

The archaeological age of KN XIV burials was determined at the Accelerator Mass Spectrometry Facility of the IsoTrace Radiocarbon Laboratory, University of Toronto, Canada, using the Libby ^{14}C half-life of 5568 years. From approximately 2700 to 2000 BC, Glazkovo people used KN XIV continuously for about 340–660 years (McKenzie, 2005). In general, the tempo of site use seems to have varied, beginning with a low-intensity early period in which approximately 15% of all of the Glazkovo burials were interred. A peak period followed that centered around 2400 BC (~3900 BP) during which 70% of all the burials were interred in fewer than 200 calendar years. Following the peak, there appears to have been another low-intensity interval of use during which the remaining 15% were buried. Weber *et al.* (2005) consider the methodological aspects of analyzing long series of ^{14}C dates derived from human bones as well as chronological patterns of Cis-Baikal cemetery use at micro-, meso-, and macroscales of analysis. From the Weber *et al.* study it was determined that radiocarbon dates derived from bone samples with less than 1% collagen are believed to be susceptible to contamination; soil acids transferred from subsurface spring runoff may have degraded the bone collagen. Charring further compromised the bone. While it was previously reported that radiocarbon dates derived from fire-affected individuals ($n = 10$) spanned the range of the entire time the cemetery was in use (Weitzel, n.d.), these dates should be reconsidered because of the low collagen yields. In a more recent study, McKenzie (2005) reported dates for 14 charred individuals: of these, only one (B38) yielded collagen greater than 1.0%, while for the other 13 the collagen yield was less than 1%. As a result of these low collagen yields and their potential for contamination, we avoided further characterizing burials affected by fire within a precise temporal framework. However, the comprehensive analysis of radiocarbon dates at KN XIV by McKenzie (2005) demonstrated very clearly that intrasite variability in mortuary practices was not related to chronological changes and, instead, must have had sociopolitical or cosmological significance. While this contrasts fundamentally with the interpretation of mortuary variability at most sites in the Baikal region over the past 50 years, where nonnormative mortuary practices have almost been exclusively attributed to changes through time, the chronological distribution of virtually every attribute of mortuary practice at

KN XIV suggests that such practices are remarkably stable in chronological terms and it appears that this is indeed the case for the use of fire.

The demographic information was gathered using standard osteological procedures on the remains that were sufficiently preserved (Lieverse, n.d.). Biological age was assigned as young adult (20–34 years), middle adult (35–49 years), old adult (50+ years), adult (20–50+ years), or undetermined (Buikstra and Ubelaker, 1994: 9). In cases where age could be further defined by dental formation and eruption, or attrition, specific age ranges were assigned (Ubelaker, 1989). The osteological sex was recorded as male, probable male, female, probable female, or undetermined. These determinations were based on a number of standard nonmetric indicators of the skull and the pelvis, as well as metric traits. As reported in a previous study (Weitzel, n.d.), the demographic information revealed that fire was used for all age groups over the age of 4. Approximately 70% ($n = 14$) of the charred individuals were adults of over 20 years, approximately 25% ($n = 4$) were under 20 years, and 5% ($n = 1$) were undetermined. The mortuary use of fire occurred more commonly in the context of adult burials. Fire was also used in both male and female burials. Approximately 45% ($n = 9$) of the charred individuals were male or probable male, 10% ($n = 2$) were female or probable female, and 45% ($n = 9$) were undetermined due to poor skeletal condition. While there were a large percentage of males versus females among the charred individuals, there were also many more males ($n = 28$) than females ($n = 5$) in total at KN XIV. In sum, the demographic data from KN XIV reveal that the construction of fires over burials was intended for both males and females and affected all age groups, from at least age of 4 to over age of 50.

The great diversity within the mortuary context made it difficult to interpret the intentions behind the use of fire, but interpretations can be aided by examining the use of fire from a bioarchaeological perspective. It is now generally accepted that activity associated with culture, in addition to the natural environment, can play a major role in the extent to which human remains are influenced between the time of death and the time of recovery (Nawrocki, 1995; Weitzel, 2005). Previous taphonomic research (Weitzel, n.d.) was developed to examine the visual characteristics associated with the use of fire during the 2001 field season. These characteristics included color and location of charring, and the completeness, articulation, and fragmentation of the skeletal remains.

METHODS

Using slight modifications to the taphonomic variables mentioned above, a method was developed for characterizing the charred burials. Variables included color and location of charring, proportion of charred/uncharred bone, articulation, fragmentation, and pathological conditions. Categories were created for each of these variables in order to identify the main sources of variation of the burials affected by fire and to classify them with the final objective of linking this information to the mortuary data to better explain Glazkovo mortuary behavior.

Colors were recorded for each skeletal element, exhibiting evidence of fire exposure, by comparison with a Munsell color chart (1992). The burials were classified into two groups on the basis of color: those that exhibited cases of charring in which the color was mostly black and that penetrated the depth of the entire bone or the bone fragment and those that not only were charred as described above but also exhibited small spots of dark gray, gray, and white – all colors representing calcination resulting from higher temperatures and/or longer durations of exposure to fire (Chapter 2).

The location of charring was recorded for each skeletal element and then for the skeleton as a whole. Two separate variables were created for characterizing the location of charring on the skeleton. The first variable described the side of the skeleton affected with categories that included: the skull only, the skull and the right side of the skeleton, the skull and the left side of the skeleton, or both the right and the left sides of the skeleton. The second variable characterized the superior or the inferior portion of the skeleton with categories that included: the portion of the skeleton superior to and including the pelvis, inferior to the pelvis, or both the superior and the inferior portions of the skeleton.

Categories were created based on the number of skeletal elements affected by fire in proportion to the number of total elements present. The burials were classified into one of the following categories: 100%, 75–99%, 50–74%, 25–49%, and 0–24% charred. For example, if a burial consisted of a portion of the cranium, dentition, right humerus, and left tibia and all were charred, the burial would be considered 100% charred. However, if only the right humerus and the left tibia were charred, then the burial would be 50% charred.

The burials were distinguished as articulated or disarticulated. This qualitative determination was made based on the extent to which all the skeletal elements of the burial were articulated or disarticulated and whether or not they were all charred. The burials with charred bones were divided between those that were generally fractured longitudinally and those that were fractured both longitudinally and transversely; fragmentation was scored as either extreme or not extreme.

Finally, categories were created based on a previous study by Lieverse (n.d), which examined the pathological conditions of KN XIV remains. These categories included the following: osteoarthritis, dental pathology (i.e., antemortem tooth loss, pariapical abscess, and periodontitis), enamel hypoplasia, trauma, and osseous coalition.

RESULTS

The results are presented in Table 11.1 and discussed in detail below.

COLOR

The charred bones were black, very dark gray, dark reddish gray, dark gray, gray, and white. Fourteen burials exhibited cases of charring in which the color was mostly black (B 25, 34, 37.1, 37.2, 38, 41, 52, 57.1, 57.2, 62.1, 62.2, 65, 66, and 82). Six burials were charred as described above but also

TABLE 11.1 Taphonomic and Mortuary Variables Associated with Charred Burials at Khuzhir-Nuge XIV

Burial No.	Color-1	Color-2	Location I-1	Location I-2	Location I-3	Location I-4	Location II-1	Location II-2	Location II-3	Percent-1	Percent-2	Percent-3	Percent-4	Percent-5
9		X		X			X						X	
24		X			X				X					X
25	X					X	X				X			
28		X		X			X					X		
29		X		X			X						X	
34	X			X				X						X
37.1	X		X				X							X
37.2	X		X				X							X
38	X		X				X							X
41	X					X	X							X
43		X				X			X		X			
52	X					X			X	X				
54		X				X			X	X				
57.1	X					X			X				X	
57.2	X					X			X		X			
62.1	X					X			X			X		
62.2	X				X				X				X	
65	X		X				X							X
66	X						X						X	
82	X													

Color: 1, charred; 2, charred and calcined. Location on skeleton I: 1, skull only; 2, skull and right side; 3, skull and left side, 4, both right and left sides. Location on skeleton II: 1, upper portion of skeleton; 2, lower portion of skeleton; 3, both upper and lower portions. Percent charred/complete: 1, 100%; 2, 75–99%; 3, 50–74%; 4, 25–49%; 5, 0–24%.

(*Continues*)

TABLE 11.1 *(Continued)*

	Articulation-1	Articualtion-2	Fragmentation I-1	Fragmentation I-2	Fragmentation II-1	Fragmentation II-2	Placement-1	Placement-2	Placement-3
9	X			X	X			X	
24	X			X	X			X	
25	X			X	X				X
28		X		X	X				X
29	X			X		X			X
34	X		X			X			X
37.1	X		X			X			X
37.2	X		X			X			X
38	X		X		X				X
41		X	X		X				X
43		X	X		X				X
52		X	X		X				X
54	X		X		X				X
57.1		X	X		X				X
57.2		X	X		X				X
62.1		X	X		X				X
62.2		X	X		X				X
65	X		X		X				X
66	X		X			X			X
82	X		X		X		X		

Articulation: 1, articulated; 2, disarticulated. Fragmentation: 1, longitudinal fractures; 2, longitudinal and transverse fractures. Fragmentation II: 1, extreme fragmentation; 2, fragmentation. Placement of burials affected by fire within site: 1, East Cluster; 2, West Cluster; 3, Center Cluster.

	Sex-1	Sex-2	Sex-3	20–50+ years	11.5–20 years	4–7.5 years	Undetermined	Pathology-1	Pathology-2	Pathology-3	Pathology-4	Pathology-5
9	X			X				X				
24			X	X								
25	X			X					X			
28		X		X								
29	X			X				X				
34	X			X				X				
37.1			X		X					X		
37.2			X		X							
38	X			X							X	
41			X				X					
43	X			X				X				
52			X	X								
54			X	X				X				
57.1		X			X							X
57.2	X			X					X			
62.1	X			X								
62.2			X		X							
65			X			X						
66	X			X					X			
82			X	X								

Sex: 1, male; 2, female; 3, undetermined. Age: 1, 20–50+ years; 2, 11.5–20 years; 3, 4–7.5 years; 4, undetermined. Pathological condition: 1, osteoarthritis; 2, dental pathology; 3, enamel hypoplasia; 4, trauma; 5, skeletal pathology.

exhibited small spots of dark gray, gray, and white (B 9, 24, 28, 29, 43, and 54). The prevalence of darker colors indicates that most of the individuals were burned at temperatures below 800°C (Ubelaker, 1989). However, it is possible that the six burials with calcined bone were the result of burning at higher temperatures or that they burned for a much longer duration.

LOCATION

Fire affected the skull alone on four of the burials (B 37.1, 37.2, 38, and 65). The right side of the skeleton and the skull were charred on four other burials (B 9, 28, 29, and 34), while the left side of the skeleton and the skull were affected on only two burials (B 24 and 62.2). Both the left and the right sides were influenced by fire in eight cases (B 25, 43, 52, 54, 57.1, 57.2, 62.1, and 66). Burials 41 and 82 were not included; burial 41 had only fragments of an unsided ulna or radius, and no. 82 included only unidentified long bone fragments. Only the upper portion of the skeleton was affected on 10 burials (B 9, 25, 28, 29, 37.1, 37.2, 38, 41, 65, and 66). The lower portion of the skeleton was affected on only one burial (B 34). The upper and the lower skeleton were affected in eight cases (B 24, 43, 52, 54, 57.1, 57.2, 62.1, and 62.2). Again, burial 82 was not included.

In general, the exposure to fire shows a distinct pattern associated with the placement of the fire over the body. The location of charring and calcination is on both anterior and posterior sides of the skeleton, usually at the superior end, particularly at the skull, and often including the dentition. Charring also is more common on the right side. This location could be a result of how fires were originally built with respect to the slope of the site and the location of the pit. Fires were probably built by individuals standing on the downhill side of the pits. Fuel was placed over the body closer to the individual building the fire, thus charring the right side of the burial more than the left.

PROPORTION OF CHARRED/UNCHARRED BONE

In descending order of frequency, the skeletal elements most likely to be preserved at KN XIV, whether or not they were charred, were the following: the skull (including dentition), followed by the right humerus, left femur, right radius, right femur, right ribs, right ulna, left ribs, left humerus, tibiae, right fibula, and left ulna. Two burials were 100% charred (B 52 and 54), three were 75–99% charred (B 25, 43, and 57.2), two were 50–74% charred (B 28 and 62.1), five were 25–49% charred (B 9, 29, 57.1, 62.2, and 66), and seven were less than 24% charred (B 24, 34, 37.1, 37.2, 38, 41, and 65). In only two of the charred burials do we see 100% of skeletal elements present with some evidence of burning. Many of the charred burials have less than 24% of their total skeletal elements affected by fire. In a previous study it was demonstrated that completeness of skeletal elements was significantly associated with charred burials (Lieverse, 1999; Lieverse *et al.*, 2000). It is possible that the absence of skeletal elements is, in some cases, a direct result of exposure to fire. However, the overall preservation of bone at KN XIV is poor, even among the uncharred burials.

ARTICULATION

Overall, twelve burials were in articulation (B 9, 24, 25, 29, 34, 37.1, 37.2, 38, 54, 65, 66, and 82) and eight burials were disarticulated (B 28, 41, 43, 52, 57-1, 57-2, 62-1, and 62-2). While previous studies (Lieverse, 1999; Lieverse *et al.*, 2000) showed that uncharred elements had a greater probability of being articulated with each other than with charred elements, surprisingly here, there are more charred burials in articulation than in disarticulation.

FRAGMENTATION

The bones in 15 burials were fractured longitudinally (B 34, 37.1, 37.2, 38, 41, 43, 52, 54, 57.1, 57.2, 62.1, 62.2, 65, 66, and 82) while only five burials showed distinct cases of both longitudinal and transverse fracture (B 9, 24, 25, 28, and 29). Fifteen burials exhibited extreme fragmentation (B 9, 24, 25, 28, 38, 41, 43, 52, 54, 57.1, 57.2, 62.1, 62.2, 65, and 82) as opposed to five that showed less than extreme fragmentation (B 29, 34, 37.1, 37.2, and 66). It should be noted that fragmentation was pervasive among all the skeletal remains at KN XIV despite the fact that there was a higher degree of fragmentation among charred individuals than among the uncharred ones (Lieverse, 1999; Lieverse *et al.*, 2000). Because most of the remains at KN XIV exhibit a large degree of longitudinal fragmentation, it would be erroneous to claim that fractures are solely the result of exposure to fire. Since unburned burials had taphonomically fractured bones, other factors must be considered when evaluating the cause of missing elements and the degree of disarticulation and fragmentation such as animal scavenging, human disturbance and/or looting, and natural decomposition through time. Transverse fractures, however, are associated with burned remains.

PATHOLOGY

Eleven of the 20 charred burials exhibited pathological conditions: five with osteoarthritis (B 9, 29, 34, 43, and 54), three with dental pathological conditions including antemortem tooth loss, periodontitis, and a periapical abscess (B 25, 57.2, and 66), one with enamel hypoplasia (B 37.1), one with healed fractures of the left second, third, and fourth metatarsals (B 38), and one with congenital fusion of the left navicular and intermediate cuneiform (B 57.1) (Lieverse, n.d.). The presence of these pathological conditions does not explain the use of fire in the grave as uncharred burials also reveal similar pathological conditions.

It should be mentioned that fire not only affected the skeletons, but also influenced other aspects of the graves, including the paving stones and the sediment (McKenzie, 2005). In some graves, it was clear that a single firing event was responsible for both the charring of the bones and the discoloration of the surrounding stones and the sediment (black and red). In other cases, it appeared that there was evidence for more than one fire affecting the grave, ranging from a few pieces of charcoal to defined patches of sediment, charcoal, and burned birch bark. In all cases where the skeletal remains exhibited charring, the surrounding sediment also exhibited a blackish discoloration, often containing charcoal and charred birch bark. It was interesting that in some cases where the skeletal charring was limited to a small area, charred sediment was found distributed throughout the pit (G 37 and 38). Occasionally, traces of fire were recorded at burial levels that did not affect the bones (G 27, 36, 47, 68, 72, 84, and 85); there were also cases in which the burial levels showed no evidence of fire, but the upper layers of the grave shaft did (G 26, 32, 34, 35, 36, 42, 44, 45, 46, 50, 51, 63, 64, 70, 73, 81, and 82). Finally, there were cases in which evidence for fire use was found beside or near the graves but not actually within the confines of the grave pits (G 9, 11, 22, 41, 42, 43, 45, 46, 51, 52, 68, 72, and 84).

CONCLUSION

The variability in fire use at KN XIV is displayed in the color and the location of charred remains, the proportion of charred/uncharred bones, articulation, fragmentation, and pathological conditions indicating that the fires are intended to be relatively small overall, i.e., of low temperatures, and are localized at the superior end of the individual (Figure 11.3). It does not

FIGURE 11.3 Burial No. 66 showing charred skeletal elements. Photo by A. Weber. (see Plate 28)

appear that individuals are meant to be fully cremated. Nor are they meant to be disturbed to any large degree since their completeness, articulations, and levels of fragmentation are similar to uncharred burials. Clearly, the mortuary practice was not designed to quickly and completely dispose of the remains through cremation. The use of fire appears to have been more symbolic in nature. The presence of fire in the grave of the skeleton that was not thermally affected also supports this idea. Whereas in some cases, charcoal may have entered the grave by natural agents, such as wind or rodents or through spaces between the stones; there may be some cases where small fires were built nearby or inside the grave but not directly on the body. Whether or not these small fires served the same purpose as those directly over the skeleton is difficult to assess. The function of fires above the burial level is even more ambiguous but may reflect instances when people revisited the graves in order to perform acts of remembrance (McKenzie, 2005).

There are a few burials that appear to be influenced by the exposure to fire more than the others (B 43, 52, 54, and 57.2; Figure 11.4). These burials have five or six of the seven traits that represent more extreme cases of fire use. These traits include calcination of bone, fire affecting the skull and both

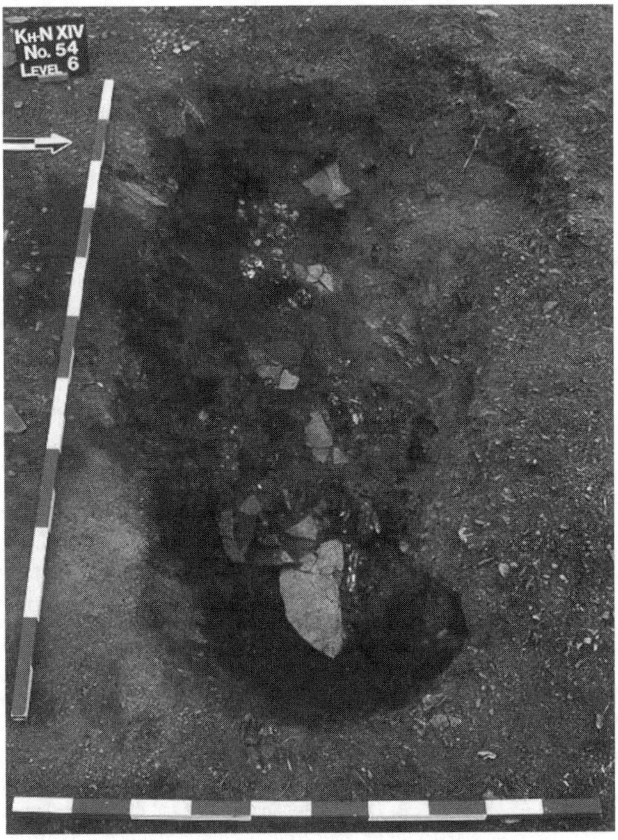

FIGURE 11.4 Burial No. 54 showing more extreme evidence of fire. Photo by A. Weber. (see Plate 29)

right and left sides of the skeleton, fire affecting both the upper and lower portions of the skeleton, 75–100% of skeletal elements present showing some evidence of fire, disarticulation, both longitudinal and transverse fractures, and extreme fragmentation overall. While they are unique in the extent to which they have been exposed to fire, these four burials are not unique with regard to where they are located or their demographic characteristics. They are all from the Center Cluster of the cemetery. Two are male or probably male and two are unidentified, with ages between 20 and 50 years.

In fact, when all of the taphonomic variables associated with fire are considered with the variables associated with the mortuary context, no clear patterns emerge and we still see a high degree of variability among charred burials with respect to where they are located within the site (outside of being located mainly at the center), chronology, and demography. Clearly the use of fire was not meant for everyone but it was not exclusive to a particular period in time, a biological age or sex, or placement other than within the center of the cemetery.

Indeed Glazkovo mortuary ritual consisted of many practices that varied in their independent expressions, not least of which was the use of fire in the grave. Faced with the event of human death, people's experiences of the world during and after death may differ individually and culturally (Parker Pearson, 1999). Perhaps this is demonstrated among prehistoric populations of the Baikal region where fire is placed over an individual as a sociopolitical or cosmological expression, which could be representative of a single individual or an entire group, and varies accordingly. Certainly, this would help explain the wide variety seen in the mortuary use of fire during the early Bronze Age in this region of Siberia.

REFERENCES

Brown, J.A. (1995). On mortuary analysis – with special reference to the Saxe-Binford research program. In, *Regional Approaches to Mortuary Analysis* (L.A. Beck, ed.), pp. 3–26. Plenum Press, New York.

Buikstra, J.E. and Ubelaker, D.H. (1994). *Standards for Data Collection from Human Skeletal Remains*. Arkansas Archaeological Survey Research Series No. 44, Fayetteville, NC.

Cannon, A. (2002). Spatial narratives of death, memory, and transcendence. In *The Space and Place of Death* (H. Silverman and D. Small, eds.), pp. 191–199. Archaeological Papers of the American Anthropological Association No. 11, Arlington, TX.

Charles, D.K. and Buikstra, J.E. (2002). Siting, sighting, and citing the dead. In *The Space and Place of Death* (H. Silverman and D. Small, eds.), pp. 13–25. Archaeological Papers of the American Anthropological Association No. 11. Arlington.

Goldstein, L.G. (1980). *Mississippian Mortuary Practices: A Case Study of Two Cemeteries in the Lower Illinois Valley*. Northwestern University Archaeological Program, Evanston.

Goldstein, L.G. (1981). One-dimensional archaeology and multi-dimensional people: spatial organization and mortuary analysis. In *The Archaeology of Death* (R. Chapman, I. Kinnes, and K. Randsborg, eds.), pp. 53–69. Cambridge University Press, Cambridge.

Lieverse, A. (1999). *Human taphonomy at Khuzhir-Nuge XIV, Siberia*. Master's thesis, University of Alberta.

Lieverse, A. (n.d.). Human osteological evidence: demography and health. In *Khuzhir-Nuge XIV Osteology: Data from a Late Neolithic/Early Bronze Age Cemetery in Cis-Baikal, Siberia (1997–2001 excavations)* (M.A. Katzenberg and A. Weber, eds.). Northern Hunter-Gatherers Research Series. CCI Press, Edmonton, Alta.

Lieverse, A., Weber, A., and Goriunova, O.I. (2000). *Variable Skeletal Condition at Khuzhir-Nuge XIV, Siberia: Taphonomic Implications*. Paper presented at Meso 2000 Conference, Stockholm, 4–8 September 2000.

McKenzie, H.G. (2005). *Mortuary Variability Among Middle Holocene Hunter-Gatherers in the Lake Baikal Region of Siberia, Russia*. Ph.D. dissertation, University of Alberta.

Munsell Color Company, Inc. (1992). *Munsell Soil Color Charts*. Newburgh, New York.

Nawrocki, S.P. (1995). Taphonomic processes in historic cemeteries. In *Bodies of Evidence: Reconstructing History Through Skeletal Analysis* (A.L. Grauer, ed.), pp. 49–66. Cambridge University Press, Cambridge.

Parker Pearson, M. (1999). *The Archaeology of Death and Burial*. Texas A&M University Press, College Station, TX.

Silverman, H. and Small, D.B. (eds.) (2002). *The Space and Place of Death*. Archaeological Papers of the American Anthropological Association, No. 11, Arlington.

Ubelaker, D.H. (1989). *Human Skeletal Remains: Excavation, Analysis, Interpretation*, 2nd edn. Taraxacum Press, Washington, DC.

Weber, A.W., McKenzie, H.G., Beukens, R., and Goriunova, O.I. (2005). Evaluation of radiocarbon dates from the Middle Holocene hunter-gatherer cemetery Khuzhir-Nuge XIV, Lake Baikal, Siberia. *Journal of Archaeological Science* **32**, 1481–1500.

Weitzel, M.A. (2005). *Human Taphonomy: Khuzhir-Nuge XIV, Siberia and Edmonton, Alberta*. Ph.D. dissertation, University of Alberta.

Weitzel, M.A. (n.d.). The mortuary use of fire at Khuzhir-Nuge XIV. In *Khuzhir-Nuge XIV Osteology: Data from a Late Neolithic/Early Bronze Age Cemetery in Cis-Baikal, Siberia (1997–2001 excavations)* (M.A. Katzenberg and A. Weber, eds.). Northern Hunter-Gatherers Research Series. CCI Press, Edmonton, Alta.

12

Putting Together the Pieces: Reconstructing Mortuary Practices from Commingled Ossuary Cremains

A. Joanne Curtin, Ph.D

Department of Anthropology, University of West Florida, Pensacola, FL

INTRODUCTION

Cremations pose challenging analytical problems to biological anthropologists due to the shrinkage, warping, and high degree of fragmentation that typify such remains. These problems are exacerbated when cremains of a large number of individuals are commingled in an ossuary context. This study presents the results of an analysis of commingled cremated remains from a prehistoric ossuary (DgRw 199-F1) on Gabriola Island, on the Northwest Coast of British Columbia, Canada. The main focus of this analysis is the elucidation of mortuary practices at the site. Several studies of mortuary practices involving cremated remains have focused on distinguishing between fleshed, green, and dry cremated bones by analyzing coloration and patterns of burning and fragmentation (e.g., Baby, 1954; Binford, 1963; Gejvall, 1963; Shipman *et al.*, 1984). This study draws on these earlier works, but adds an additional dimension to the analysis: the spatial patterning of burned bone fragments.

BACKGROUND

DgRw 199-F1 is one of more than 50 burial features located among a cluster of huge sandstone boulders at the foot of an escarpment that runs along the southwest shore of Gabriola Island (Wilson, 1987; Curtin, 1991; Skinner, 1991). The burials appear to have been surface interments that were placed in crevices and caves formed beneath and between the fallen rocks. When first discovered, the burials were assumed to be late prehistoric or protohistoric in age, based on the current understanding of mortuary practices in this region of the Northwest Coast. Until about 800–500 years ago, the typical burial pattern was flexed inhumation in midden deposits at village sites. Sometime after 800 BP, however, for reasons that are still not well understood, there was a dramatic shift from inhumation to above-ground disposal in caves, grave houses, trees, and mortuary poles (Borden, 1970; Cybulski, 1978, 1992).

The presumption that these burials were relatively recent was contradicted, however, by radiocarbon dates obtained on bone collagen from skeletal elements collected from the site's surface (Skinner, 1991). These dates ranged from approximately 2170 ± 70 to 2760 ± 60 years BP (uncalibrated), placing the burials in the Marpole and Locarno periods of South Coast prehistory (Matson and Coupland, 1995). In an attempt to reconcile these unexpectedly early dates with the current understanding of the culture history of the Northwest Coast, it was suggested that the human remains might represent ancient primary midden interments that had been dug up sometime in the historic period, afforded a mass cremation, and then disposed of in the rock features (Wilson, 1987). Alternatively, it was suggested that the radiocarbon samples might have been contaminated, yielding anomalously old dates (Cybulski, 1992, personal communication).

In an attempt to better understand mortuary practices at the site, and by extension, on the southern coast of British Columbia, additional field work was undertaken on Gabriola Island in 1992, at which time four of the burial features were completely excavated and all human remains removed. This chapter presents the results of the analysis of the human remains from one of these features, DgRw 199-F1.

DgRw 199-F1

DgRw 199-F1, the largest of the 50 burial features situated along the Gabriola Island escarpment, is an 8×2-m cave-like recess located beneath two massive sandstone blocks (Figure 12.1). Initially the feature was believed to contain the remains of seven individuals, based on the number of skulls visible on the floor of the cave when the site was first recorded by Wilson (1987). In 1992, in response to continued vandalism at the site, a project

FIGURE 12.1 DgRw 199-F1 located beneath massive sandstone slabs.

was mounted to recover all of the human remains and the associated cultural materials from this feature. The cave floor was divided into 15, 1 × 1-m excavation units (Figure 12.2), and the entire feature was excavated to sterile deposits.

These excavations revealed that the burials were not limited to visible scattered surface remains. The floor of the cave actually consisted of densely packed and fragmented burned bones, to a depth of approximately 10 cm in the eastern, less accessible part of the cave and up to 55 cm in the western half. Figure 12.3 illustrates the density of the deposits; all of the bones pictured were recovered from a single 5-cm level of excavation unit 13. In total,

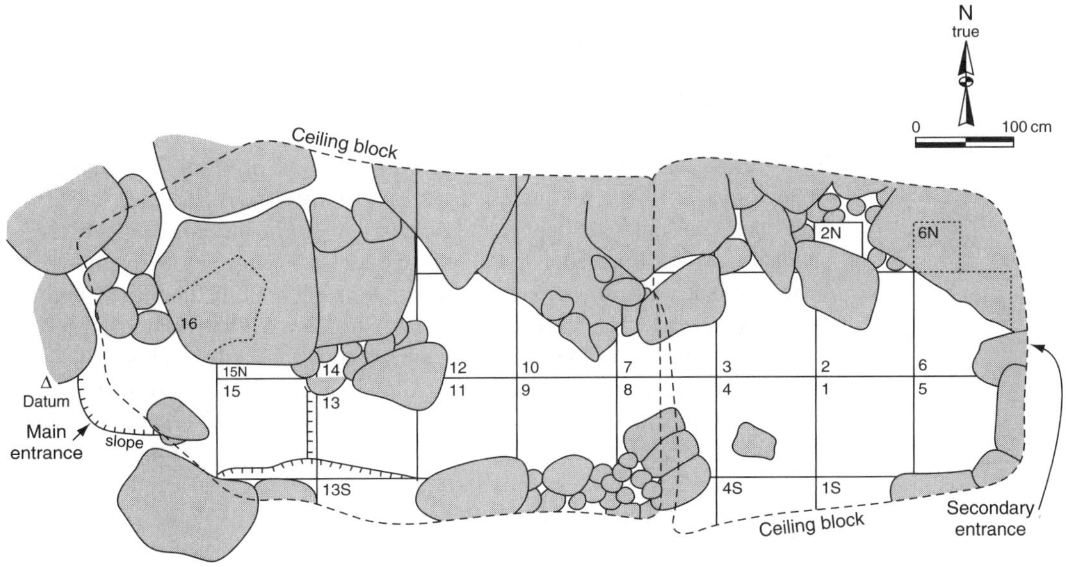

FIGURE 12.2 Plan view of feature 1 showing excavation units.

FIGURE 12.3 Bones recovered from E.U. 13, level 4, illustrating the density of deposits. (see Plate 30)

159,323 human teeth, bones, and bone fragments were recovered from the feature, representing a minimum of 118 people, ranging in age from newborns to old adults. In addition to the human remains, the excavations recovered a substantial number of faunal remains (fish, mammal, and shellfish), most of which may represent food offerings to the dead, and 148 artifacts made of antler, bone, stone, and wood.

METHODS OF ANALYSIS

The recovered human remains were washed and cataloged prior to data collection. Information recorded in the specimen catalog included three-dimensional field provenience (excavation unit, quadrant, and 5-cm arbitrary level), identification (element, side, and portion), demographic characteristics (age and sex), condition/preservation (poor, fair, or good), postmortem modifications (burning, tool marks, animal chewing, and rodent gnaw marks), and antemortem conditions (discrete traits, anomalies, and pathological conditions). The presence and degree of burning was assessed primarily on the basis of bone coloration, which allowed separation of specimens into four categories: unburned (no apparent heat-related changes), slight burning (light brown or reddish discoloration, often localized), moderate burning (more extensive dark brown or black discoloration), and severe burning (calcined bone, white, gray, or blue-gray in color, often warped and shrunken). Observations were also made on the nature of the fracture patterns on the burned remains (e.g., superficial versus deep; curvilinear versus linear). Other bone modifications (tool marks and chewing/gnaw marks) were identified using the criteria established by White (1992). Observations were made using a 10-power hand lens under strong natural light; where the morphology of the marks was ambiguous, they were further examined under a dissecting microscope.

Not all of the data sets could be collected for every specimen. More than half of the specimens recovered (51.7%) were too small to be identified to the skeletal element; these were merely counted and bagged according to the provenience. Other specimens could be identified only within broad anatomical categories, such as large-diameter long bone (humerus, femur, and tibia) and small-diameter long bone (radius, ulna, and fibula). Demographic data collection was equally problematic. An attempt was made to assign specimens to broad age categories (infant, child, juvenile, young adult, middle adult, and old adult), but in practice it was often possible only to distinguish immature from adult. Sex could be determined only for diagnostic elements (pelvis and cranium) or for specimens at the extreme ends of the range of size.

Once the human remains were assessed and cataloged, they were sorted according to the skeletal element, and all broken edges were compared for possible articulation with other fragments. Pieces found to conjoin were reconstructed using water-soluble glue, and a record was kept of the number of articulating pieces and the original provenience of each piece of a conjoined set. This information was used to assess the degree of horizontal and vertical dispersal of fragments from a single element. A very rough method of determining the degree of dispersal of fragments of the same element was devised, as follows. Conjoined sets whose constituent pieces all came from the same

$50 \times 50 \times 5$-cm provenience unit were assigned horizontal and vertical dispersal scores of 1 ($H = 1$, $V = 1$); these numbers were summed to provide a total scatter (TS) score of 2. Similarly, sets whose members were found at the same level of two adjacent unit quadrants were scored, $H = 2$, $V = 1$, TS = 3. Calculated dispersal scores were then used to evaluate mortuary behavior and taphonomic processes within the burial feature.

RESULTS

The human remains from DgRw 199-F1 appear to be secondary burials. None of the elements was found articulated in anatomical position, and the remains of different individuals were thoroughly commingled. The skeletal assemblage is characterized by extreme fragmentation: 52% of the recovered specimens (n = 82,330) are less than 1 cm in maximum diameter, often too small to be identified to the skeletal element. Some of the fragmentation is no doubt due to trampling by visitors to the cave, both pothunters and tourists, but most is probably attributable to the cremation process. The majority (56%) of the human bones collected from the site exhibit some evidence of burning, ranging from slight, localized discoloration, to severe calcination (Figure 12.4). Burning is strongly correlated with age at death: 82% of infant remains appear unburned compared to 50% of child remains and 13% of adults. Only 127 specimens exhibited evidence of animal modification in the form of rodent gnaw marks or carnivore chewing. Cut marks were observed on additional 94 specimens, predominantly cranial fragments and long bone joint margins.

The pattern of burning observed at the site is typical of the cremation of fleshed remains or green bones (Binford, 1963; Buikstra and Swegle, 1989). Burned and fragmented elements exhibit deep transverse splitting and curvilinear transverse fractures, often ending in longitudinal hinge terminations, rather than the straight longitudinal fractures and superficial checking, more typical of dry, defleshed cremains. Many heavily burned (calcined) specimens are significantly warped, which prevented complete reconstruction of articulating fragments. Cranial fragments often exhibit exfoliation of the outer table of the bone, which is seldom seen in dry cremated bones Finally, only one of the specimens recovered from DgRw 199-F1 displayed the distinctive coloration

FIGURE 12.4 (a) Lightly burned skull, reconstructed from 64 pieces. (see Plate 31a)

FIGURE 12.4 (b) Calcined long bone fragment, reconstructed from six pieces. (see Plate 31b)

that Buikstra and Swegle (1989) identified as a signature of burned dry bones: tan to light brown outer cortex overlying black or gray cortex and trabeculae.

In the refitting exercise, a total of 11,952 fragments were fitted to other pieces to form 3,435 reconstructed 'sets' of 2–99 pieces. Ninety percent of the reconstructed sets comprise five or fewer pieces, but a small number of sets ($n = 13$) are very large, with more than 50 conjoined pieces. All of the very large sets consist of reconstructed crania.

In general, there appears to have been very little scattering of broken pieces of the same bone. In more than 80% of the conjoined sets, all members came from the same or adjacent provenience units ($\mathrm{TS} = 2 - 3$), and fewer than 1% showed substantial dispersal of constituent fragments (TS>10) (Figure 12.5). Vertical and horizontal dispersal scores are very similar, although fragments are more likely to be dispersed vertically (mean $V = 1.5$) than horizontally (mean $\mathrm{H} = 1.3$); 94% of conjoined sets came from the same or adjacent quadrants, whereas 87% are from the same or adjacent levels (maximum vertical dispersal ($V = 10$) is less than maximum horizontal dispersal ($H = 15$) because the former measure is limited by the depth of the cultural deposits, which is less than their horizontal extent).

Thirty conjoined sets with high TS scores (TS>10) were examined in greater detail for the information they could provide about postdepositional taphonomic processes in the burial cave. These 30 sets can be grouped into three categories, defined by their dispersal values and the inferred agent of dispersal:

Type 1: High horizontal and low vertical dispersal values. Many of these sets exhibit evidence of carnivore chewing, and it is likely that their dispersal is the result of postdepositional disturbance by animals.

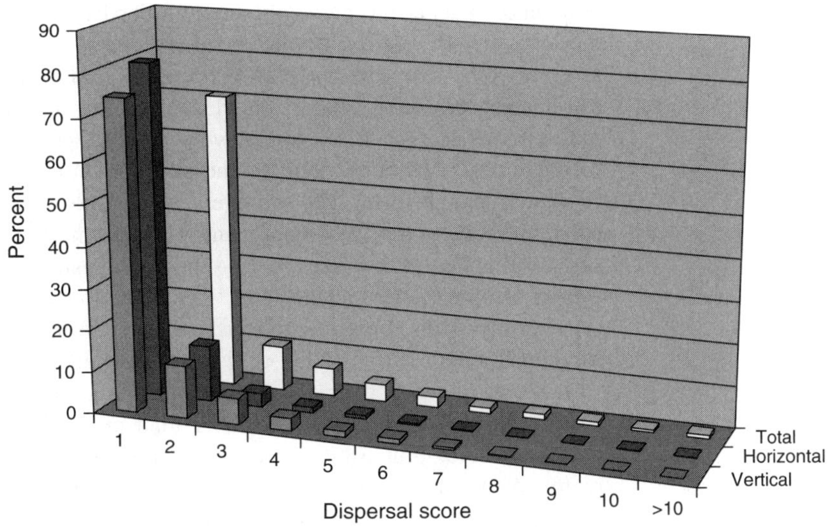

FIGURE 12.5 Distribution of dispersal scores of conjoined sets. Note that none of the total scores are '1' because a minimum total score (i.e., horizontal + vertical score) is '2'.

Type II: High vertical and low horizontal dispersal values. The dispersal of these sets appears to be the result of natural downward filtering of fragments through cracks in the boulder-strewn cave floor.

Type III: High vertical and high horizontal dispersal values. Some members of these sets were found at the bottom of the excavated pits that dotted the floor of the cave, while others were scattered some distance away on the current ground surface. These dispersal scores are suggestive of the recent disturbance from pits dug into deposits by pothunters searching for artifacts.

One additional line of evidence was examined in the analysis of skeletal dispersal within the burial feature. Although none of the skeletal elements was articulated in anatomical position, it was occasionally possible to identify several elements from a specific individual, usually on the basis of unusual size, age, or pathological characteristics. When this was possible, a pattern similar to the conjoining exercise was revealed: most of the elements from a single individual were found in the same or adjacent provenience units.

MORTUARY BEHAVIOR RECONSTRUCTED

The unusual context and unexpected early dates obtained on cremated bones from DgRw 199-F1 raised questions as to the nature and origins of these deposits. Specifically, it was suggested that either the radiocarbon dates were incorrect, and the burials were much more recent than indicated, or that the remains represented ancient burials that were exhumed, cremated, and deposited in the cave sometime in the recent past. The current analysis does not support either scenario. Two more radiocarbon dates were obtained on bones from the cave and were submitted to a different lab for analysis. These new dates bracket the dates obtained earlier by Skinner (1991), and support the antiquity of the burial site. Artifacts recovered from the burial feature also are consistent with the radiocarbon dates. Burning patterns indicate that the bodies were cremated in the flesh, and the presence of cut marks near major joints suggests that some were at least partially dismembered prior to cremation. Although none of the skeletal remains was found articulated in anatomical position, the relatively low dispersal rate of fragments from the same bone and of elements from the same individual (where discernible) suggest that the bodies were processed and deposited individually, rather than as part of a mass cremation event. Postdepositional disturbance from animals scavenging, from subsequent interments, and from modern pothunting may account for the relatively few conjoined sets with high dispersal scores.

It seems clear from this analysis that prehistoric mortuary rituals on the Northwest Coast were more variable than previously assumed and warrant more careful examination. In particular, the role of cremation in the broader realm of mortuary practices needs to be elucidated. Burned human remains are an unusual occurrence in prehistoric midden sites on the Northwest Coast (although occasional examples have been reported from at least eight sites in the region), but become increasingly common in the late prehistoric period, particularly in association with large earthen burial mounds/cairns. In most cases, whether ancient or protohistoric, the burned bones have not been subjected to careful analysis, so it is difficult to determine not only the demographic characteristics of the cremated individuals, but also the context (deliberate or incidental to other activities), nature (fleshed, green, or dry bone), and extent of burning. The careful analysis of cremated bones is labor-intensive and requires a major investment of time, but the benefits to our understanding of burial ritual cannot be overemphasized.

REFERENCES

Baby, R.S. (1954). *Hopewell Cremation Practices*. Papers in Archaeology No. 1. Ohio Historical Society, Columbus, OH.

Binford, L.R. (1963). An analysis of cremations from three Michigan sites. *Wisconsin Archaeologist* **44**, 98–110.

Borden, C.E. (1970). Culture history of the Fraser-Delta region: an outline. *B.C. Studies* **6 & 7**, 95–112.

Buikstra, J.E. and Swegle, M. (1989). Bone modification due to burning: experimental evidence. In *Bone Modification* (R. Bonnichsen and M.H. Sorg, eds.), pp. 247–258. Center for the Study of First Americans, Orono, ME.

Curtin, A.J. (1991). *Gabriola Island Burial Recovery Project: Report on the 1989 Field Season*. Unpublished permit report #1980–92, submitted to the Archaeology Branch, Victoria, BC.

Cybulski, J.S. (1978). *An Earlier Population of Hesquiat Harbour, British Columbia*. British Columbia Provincial Museum Cultural Recovery Papers No. 1, Victoria, BC.

Cybulski, J.S. (1992). *A Greenville Burial Ground*. Archaeological Survey of Canada Mercury Series Paper No. 146. Canadian Museum of Civilization, Ottawa.

Gejvall, N.-G. (1963). Cremations. In *Science in Archaeology* (D.R. Brothwell and E.H. Higgs, eds.), pp. 379–390. Praeger, New York.

Matson, R.G. and Coupland, G. (1995). *The Prehistory of the Northwest Coast*. Academic Press, San Diego, CA.

Shipman, P., Foster, G., and Schoeninger, M. (1984). Burnt bones and teeth: an experimental study of color, morphology, crystal structure, and shrinkage. *Journal of Archaeological Science* **11**, 307–325.

Skinner, M. (1991). *Osseous Treponematosis of the Americas: A Review with New Evidence from Gabriola Island (DgRw 199), BC, Canada*. Unpublished permit report #1988-1, submitted to the B.C. Heritage Conservation Branch, Victoria.

White, T.D. (1992). *Prehistoric Cannibalism at Mancos 5MTUMR-2346*. Princeton University Press, Princeton, NJ.

Wilson, I.R. (1987). *The False Narrows Bluffs Archaeological Investigations on Gabriola Island*. Unpublished permit report #1987–40, submitted to the B.C. Heritage Conservation Branch, Victoria, BC.

13

A Taphonomic Analysis of Human Cremains from the Fox Hollow Farm Serial Homicide Site

Amanda Baker Bontrager, M.S. and Stephen P. Nawrocki, Ph.D., D.A.B.F.A.

University of Indianapolis, Archeology and Forensics Laboratory, Indianapolis, IN

INTRODUCTION

In June 1996, the University of Indianapolis Archeology and Forensics Laboratory was asked to conduct the recovery of human remains from Fox Hollow Farm, an 18-acre estate located north of Indianapolis in Westfield, Hamilton County, Indiana. The bones of at least 11 homicide victims were discovered in two main areas of the property: a wooded area directly behind the main residence (Area 1), and within a seasonal drainage ditch in the woods on the western border of the property (Area 3) (Nawrocki *et al.*, 1998; Nawrocki and Baker, 2001).

Unfortunately, little is known about the specific events that led to the deaths and eventual disposal of these individuals. The suspected assailant committed suicide shortly after the remains were discovered on his property, leaving the investigators and anthropologists to reconstruct the details. It is believed that he had picked up young adult males in Indianapolis bars and had strangled them in his home while his family was out of town (Weinstein and Wilson, 1998). No evidence of traumatic injury, dismemberment, or deliberate subsurface burial was detected at the site.

The 413 bones and teeth found in Area 3 were unburned and largely intact but were commingled and scattered throughout the drainage ditch, having been subjected to fluvial transport over a distance of 100 m or more. Genetic testing of 17 bones and teeth from the ditch indicated the presence of at least five individuals, four of whom have been positively identified via DNA and/or dental analysis. These four were reported missing in 1993 and therefore had decomposed over a 3-year period.

In stark contrast to Area 3, most remains recovered from Area 1 were burned and heavily fragmented. The area is moderately wooded and slopes downhill away from the residence, descending 3 m in elevation over a distance of 20 m. A narrow channel cuts through the center of Area 1, diverting rainwater from the gutter system of the residence. While the burning events appear to have taken place in a centralized spot near the top of the slope, the remains were distributed downhill by rainwater into a delta covering 80 m^2 of sloping forest floor. A shallow basin and large trees near the bottom of the slope served to trap bone fragments and allowed them to precipitate out of the flow. A total of 5600+ bone fragments, 133 complete bones, and 45 teeth and tooth fragments were recovered from Area 1.

Genetic testing of 13 complete bones and teeth from Area 1 indicated the presence of at least five individuals, three of whom have been positively identified via DNA and/or dental analysis. These three were reported missing in 1994 and therefore had been dead for about 2 years at the time of discovery. Their location immediately behind the residence suggests that the assailant was becoming more comfortable with or more daring in his activities over time, particularly since his wife and children were living there during the killings.

A few lines of evidence indicate that at least some of the victims deposited in Area 1 were allowed to decompose before being burned (Baker, 2000, 2004; Baker Bontrager and Nawrocki, 2005). Family members reported that they had stumbled upon an intact, unburned skeleton on the surface of Area 1 in the winter of 1994–1995. The assailant explained that it was an anatomical teaching skeleton that he had dumped in the woods. Once given this explanation, the family temporarily forgot about it. The osteological analysis also provided hints that some victims were burned in a skeletonized state. The majority of anterior teeth (incisors and canines) recovered from Area 1 were unburned, suggesting that they had fallen free of the jaws before the skulls were burned. In contrast, all multirooted molars, which would have resisted falling out for a longer period of time, were heavily burned. Finally, of the 133 complete bones found in Area 1, nearly all were small bones of the hands and feet, and most of these were unburned or only lightly singed. If the assailant had allowed his victims to decompose and then raked the bones together into piles to burn (perhaps inspired by his family's discovery of one of his victims), small bones of the extremities might have been missed and would have remained peripheral to the flames.

Other issues complicate the interpretation of Area 1. First, it is possible that some of the assailant's last victims were burned immediately after being killed rather than being allowed to decompose first. Second, fragments from earlier victims might have been subjected to repeated rakings and burnings as later victims were added to the pile. Third, the wash of water through the channel during rainstorms swept the burned fragments downhill, churning and mixing them with leaf litter, ash, and sediment. Finally, the process of excavation, screening, and cleaning might have increased the degree of fragmentation.

The analysis of the burned remains was complicated by these various behavioral and taphonomic factors. Questions that the analysts considered included the following: Is it possible to distinguish between victims that were burned fresh versus those that were burned in a skeletonized or partially decomposed state? How does fluvial transport alter burned fragments? How are the cremains from Fox Hollow compared to burned remains from other forensic and archeological contexts? While careful examination of this particular assemblage may provide information that can assist in the interpretation of specific assailant behaviors, detailed scientific study also offers an opportunity to examine how numerous intersecting taphonomic processes affect a skeletal assemblage over a relatively short period of time.

THE STUDY SAMPLE

FOX HOLLOW FARM

During the excavation of Fox Hollow, nineteen 2 × 2-m grid squares were established in Area 1, and a grid-specific provenience was given to each fragment. From the 5651 bone fragments recovered (excluding complete bones and teeth), a random sample was taken for detailed analysis. The fragments from each archeological unit that contained over 100 fragments were spread evenly over a 30 × 30–in. piece of paper, which had on it 3 × 3-in. squares numbered from 00 to 99. Using a table of random numbers, 100 fragments were randomly selected (Greene and Schmidt, 2000). No fragments were chosen from the 11 grid units that produced less than 100 fragments (none of these units held more than 12 fragments). Therefore, a total of 800 fragments (8 units × 100) were selected for detailed analysis. However, eight of these fragments were later removed from the study sample because it was determined that they were likely broken by carnivore chewing and not by burning. Therefore, a total of 792 burned fragments were analyzed.

The remains from Fox Hollow Farm were compared to five other burned specimens or assemblages, including two recent forensic cases, an Early Archaic prehistoric cremation, a commercial human cremation, and a commercial dog cremation. Depending on how each specimen had been recovered and subsequently curated, slight alterations of the sampling process were required in order to obtain representative samples of fragments. Additional information on sampling procedures and assemblage characteristics can be found in Baker (2004).

LAWRENCE COUNTY, PENNSYLVANIA

This assemblage, excavated in 1987, represents the remains of a 16-year-old girl who had gone missing in 1965. An informant told police that the body was burned in a home after her death and then sometime after the initial burning the assailant returned to the property and set the house ablaze. The landowner had subsequently bulldozed dirt and burned debris into the open foundation where the house had once stood. During the excavation, a backhoe was used to remove three to four feet of overburden. At approximately 1.5 feet above the basement floor, heavy excavation was replaced by trowel and shovel work. Bones were discovered in the center of the structure. Due to the poor condition of the skeletal material, the fill from this area was taken back to the laboratory in its entirety (Dennis Dirkmaat, 2002, personal communication).

In all, the Lawrence County assemblage includes over 2300 bone fragments. None of the bones is complete, most are extensively warped and most are blue-gray to white in color. The large concentration of remains in one area suggests that they were not scattered or damaged during the bulldozing of the dwelling, although they may have been affected by compaction from the overlying debris during their 22-year interment.

The senior author examined the Lawrence County remains at Mercyhurst College in Erie, Pennsylvania, in September 2002. They had been sorted into plastic trays and bags. All of the larger, identifiable fragments were scored ($n = 184$). The remaining unidentified fragments had been divided into 10 plastic trays. Ten fragments were randomly sampled from seven of these trays ($n = 70$) using the same methods employed for the Fox Hollow assemblage. Two of the plastic trays were excluded because they did not contain even 10 fragments. Another tray was excluded because all of the fragments were fused to the pieces of melted plastic and glass. In all, 254 fragments were analyzed.

POTTER COUNTY, PENNSYLVANIA

This assemblage, recovered in 1995, represents the remains of a gracile male, aged 20–50 years. The victim had died in a house fire that may have been started by a wood-burning stove (Dennis Dirkmaat, 2002, personal communication). Fragment colors range from black to blue-gray to white. Unlike the Lawrence County remains, this assemblage contains many large, diagnostic fragments, including portions of the right shoulder girdle and thoracic spine encased in dried soft tissues.

The senior author examined the Potter County remains at Mercyhurst College in Erie, Pennsylvania, in September 2002. The fragments had been sorted into plastic bags. As with the Lawrence County case, all of the larger, identifiable fragments were scored ($n = 160$). In addition, there were six bags containing unidentified fragments. Ten fragments were randomly chosen from each bag ($n = 60$). In all, 220 fragments were analyzed.

THE JERGER SITE

The Jerger site, located in the White River Valley in Daviess County, Indiana, is an Early Archaic mortuary site (see Chapter 14). Local residents discovered cremated human bones, projectile points, and prehistoric cultural materials on the surface of a cultivated field. The Indiana Department of Transportation excavated the site 5 years after it had been discovered. By this time the debris field had grown to approximately 6000 square feet (Tomak, 1979).

The senior author examined the Jerger remains at the University of Indianapolis, where they are still undergoing analysis (Greene and Schmidt, 2000). They had been sorted into numerous plastic bags. The remains are commingled and range from unburned to completely calcined. One representative bag from the plow zone was chosen and the fragments were spread over the number grid, 100 being randomly selected for analysis.

COMMERCIAL HUMAN CREMATION

These remains were cremated and mechanically pulverized by an Indiana crematorium in 1987 and were returned to the funeral director for interment. The specimen was later donated to the University of Indianapolis for curation and study. The remains arrived at the laboratory in a cylindrical metal can.

According to the death certificate, this adult male died at the age of 63 years and weighed 190 pounds. The remains were highly fragmented and extensively altered by the heat, with colors ranging from blue/gray to white. In addition, yellow and pink colors were observed in this sample, likely resulting from metals in or around the bone at the time of cremation (Dunlop, 1978). The remains had been sorted with geological sieves. Three separate fractions were obtained: ash (passing through a #20 sieve), small fragments (caught by the #20 sieve), and large fragments (caught by a #6 sieve). Only the fragments caught by the #6 sieve were included in this study. The smaller fractions were not examined due to their extremely small size (fragments of this size were not collected at Fox Hollow). The fragments were spread over the number grid and 100 were randomly selected for analysis.

Commercial Dog Cremation

This specimen, of *Canis familiaris*, was commercially cremated in New York State but not mechanically pulverized. The specimen was donated to the University of Indianapolis in 1996 and had been curated since then in its original styrofoam shipping container. It has been used for anatomical study and teaching. The fragments were extensively burned, and most were heavily calcined and fragile. Some likely broke during shipping, although many fragments were large enough to retain diagnostic morphological features. The fragments were spread over the number grid and 100 were randomly selected for analysis.

SCORING METHODS

All fragments from each of the six assemblages were assigned to one of the 12 bone categories based on morphology (Table 13.1). Two of these categories

TABLE 13.1 Bone Categories

1	Cranium and mandible
2	Vertebrae
3	Thorax (ribs and sternum)
4	Clavicle
5	Scapula
6	Upper extremity (humerus, radius, and ulna)
7	Hand
8	Pelvis and sacrum
9	Lower extremity (femur, tibia, fibula, and patella)
10	Foot
11	Unidentifiable cortical bone fragment
12	Unidentifiable trabecular bone fragment

The numbering system given here differs slightly from that used by Baker (2004).

(numbers 11 and 12) represent fragments of either cortical or trabecular bone that could not be assigned to a specific element. The maximum length of each fragment was taken to the closest tenth of a millimeter with sliding calipers. Then each fragment was given a score for color, fracturing, and warping. Additional information on scoring methods can be found in Baker (2004).

COLOR

When burned, bones go through characteristic color changes depending on the duration of heat exposure, the fire's temperature, and the presence of soft tissues (Binford, 1963; Buikstra and Swegle, 1989; Correia, 1997; see Chapter 2). In the current study, the color of the outer surface (or the ectocranial surface, in the case of vault bone) of each fragment was scored as 'unburned' (no change in color), 'lightly burned' (brown or black, including 'smoked' fragments), or 'heavily burned' (blue-gray or white, also known as 'calcined'). These categories reflect increasing exposure to higher temperatures, with white indicating the greatest exposure (Ubelaker, 1999). Calcined bones no longer contain organic components, such as collagen, and so bones become very brittle.

The inner (or endocranial) surface of each fragment was scored separately in the same way as the outer surface. The term 'inner surface' is used here to indicate either the exposed medullary (or diploic) surface of the cortex or the endocranial cortical surface of the intact cranial vault bone.

FRACTURES

Heating the bone dehydrates and eventually destroys collagen, resulting in the loss of elasticity and tensile strength. As a result, the bone's structural integrity is compromised. As bones begin to shrink and warp, they fail or fracture along the lines of greatest stress. Published rates of shrinkage of fresh bones vary from 0% to 25% at approximately 800°C (Bradtmiller and Buikstra, 1984; Buikstra and Swegle, 1989; Correia, 1997; Herrman and Bennett, 1999; Shipman *et al.*, 1984; Ubelaker, 1999). With such extreme changes in shape and size, the bone is literally ripped or torn apart from within. Theoretically, variability in temperature, flame exposure, moisture content, soft tissue coverage, and cross-sectional diameter within and across the burning body would tend to exacerbate fracturing, as some areas of a bone may not be expanding or contracting at the same rate as immediately adjacent areas.

Herrman and Bennett (1999) summarize five distinct fracture patterns in burned bones. 'Longitudinal fractures' travel along the long axis of the bone and vary in depth. 'Curved transverse' or 'thumbnail fractures' usually appear in a nested or a stacked fashion, one curvilinear fracture within another, and have been associated with the shrinkage of soft tissues during burning. 'Straight transverse' or 'step fractures' usually extend from the edges of a longitudinal fracture, against the grain of the bone. 'Patina' usually affects the outer layer of the cortical bone and is typically found on the epiphyses, with an appearance that has been likened to that of an aged, cracked oil painting. 'Delamination' is a peeling or flaking of the cortical bone away from the underlying spongy bone, occurring most commonly in the cranial bones and at the epiphyses.

In the current study, the predominant fracture type evident on the outer surface of the body of each fragment was recorded. Because fractures to a fragment's edge can be created by postburning processes (e.g., soil compaction, raking, and transport), edge fractures were not scored. If more than one type of fracture was present on the outer surface of a fragment, only the more extensive or more obvious type was scored.

WARPING

In the current study, warping was assessed by carefully examining the contours of each fragment. If the natural contours appeared to be twisted or deformed to any degree, warping was scored as 'present.' Warping is more easily detected in large fragments and in portions of the skeleton that have thinner cortices.

Fracturing and warping can offer some evidence regarding the condition of the remains at the start of the burning event. Most commonly, fracture patterns are used to determine whether the bones were burned with a considerable amount of soft tissue, fats, and moisture still present (a 'wet' or 'green' state) or whether they were burned essentially devoid of soft tissue and fats (a 'dry' state) (Baby, 1954; Binford, 1963; Buikstra and Swegle, 1989; Thurman and Wilmore, 1980). In general, remains burned in a wet/fleshed state should exhibit highly varying or dichotomous color patterns, deep fracturing, delamination, and pronounced warping. Conversely, bones burned in the dry/decomposed state usually exhibit very little color variation on the outer surface, superficial fracturing, and little, if any, delamination or warping (Buikstra and Swegle, 1989; Correia, 1997; Nawrocki, 2003).

RESULTS

BONE CATEGORIES

In general, there are similarities between the six samples with respect to bone categories present (Table 13.2). For the majority (4/6) of the samples, the greatest number of fragments were assigned to bone category 11 (unidentifiable cortical bone). Jerger has the greatest percentage (81%) of its sample assigned to this category. In addition, Fox Hollow and the two

TABLE 13.2 Percentage of Fragments by Bone Category for Each of the Six Samples

	Bone category											
Sample	1	2	3	4	5	6	7	8	9	10	11	12
Fox Hollow	9	5	4	1	2	4	2	1	3	2	54	15
Lawrence	35	22	11	0	0	2	2	2	2	1	21	4
Potter	26	9	4	1	1	3	6	7	8	5	24	9
Jerger	12	3	1	0	0	0	0	0	0	0	81	3
Commercial human	5	3	0	0	0	0	0	0	0	0	76	16
Commercial dog	1	11	11	0	0	1	1	0	0	2	58	15

Bone categories are listed in Table 13.1. Percentages are rounded to the closest whole number and therefore may not sum to 100.

commercial cremations each have at least 50% of their fragments in category 11. Conversely, for both the Lawrence and the Potter samples, bone category 1 (cranium/mandible) is the most common fragment type at 35% and 26%, respectively, although category 11 is still notable at 21% and 24%.

All of the samples show an overall paucity of identifiable appendicular elements (categories 4, 5, 6, 7, 8, 9, and 10). In both the commercial human and Jerger samples, no appendicular elements were identified. Only 4% of the commercial dog cremation fragments were assigned to appendicular bone categories. In the Lawrence, Fox Hollow, and Potter samples, 9%, 15%, and 31% of the fragments were appendicular elements, respectively. In the Lawrence and Potter samples, the lower extremity is slightly better represented than the upper, but the situation is reversed at Fox Hollow. Vertebra and thorax fragments (categories 2 and 3) are fairly well represented in the Lawrence (33% combined) and commercial dog (22% combined) samples, but are less frequent in the other samples (3% to 13%).

The relative prevalence of specific bone categories will be dependent on a number of factors, not all of which are taphonomic in nature. For example, certain areas of the skeleton (such as the pelvis, cranium, and vertebral column) possess a substantial surface area and/or volume and therefore would naturally be expected to produce more fragments than, say, the hands (holding fragment size constant). Other areas of the skeleton (such as the larger limb bones) contain a significant amount of dense, strong cortical bone that would, therefore, predispose them to better survival and disproportional representation, compared to, say, the largely spongy sacrum.

One would expect there to be a non-random association between certain categories. For example, as the limbs become more fragmented and less identifiable, the number of unidentified cortical bone fragments should increase. For the most part, this inverse relationship is exactly what is observed here. Combining all the fragments from dense tubular long bone (categories 4, 6, 7, 9, and 10) and comparing the sums with category 11, we see that only the Lawrence sample does not fit the trend perfectly (Table 13.3).

FRAGMENT LENGTH

Of the six samples, the commercial human cremation has the smallest mean fragment size, at 12.7 mm, followed by Jerger (14.4 mm), Fox Hollow (19.4 mm), and the commercial dog cremation (20.8 mm) (Table 13.3). The

TABLE 13.3 Fragment Percentages and Lengths

Sample	Percentage of identifiable long bone fragments	Percentage of unidentifiable cortical bone fragments	Mean fragment length
Jerger	0	81	14.4
Commercial human	0	76	12.7
Commercial dog	4	58	20.8
Lawrence	7	21	22.5
Fox Hollow	12	54	19.4
Potter	23	24	33.4

Lawrence and Potter samples have the largest mean fragment sizes, at 22.5 and 33.4 mm, respectively.

A number of *t*-tests were used to compare the Fox Hollow fragment lengths to the other samples. Each comparison was constructed under the assumption that there was no statistical difference between the Fox Hollow sample and the opposing sample. Four out of the five comparisons made with Fox Hollow show statistically significant differences at $p < 0.05$ level ($t = 4.18 - 19.14$, d.f. $= 890 - 1044$). The only comparison in which the null hypothesis could not be rejected is between the Fox Hollow and the commercial dog cremation, which are the closest in average fragment size ($t = 1.64$, d.f. $= 890$). Potter is the most different from Fox Hollow and has, in fact, far larger fragments than any of the other specimens. The Potter specimen was ruled an accidental burning and the individual was found in rough anatomical position. The remains were not significantly disturbed after the burning event, while postmortem taphonomic and/or funerary processes altered most of the other specimens.

There is a trend for the percentage of unidentifiable cortical bone fragments to increase as the mean fragment size decreases, although the association is not perfect (Table 13.3). The smaller the fragment is, the more difficult it is to identify. This trend is not evident, however, for unidentified trabecular fragments, perhaps because burned epiphyses continue to retain diagnostic features that allow them to be identified even as they become smaller.

COLOR

All six samples are significantly damaged by fire; however, there are differences in the degree to which this damage is expressed (Figures 13.1 and 13.2). Both the commercial human and dog cremations show the greatest degree of fire alteration, with nearly all fragments being heavily burned (blue-gray to white) on both surfaces. The Jerger and Lawrence samples also have a large proportion of fragments scored as heavily burned, reaching or exceeding 80% for both surfaces. The Potter and Fox Hollow samples show the lowest degree of burn damage, with percentages of heavily burned fragments near 70% for the outer surface and 60% for the inner surface.

Interestingly, the lack of burn damage in the Fox Hollow sample, as reflected in the color scores, stands in stark contrast to the sample's ranking by mean fragment length. Potter has the largest fragments and displays the least burning, and so one would expect the Fox Hollow fragments to be larger than they are. This disparity may indicate that the Fox Hollow assemblage was subjected to significant postburning fracturing in the environment.

There is an inverse relationship between the burn categories. For example, Fox Hollow and Potter display the largest proportions of lightly burned (brown to black) fragments, in both cases approaching 30% on the outer surfaces and 40% on the inner surfaces, and they show only moderate levels of heavy burning. Lawrence and Jerger are intermediate with respect to the proportion of lightly burned to heavily burned fragments, while the commercial cremations show the lowest proportions of light burning. This inverse relationship is predictable in that as the sample is exposed to increasing levels

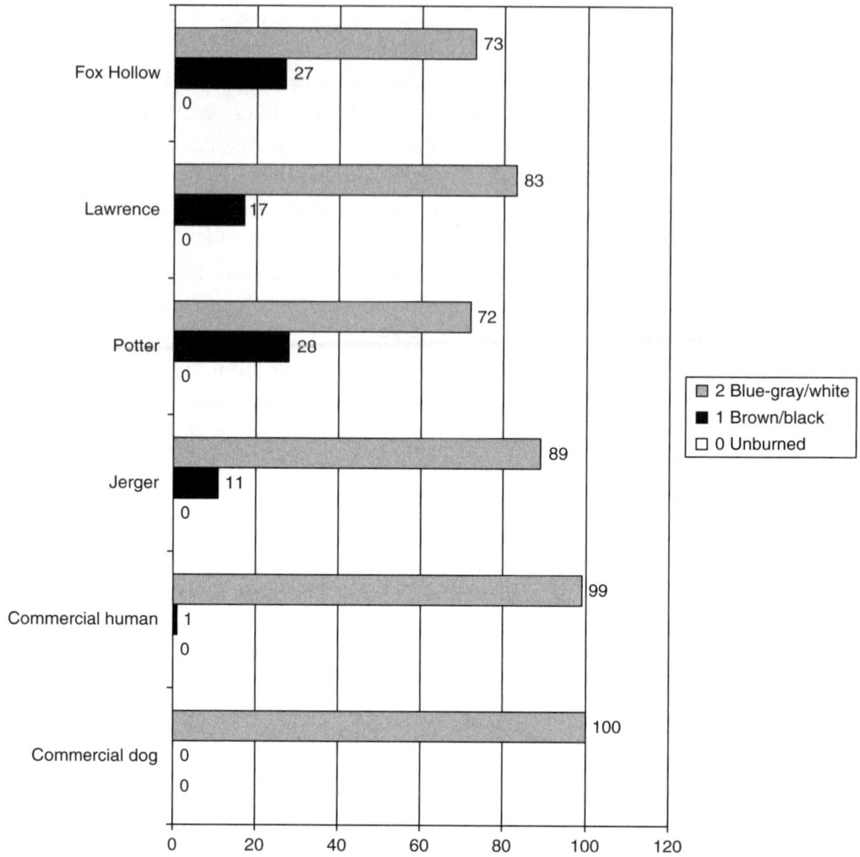

FIGURE 13.1 Percentages for outer surface color scores, by sample.

of heat, more and more fragments are converted from the lightly burned category to the heavily burned category. Theoretically, these fragments will also be getting smaller in the process, although our data are not precise enough to demonstrate this relationship conclusively.

It should be noted that burning is generally less marked on the inner surfaces compared to the outer. This phenomenon may be simply a function of exposure. As flesh and soft tissues are removed from the bone surface by fire, the underling outer bone is exposed. Until the bone fractures, however, the inner surface of the bone is partially protected from the heat. The outer bone surface is subjected to more heat for a greater duration of time and therefore it incurs more heat damage than the inner surface.

FRACTURING AND WARPING

In all six of the samples there are examples of each of the five fracture types (Table 13.4 and Figure 13.3). However, each fracture type varies in frequency between the samples. For Fox Hollow, the commercial human cremation, and Jerger, the majority of the fragments (60–70%) do not display surface fracturing. The Lawrence, Potter, and commercial dog cremation samples,

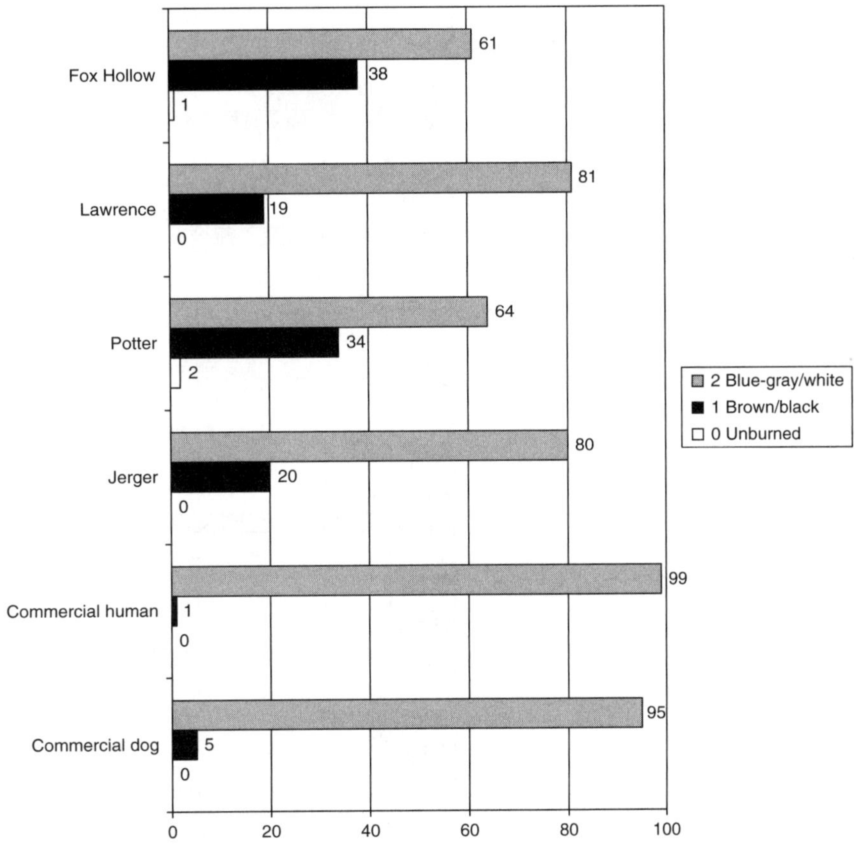

FIGURE 13.2 Percentages for inner surface color scores, by sample.

on the other hand, have greater proportions of scorable fracture types, with 61%–93% scored as present.

It is clear that the scorability of surface fracturing is related to the overall fragment size. The samples with the lowest mean fragment length (commercial human cremation, Jerger, and Fox Hollow) have the largest percentages of fragments with no surface fracturing. Table 13.4 shows that this inverse relationship is nearly perfect across samples. It is likely that surface fractures become enlarged as they become eroded, transported, or commercially

TABLE 13.4 Percentages of Fragments with Each Fracture Type, Arranged by Mean Fragment Length

Sample	Mean length	Delamination	Patina	Curvilinear	Longitudinal	Transverse	None	Warping
Potter	33.4	11	27	12	1	40	7	27
Lawrence	22.5	8	34	6	4	37	13	19
Commercial dog	20.8	1	21	9	8	22	39	20
Fox Hollow	19.4	2	4	1	1	27	66	2
Jerger	14.4	1	17	5	2	15	60	6
Commercial human	12.7	2	8	8	1	11	70	6

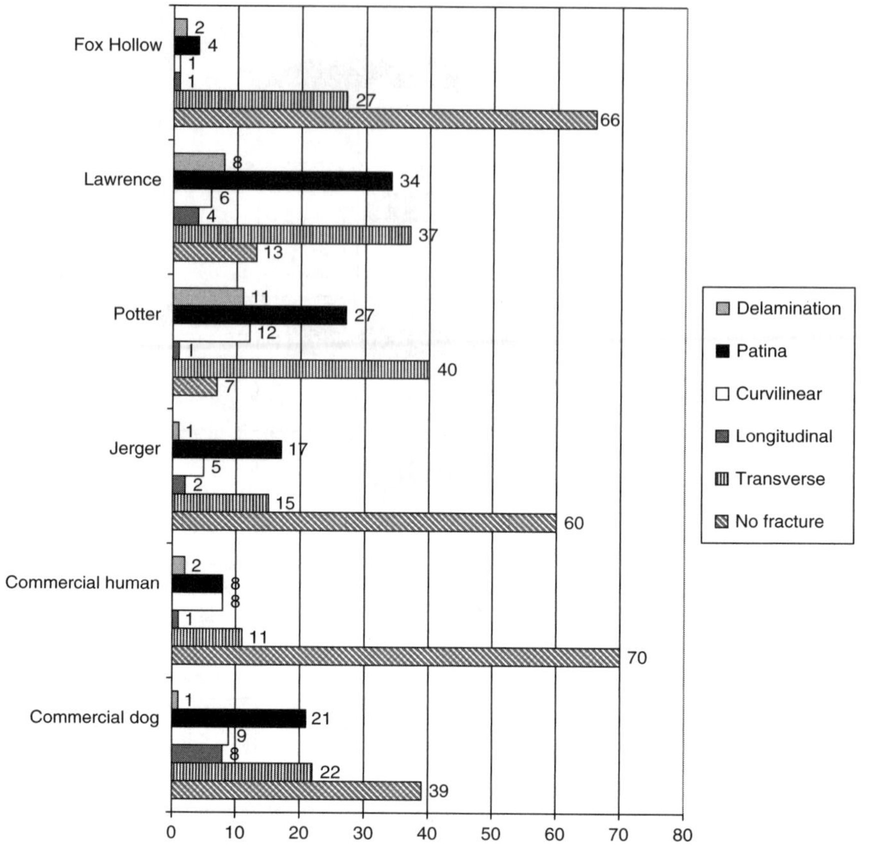

FIGURE 13.3 Percentages of fragments with each fracture type, by sample.

pounded, breaking larger fragments into smaller pieces and thus removing the signs of fracture lines from their surfaces.

Delamination is found in small proportions in all six of the samples. Potter and Lawrence have the greatest amount of delamination (11% and 8%, respectively). However, both samples have a greater proportion of fragments identified as cranial, and both have large fragment sizes. Therefore, high delamination scores may simply be a reflection of the availability of large cranial fragments, which are more likely to display delamination than other skeletal elements.

The Potter, Lawrence, and commercial dog samples have large proportions of fragments (21–34%) displaying patina. It is no coincidence that these samples also have the largest mean fragment size (Table 13.4). It appears that patina fracturing is easily masked by postburning events and is more likely to be found on bones that have been burned but not subjected to excessive fragmentation.

On the other hand, there is no simple relationship between curvilinear fractures and fragment size or between longitudinal fractures and fragment size. However, the percentages are very low for both types of fractures in most samples, so any conclusions would be tentative.

Transverse fracturing displays a strong direct relationship with fragment size (Table 13.4). The samples with the largest fragments (Potter and Lawrence) have the greatest percentages of transverse fractures. The commercial human sample, with the smallest fragments, displays the lowest percentage of transverse fractures.

The majority of the fragments from the six samples did not show signs of warping, with the highest values being of Potter (27%), the commercial dog cremation (20%), and Lawrence (19%). Fox Hollow, the commercial human cremation, and Jerger displayed nominal amounts of warping, ranging from 2% to 6% (Figure 13.4). As shown in Table 13.4, there is a fairly direct relationship between the presence of warping and the fragment size.

It is clear that the presence of many if not most types of fracturing and warping is strongly associated with the fragment size. Unfortunately, most studies of burned and cremated bones have not rigorously controlled for fragment size when examining fracture characteristics. To the extent that fracture patterns may be able to reveal information about the condition of the bone at the time of burning, skeletal biologists must be cautious not to overinterpret remains that have been heavily modified by postburning processes or that have been extensively burned for long durations.

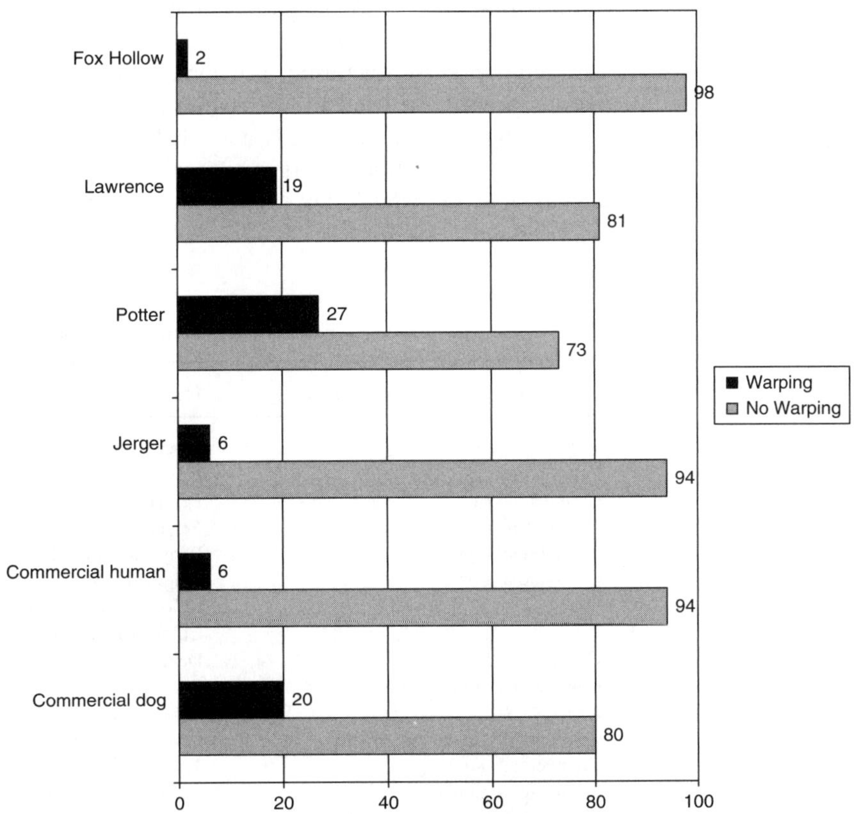

FIGURE 13.4 Percentages of fragments displaying warping, by sample.

SUMMARY AND CONCLUSIONS

The Fox Hollow Farm assemblage presented numerous difficulties for the forensic anthropologists involved in the recovery and analysis. The current study was conducted to help shed light on the behavioral (assailant-induced) and environmental factors that altered the remains in the postmortem interval. Our goal was to determine whether the physical characteristics of the cremains could indicate if at least some of the victims were burned in a fleshed or dry state even though the fragments had been transported and buried subsequent to burning. We compared the Fox Hollow remains to five other samples of burned bones, at least four of which were known to have been burned or cremated in a fleshed state. These five samples experienced different post-burning histories, including plowing, excavation, commercial pulverizing, and anatomical study. Some were collected quickly after death, while others had been buried for different periods of time.

Table 13.5 lists the sample or samples that most closely resemble Fox Hollow for each of the categories of observation. Some of the comparisons are not particularly informative because of low frequencies of certain traits in some categories or samples. For example, while Fox Hollow and the commercial human cremation both have the same percentage of delaminated fragments (2%), the commercial dog and Jerger samples are also very close (1% each). A similar problem exists for longitudinal fractures. Most of the remaining comparisons, however, have more validity. Fox Hollow is most similar to the commercial dog cremation five times, in bone category percentages, fragment size, and transverse fracturing. The commercial human cremation also makes five appearances on the chart, with patina and percentage of fragments with no body fractures being particularly notable. Potter appears four times (including both color observations), Jerger appears three times (including curvilinear fractures), and Lawrence fails to appear on the list at all.

TABLE 13.5 Comparisons of Fox Hollow with the Other Five Samples

Fox Hollow most closely resembles …	In…
Jerger	Percentage of cranial fragments
Potter	Percentage of vertebra and thoracic fragments
Commercial dog	Percentage of long bone fragments
Commercial dog	Percentage of unidentified cortical bone
Commercial dog	Percentage of unidentified trabecular bone
Commercial dog	Fragment size
Potter	Color of outer surface
Potter	Color of inner surface
Commercial human	Delamination
Commercial human	Patina
Jerger	Curvilinear fractures
Commercial human and Potter	Longitudinal fractures
Commercial dog	Transverse fractures
Commercial human	Percentage with no fractures
Commercial human and Jerger	Warping

It is interesting that the Fox Hollow remains are most similar to the commercial cremations and have less in common with the forensic cases. The Fox Hollow assailant is known to have processed multiple victims in Area 1, and it is possible that the remains of at least some of the victims were burned on more than one occasion. However, the fragment color indicates that the specimens are actually less heavily burned than seen in the two commercial cremations. The effects of water transport, burial, and archeological excavation likely reduced fragment size at Fox Hollow beyond what would be expected for forensic contexts where the victim typically burns in a short-lived campfire or house fire. Therefore, fragment color may be a more effective and less biased indicator of accumulated thermal trauma than fragment size, allowing the investigator to differentiate deliberate cremations from more superficial or accidental burning in those instances where the postmortem history of the remains is poorly understood. On the other hand, finding highly fragmented bones that are just barely burned probably indicates that forces other than just thermal energy caused the fracturing.

The Fox Hollow remains do display fractures that are associated with burning of fresh or fleshed remains, including deep transverse, longitudinal, and curvilinear fractures, delamination, and warping. However, the occurrence of these fracture types in the Fox Hollow sample is considerably lower than that displayed by most of the other samples. By itself, this evidence would seem to suggest that a larger number of Fox Hollow victims were burned in a dry state than in a fleshed state. Indeed, archeological evidence and witness statements seem to corroborate this view. However, many of the fragments at Fox Hollow are very small and thus do not present many surface fractures for analysis. Extensive cremation and fragmentation of fragile burned bone will effectively remove much of the evidence used by the anthropologist to infer the original condition of the remains. Therefore, the post-burning taphonomic history of the remains must be taken into account, and assumptions based solely on the laboratory analysis of the remains may be misleading.

ACKNOWLEDGMENTS

We would like to thank Dr. Dennis Dirkmaat of Mercyhurst College and Dr. Christopher Schmidt of the University of Indianapolis for access to their collections. Dr. Gregory Reinhardt of the University of Indianapolis provided valuable guidance and suggestions early in the formation of this study. Finally, we are grateful to the editors of this volume for including us in their project.

REFERENCES

Baby, R.S. (1954). *Hopewell Cremation Practices*. Ohio Historical Society Papers in Archaeology, No. 1.

Baker, A. (2000). *Analysis of Human Cremains from the Fox Hollow Farm*. Paper presented at the 7th Midwest Bioarcheology and Forensic Anthropology Association Conference, Columbia, MO.

Baker, A. (2004). *A Taphonomic Analysis of Human Cremains from the Fox Hollow Farm Serial Homicide Site*. Master's thesis in Human Biology, University of Indianapolis.

Baker Bontrager, A. and Nawrocki, S.P. (2005). A taphonomic analysis of burned remains from the Fox Hollow serial homicide site. *American Journal of Physical Anthropology* **Supplement 40**, 68–69.

Binford, L.R. (1963). The analysis of cremation from three Michigan sites. *The Wisconsin Archeologist* **44**, 98–110.

Bradtmiller, M.A. and Buikstra, J.E. (1984). Effects of burning on human bone microstructure: a preliminary study. *Journal of Forensic Science* **29**, 535–540.

Buikstra, J.E. and Swegle, M. (1989). Bone modification due to burning: experimental evidence. In *Bone Modification* (R. Bonnichsen, and M.H. Sorg, eds.), pp. 247–258. University of Maine, Orono.

Correia, P.M. (1997). Fire modification of bone: a review of the literature. In *Forensic Taphonomy: Postmortem Fate of Human Remains* (W.D. Haglund and M.H. Sorg, eds.), pp. 275–293. CRC Press, New York.

Dunlop, J.M. (1978). Traffic light discoloration in cremated bones. *Medical Science Law* **18(3)**, 163–173.

Greene, T.R. and Schmidt, C.W. (2000). Analysis of cremated human remains from an Early Archaic mortuary from Southern Indiana. *American Journal of Physical Anthropology* **Supplement 30**, 166.

Herrman, N.P. and Bennett, J.L. (1999). The differentiation of traumatic and heat-related fractures in burned bone. *Journal of Forensic Science* **44**, 461–469.

Nawrocki, S.P. (2003). *Experimental Burning of a Dry Human Cranium.* Paper presented at the 10th Midwest Bioarcheology and Forensic Anthropology Association Conference, Chicago, IL.

Nawrocki, S.P. and Baker A. (2001). Fluvial transport of human remains at the Fox Hollow serial homicide site. *Proceedings of the American Academy of Forensic Sciences* **7**, 246–247.

Nawrocki, S.P., Schmidt, C.W., Williamson, M., and Reinhardt, G. (1998). Excavation and analysis of human remains from the Fox Hollow serial homicide site, Hamilton County, Indiana. *Proceedings of the American Academy of Forensic Sciences* **4**, 205–206.

Shipman, P., Foster, G., and Schoeninger, M. (1984). Burnt bones and teeth: an experimental study of color, morphology, crystal structure and shrinkage. *Journal of Archeological Science* **11**, 307–325.

Thurman, M.D. and Wilmore, L.J. (1980). Replicative cremation experiment. *North American Archaeology* **2**, 275–283.

Tomak, C.H. (1979). Jerger: an Early Archaic mortuary site in Southwestern Indiana. *Proceedings of the Indiana Academy of Science* **85**, 62–69.

Ubelaker, D.H. (1999). *Human Skeletal Remains: Excavation, Analysis, Interpretation*, 3rd edn. Taraxacum, Washington, DC.

Weinstein, F. and Wilson, M. (1998). *Where the Bodies are Buried.* St. Martin's Paperbacks.

14

Early Archaic Cremations from Southern Indiana

Christopher W. Schmidt, Ph.D.[1], Curtis Tomak, M.A.[2], Rachel A. Lockhart, B.S.[3], Tammy R. Greene, Ph.D.[4], and Gregory A. Reinhardt, Ph.D.[5]

[1]*Department of Anthropology, Indiana Prehistory Laboratory, University of Indianapolis, Indianapolis, IN*
[2]*Indiana Department of Transportation, Indianapolis, IN*
[3]*Department of Anthropology, Indiana Prehistory Laboratory, University of Indianapolis, Indianapolis, IN*
[4]*Department of Biology, Arkansas State University, Beebe, AR*
[5]*Department of Anthropology, University of Indianapolis, Indianapolis, IN*

INTRODUCTION

Although the early inhabitants of North America are popularly characterized as culturally simplistic, recent archeological evidence indicates that they actually were making a wide variety of artifacts out of both local and exotic materials, often living semisedentary lives, and trading or traveling over great distances (e.g., Anderson, 1998; Kimball, 1998; Sassaman, 1998; Deller and Ellis, 2001). Moreover, their mortuaries show that as early as 9000 years ago American Indians were investing significant energy in the burial rites of their dead. The current study investigates one of these early Holocene cemeteries in an effort to archeologically and osteologically understand the mortuary practices of early inhabitants of eastern North America.

Jerger is an Early Archaic burial site located in southern Indiana; situated on a sand ridge, it overlooks what was probably a marshy area (Tomak, 1979). The Jerger site, and its nearby companion cemetery, the Steele site – along with habitation sites in Greene, Daviess, and Knox counties in the valley of the West Fork of White River in southwestern Indiana – form the basis of the Jerger Phase (Tomak 1983, 1991). This is an Early Archaic archeological tradition, which is thought to date between about 10 000 YBP and 7000 YBP. A diagnostic artifact of this occupation is the bifurcated-base Jerger point. This occupation was first identified over 35 years ago (Tomak, 1970), but no cemeteries were known at that time.

Residents of Daviess County found the Jerger site in a cultivated field. The materials they collected came from an area of archaeological debris, reportedly about 770 m^2 in size, and included cremated human bones, red ochre, stone points, bifacial chert objects, drills, and chert debris. In addition, they observed five dark areas within the limits of the debris scatter. The dark areas appeared to be features from which the materials had originated. Similar surface materials indicated that another, seemingly isolated, feature (Feature 1) occurred about 31 m away.

From 1977 to 1979 one of the authors (C.T.) investigated approximately 154 m^2 of the approximately 1846-m^2 site. He determined features 1, 3, 9, and 13 to be burials, although upon excavation it was discovered that little remained of them. The remaining portion of the largest burial (Feature 3) was 1.8 m long, 1.4 m wide, and 0.1 m deep. The fill of the burial features was dark and contained cremated bones and other materials like those from the plowzone and the surface of the site. The artifacts from the burial features were fractured and/or discolored by heat, presumably from the crematory fires. No *in situ* burned sediment was found, indicating that cremation had not taken place in the features but had occurred at some other location.

Almost all of the projectile points are heat-fractured, and many are heat-discolored. The points have a bifurcated base and are similar to MacCorkle points and other Early Archaic bifurcates; here they are referred to as Jerger points (Justice, 1987). A few drill fragments were recovered, and they also exhibit heat damage and alteration. The site discoverers found four of them in the area of Feature 1. Two consist of shaft fragments, and two are nearly complete. One of the latter has a barbed shoulder and small portion of the base intact; it was at least 60 mm long. The other nearly complete drill has a greatly expanded and rounded base.

At least 14 fragments constituting the rounded ends of some type of bone tool, with a flattened cross section, have been recovered. Their colors indicate that they were exposed to fire. Some similar specimens, but broken at both ends, were also found. They might be midsections of the artifacts with the rounded end. Also found were fragments of other kinds of burned bone artifacts whose functions have not been determined; some appear to be the remains of awl-like artifacts. At least two-dozen carnivore canines, which were perforated through the root, were found as well. They were fractured and discolored by heat, and some exhibit polish. Other finds included numerous burned and unburned animal bone fragments from deer, raccoon, dog or coyote, turtle, and turkey (William R. Adams, archeozoologist at Indiana University, personal communication).

Of great interest is the recovery of five heat-altered marine shells from Feature 3. The largest shell is 8 mm long and about 4 mm wide. All of them have a broken spire, and they may have been used as beads. William K. Emerson, Curator of Mollusks at the American Museum of Natural History, identified them as *Olivella* sp., shells having a natural range extending from North Carolina to Florida and southward. This identification supersedes a previously reported identification as *Detracia* (Tomak, 1979).

Little archaeological material occurs at the Jerger site excepting that originating in the mortuary features. No other kind of aboriginal feature has been found at the site, and it seems that the location was used mainly for funerary purposes. The materials from the plowzone are like those found in the features, and the materials from both contexts can be associated on the basis of having been heat-altered.

The known Jerger Phase cemeteries were in special mortuary locations. There is no evidence that Jerger people used these sites for any other purposes, and there is little indication that these sites were used much by any other American Indian group. The cemeteries are not particularly close to any of the more productive habitation sites but are reasonably near some of them.

OSTEOLOGY

The primary effort of the human osteological component of the Jerger analysis focuses on the following questions: How many people are present? What is the extent to which the remains are burned? How are the remains distributed among the pits? Is there evidence for the circumstances of the burial?

A total of 10 852 g of human bone fragments were recovered from Jerger. The fragments numbered in the tens of thousands and, of these, 2623 bones and 270 dental fragments were analyzed for the current study. They represent materials collected from the features and the plowzone. The excavators noted that the plow brought materials to the surface but did not appear to drag the bone fragments from their subsurface origins. Thus, a relative provenience of the plowzone material was available, even though these remains were clearly disturbed. In addition to identifying each fragment, data collected included the maximum length of each fragment and evidence of burning such as bone color, fracture pattern (texture), and bone warpage or delamination.

SAMPLING

The remains were curated in dozens of paper sacks, each labeled with a unique field specimen number. The sack contents were sorted for diagnostic bone fragments, teeth, animal bones, and chert flakes. All of the teeth and animal bones were removed and studied. However, there were far too many bone fragments to study the entirety. Once we removed the very few diagnostic fragments (mainly elements such as distal phalanges, superciliary arches, and other recognizable skull fragments), the remaining bone fragments were placed onto a grid of 15 numbered squares, each approximately 5 cm across following Greene and Schmidt (2000). A table of random numbers was used to determine the three grid squares from which the bones would be studied. It is worth noting that the bone fragments were removed from the bag manually, making sure to collect the smaller-sized fragments from deep in the bag and the larger fragments near the top of the bag. This helped to mix fragments that had become size-sorted while the remains were curated and ensured that each grid square included fragments of all sizes. The bones that were selected for study were rinsed with tap water and allowed to dry, then counted, weighed, and measured.

INVENTORY

All told we analyzed, 2893 bone and dental fragments for the current study. Of the 270 dental fragments in this sample, 71 were identified. Of those, 54 were from adults and 17 were from subadults. The identifiable teeth from adults are represented by all four tooth classes. Molars comprise 50.0%, premolars 35.2%, canines 11.1%, and incisors 3.7% of the total number of identifiable teeth. A disparity exists between left- and right-side elements. The left side yielded 18 teeth and the right side, 36. The ratio of maxillary to mandibular teeth is more even. There are 17 maxillary and 10 mandibular molars; 9 maxillary and 10 mandibular premolars; 4 maxillary and 2 mandibular canines; and 0 maxillary and 2 mandibular incisors, giving a total of 30 maxillary and 24 mandibular adult teeth.

The subadult teeth consist of 9 deciduous premolars and 8 developing adult teeth (3 canines, 2 premolars, and 3 molars). The deciduous premolars include 1 upper left first premolar, 1 upper left second premolar, 3 upper right second premolars, 2 lower left first premolars, and 2 lower right first premolars. The right and left teeth are almost even in number; there are 9 from the left and 8 from the right (Table 14.1).

Of the 2623 bone fragments, 22.8% were cranial, 4.9% were vertebrae or ribs, 0.9% were clearly upper extremity, 0.9% were clearly lower extremity, and 70.5% were unidentified postcranial long bone fragments. No bones were clearly identifiable as subadult and only two superciliary arch fragments gave any insight into sex.

MNI

The minimum number of individuals (MNI) at Jerger is 14. This was determined via a dental inventory because virtually no skeletal elements were complete enough for identification. By counting repeating teeth and controlling for developmental stages, we determined that at least 5 adults and 9 subadults were present. The subadults range in age from around birth to roughly 15 years (Table 14.2). Four fully erupted, fully developed maxillary right first, second, and third molars represent the four adults (see Table 14.3).

The total number of recovered dental elements is far less than 448 or so complete teeth that one would expect to find from 9 children and 5 adults. Taking into consideration the ages of the children represented in the Jerger assemblage, there should be 288 complete and developing deciduous and permanent teeth from the children, and 160 teeth from the adults. The missing teeth perhaps remain at the original spot where the actual cremations took place, but it is also possible that they were so fragmented they were not recovered with the standard archeological methods employed at the time of excavation.

TABLE 14.1 Frequencies of Left and Right Teeth

Subadult		Adult	
Left	Right	Left	Right
9	8	18	36

TABLE 14.2 Subadult Age Groups Represented by At Least One Individual

Age	MNI	No. of teeth	Teeth present
0	1	1	dLP^1, 1/4 cr
0–6 months	2	5	dRP^2, 1/4 cr; dRP_1, 1/2 cr; dLP_1, 1/2 cr; dRP_1, 1/2 cr; dLP_1, 1/2 cr
6–9 months	1	1	dRP^2, 1/2 cr
9–12 months	1	3	dLP^2, 3/4 cr; dRP^2, 3/4 cr; LM^1, 1/4 cr
18 months	1	1	LM^1, 1/2 cr
5–6 years	1	4	RP_1, 1/2 cr; RP_2, 1/2 cr; LC^1, 3/4 cr; RC^1, 3/4 cr
10–12 years	1	1	LC^1, 1/2 a
15 years	1	1	LM^3, 1/4 r

d, Deciduous; L, left; R, right; C, canine; M, molar; P, premolar. Superscript indicates the tooth is maxillary, subscript indicates the tooth is mandibular.

Developmental stages follow Moorrees *et al.* (1963). cr, crown; r, root; a, apex. Thus, 1/4 cr means the crown is one-quarter developed.

TABLE 14.3 MNI by Tooth Type

Maxillary teeth		Mandibular teeth	
Tooth	MNI	Tooth	MNI
RM3	5	RM3	4
RM1	4	LP3	4
RM2	4	RP3	3
RP3	4	RM1	3
LP3	3	LM3	3
RC	3	LP4	2
LM3	2	RP4	1
RP4	2	RI2	1
LM1	1	RI1	1
LM2	1	LC	1
LC	1	RC	1
Total	30		24

R, right; L, left; M, molar; P, premolar; C, canine; I, incisor.

DEMOGRAPHY

The age distribution includes individuals from nearly all stages of life. There are newborn and very young children, toddlers, and two older children, one of which is a teen. At least one adult has modest dental macrowear, while three others have extensive wear. Although specific ages could not be assigned for each adult, the range in dental wear indicates individuals from earlier and later stages of adulthood.

Sex was indeterminate for most of the bones, although a few superciliary arch fragments provided some indication of dimorphism. One particularly robust fragment suggests the presence of at least one male. Another superciliary arch fragment is less robust, perhaps representing a female, but it is impossible to draw a conclusion because of the very small size of the fragment. Sex could not be gleaned from the teeth because most were missing their crowns, and nearly all were fractured and shrunken from the heat.

Ancestry is presumed to be Native American because of the context and the associated artifacts. In addition, two unerupted canine crown fragments had evidence of shoveling and the teeth overall had an appearance consistent with American Indian morphology (e.g., significant macrowear, bifurcated upper first premolar roots, premolar odontomes, etc.).

EXTENT OF BURNING

The Jerger remains were extensively thermally modified. Around 98% of all bone fragments were burned, as were all of the teeth. Over 82% of the burned bones had been heated to a point at or near calcination. This means the bones had turned bluish-gray or white, colors primarily resulting from the combustion of the bone collagen. The remainder of the burned bones had turned brown or black, results of incomplete combustion of the collagen.

The fragments are remarkably small, averaging only 12.8 mm in maximum diameter. The cranial and postcranial fragments are about the same size, with the largest average sizes not surprisingly being attributed to the elements that could be identified as either upper or lower extremity. The upper extremity fragments average 17.1 mm while the lower extremity fragments have a mean of 20.1 mm (Table 14.4).

Some of the bone fragments looked curled, or warped, as a result of heat exposure. In cranial bones, the intense heat can lead to a separation of the inner and outer cortical tables, a condition known as delamination. Both warpage and delamination seem to be associated with moisture in the bones and when present on burned remains may indicate that at least some soft tissues or bodily fluids are still present at the time of the fire. About 22% of the cranial fragments show some sign of delamination. However, only 3.6% of the postcranial elements (67 of 1881) show any warpage. At

TABLE 14.4 Summary Data Describing the Extent of Burning

Element	Mean max L	s.d.	Percentage Calcined	Percentage Unburned	Percentage Warped
Cranial	13.0 (597)	3.55	84.9 (594)	0.1 (594)	21.6 (593)
Verts/ribs	15.0 (130)	4.93	59.2 (130)	9.2 (130)	3.9 (129)
Upper extremity	17.1 (24)	5.66	87.5 (24)	8.3 (24)	12.5 (24)
Lower extremity	20.1 (24)	8.14	80.0 (24)	0.0 (24)	12.5 (24)
Unidentified, postcranial	12.4 (1848)	4.63	83.6 (1848)	1.7 (1848)	3.3 (1704)
Totals	12.8 (2623)	4.58	82.7 (2620)	1.9 (2620)	7.9 (2474)

max, Maximum; L, length; s.d., standard deviation.

present it is difficult to reconcile this disparity. Perhaps crania are simply more likely to delaminate whether or not significant fluids are still in the body. Another possibility is that the postcranial fragments are so small that they simply do not show the warpage that may have been evident before their final fragmentation.

The degree of burning among the Early Archaic Jerger remains is unexpected. By comparing the Jerger remains to those from other sites, one sees that the Jerger cremains are highly fragmented. Two of us (T.G. and C.S.) compared Jerger cremation fragments to an assemblage of cremated bones from a Middle Woodland site in western Indiana (*ca.* 1500 YBP) and found that the Middle Woodland cremations were on average larger and less calcined than those from Jerger. The average length for the Middle Woodland fragments was 47.2 mm and the percentage of calcined bone was 48 (Greene and Schmidt, 2000). This is especially interesting because cremation is such a widespread and apparently significant practice during the Middle Woodland period (Brose and Greber, 1979), yet the Jerger material is more heavily burned.

Some of the Jerger remains were used in the Bontrager and Nawrocki study (Chapter 13). In their comparisons of forensic, archeological, and modern cremations, only the cremains from a modern commercial cremation (a process that includes pulverization) had fragments as small as those from Jerger. Likewise, Curtin (Chapter 12) provides average burned-fragment sizes, from a Native American cremation in the American northwest, that are far larger than what is reported here for Jerger.

DISTRIBUTION

At least four individuals came from Feature 3, a young adult, an old adult and at least two of the children (ages of 0–6 months and 5–6 years). Feature 1 also has evidence of more than one person. Most of the individuals who were accounted for, however, came from the plowzone material, so it is unclear if they came from additional single or multiple interments. It is possible that the remains of a particular individual came from two separate features. Two remarkably similar developing canines, both very large and starkly blue in color, were found in the plowzone, but several meters apart from each other. If our understanding of the plow damage at the site is correct, this implies that at least two of the features are immediately contemporary. Just as there is no apparent segregation by age, there is none by bone type. Each of the bone-producing features had cranial and postcranial elements. Unfortunately, this determination cannot be made to a very precise level, since we were not specifically able to identify most postcranial elements.

MORTUARY TREATMENT

The cemetery includes individuals of all ages, and probably both sexes buried together. The grave goods suggest a significant effort in the mortuary practice. The mass of chert flakes alone rivals the mass of bone fragments and a great deal of red and yellow ochre is present. The drilled animal teeth have been recovered from or around nearly every feature, and at least five came directly from Feature 1. This level of ceremonialism is consistent with groups that are semisedentary (e.g., Charles and Buikstra, 1983) and is in many ways far more complex than what has been described for the Middle and Late Archaic (7000–2500 YBP) (e.g., Muller, 1986). As noted earlier, the cremations themselves are far more heavily burned and fragmented than what is typically seen in the Early and Middle Woodland cremations (2500–1500 YBP), which are associated with ceremonial centers, increased sedentism, and horticulture (Brose and Greber, 1979). This is not to say that in total Jerger society was as complex as those of the Middle Woodland. However, there is increasing evidence that significant mortuary complexity arose well before the Woodland Period. For example, marked cultural and mortuary complexity (which happens to include distinct cemeteries and cremation) has been documented in Louisiana at sites around and including Poverty Point, which are over 4000 years old (e.g., Wagner Mires, 1991). The amazing preservation of artifacts and human remains at the Early to Middle Archaic Windover site in Florida also shows tremendous cultural complexity at surprisingly early dates (Doran, 2002). Likewise, the mortuary evidence coming from the Early Archaic people of Indiana indicates well-established and robust funerary ceremonialism early in the prehistory of the Eastern Woodlands, a phenomenon that predates other social complexities (such as the construction of earthworks, permanent villages, and a dependence upon domesticates) by several millennia.

While the quantity of grave goods at Jerger is impressive, the distribution of these artifacts is not well understood. It is possible that the grave goods were associated with all of the individuals. It is also possible that they were actually meant only for the children, but ended up being associated with the adults as well because of the cremation process. Having highly decorated children is common in Middle and Late Archaic burials from the Green River region in Kentucky. At these sites certain subadults were copiously covered with shells while most adults had few if any grave inclusions (i.e., Webb, 1946). Presently, we do not know if the Jerger grave goods are intended for everyone or if there are differences in the distribution based upon age or other factors like sex and social status.

The complex ceremonialism may have included a specific, consistent orientation of the bodies before they were burned. The adults, for instance, may have been placed on their left sides, resulting in the poor representation of left-side teeth. The elements nearer to the ground may have become obscured by ash and not been collected when the cremains were taken from their point of burning to their burial pits. It is not clear if the burial features represent a single mortuary event, or if the features indicate a series of burial events. What is clear is that the Jerger burials are within a distinct cemetery, which suggests some level of corporate 'ownership' of the area by the Jerger people (see Charles and Buikstra, 1983).

Moreover, it is possible that the remains are so fragmented because they were burned more than once (i.e., Beck, 2005). In fact, the size of the mortuary features and the frequency of artifacts are consistent with 'mourning ceremonies' that Beck (2005) describes. If the Jerger people were setting aside lands specifically for interring their ancestors (as opposed to burying them within a habitation area), then such veneration is not unexpected (e.g.,

Byers, 2005; Duncan, 2005). However, the overall volume of bone per feature is more consistent with single burning events. More work is necessary to resolve this particular issue.

DISCUSSION

The southern Indiana Jerger site provides evidence for the early occurrence of mortuary ceremonialism in the eastern United States (Tomak, 1991). Its graves occur in a special cemetery area; the cremation, red ochre, burned artifacts, and quantities of artifacts indicate a great deal of attention, if not special treatment, from the Jerger people to their dead. The presence of marine shells is significant and indicates procurement of exotic items from a distant source and the use of these items as burial accompaniments at an early date in the eastern United States.

Cemeteries like the Jerger site are rare. Not far from Jerger is the Steele site, but it is yet to be studied osteologically. Besides Jerger and Steele, there is only one other site known to the authors that contains cremations accompanied by heat-altered artifacts and Early Archaic bifurcated base points. This site is McCullough's Run (Cochran *et al.*, 1997), which is also in Indiana. It is located in Bartholomew County in the valley of the East Fork of White River, about 75 miles northeast of Jerger. A number of Early Archaic burial features at McCullough's Run contained cremated bones and heat-altered artifacts. Three of the features contained bifurcated base points that are reported to be St. Albans and MacCorkle points. Radiocarbon dates from two features with bifurcates are 8630 YBP and 9045 YBP (calibrated). Two other features produced dates of 6340 YBP and 8415 YBP (calibrated). Although it may just be coincidental, there is a basic similarity in the locations of the Jerger site and the McCullough's Run site. Both are on a southeastern extension of the southern end of a ridge northwest of a sizeable creek. The Steele site, however, is on a northern projection of upland south of a sizeable creek.

McCullough's Run produced 14 232 g of cremated human bones and evidence for 18 people (8 adults, 8 subadults, and 2 who were unidentifiable). Ten features produced human remains in a condition similar to that seen at Jerger. Approximately 92% of the burned fragments were calcined or nearly calcined. Less than 1% of the fragments were unburned (Cochran *et al.*, 1997). Like Jerger, McCullough's Run had pits with multiple individuals, including two that had remains from both adults and subadults (Cochran *et al.*, 1997). The ratio of grams of cremains per individual is similar between the sites. For McCullough's Run, the average amount for each person was 790.6 g, and for Jerger, 775.3 g; the modest difference may reflect the higher percentage of subadults at Jerger. Perhaps the only notable distinction between the two cemeteries is in the degree of bone fragmentation. The McCullough's Run MNI was determined using the presence of external auditory meati and odontoid processes of the second cervical vertebra. These were almost never found at the Jerger site, which must be seen as a peculiarity because the petrous portion of the temporal bone (of which the external auditory meatus is a part) is very dense and is usually a good bone for determining MNI, especially in instances of highly commingled bones (e.g., Schmidt and Larsen, 2002).

The present archeological evidence does not indicate that southern Indiana was inhabited by large numbers of Jerger people or that they used it heavily. Three small cemeteries and a fair number of habitation sites have been identified, but none of the latter is particularly substantial. Furthermore, recognizable Jerger artifacts are not common in private collections from the area. For example, from a sample of over 2500 identifiable points collected within the geographic extent of the Jerger phase, only 38 of them were Jerger points (Tomak, 1970, 1987). The Jerger people not only utilized local resources, but also exploited chert sources from along the Ohio River, at least 50 miles away to the southeast. The *Olivella* beads and the similarity of the Jerger points to MacCorkle indicate long-distance trade, travel, or both with groups in the mid-South (Tomak, 1970, 1980, 1987).

CONCLUSION

The Jerger cemetery provides cultural insight of some of the earliest inhabitants of Indiana and indicates a cultural complexity that would go unnoticed in the absence of osteological study. The extraordinary level of cremation suggests a significant effort to reduce the remains to very small fragments, through at least one episode of burning, and perhaps more than one. The ceremonialism does not seem to be limited to certain individuals, as all of the remains seem to be treated similarly.

The 14 people interred at Jerger were accompanied by a variety of grave goods and were cremated to a degree that exceeds most other prehistoric cremations. The remains represent people of all ages, buried within the confines of a distinct cemetery, with adults and subadults sharing graves. The cemetery suggests that the Jerger people recognized corporate control over those areas where their kin were buried and that they placed a certain level of emphasis on adorning their dead. This mortuary contradicts any notion that mortuary complexity steadily increased through time in the Eastern Woodlands. In fact, by 9000 years ago, at least in Indiana it was already quite elaborate.

REFERENCES

Anderson, D.G. (1998). Models of Paleoindian and Early Archaic settlement in the lower southeast. In *The Paleoindian and Early Archaic Southeast* (D.G. Anderson and K.E. Sassaman, eds.), pp 29–57. The University of Alabama Press, Tuscaloosa, AL.

Beck, L.A. (2005). Secondary burial practices in Hohokam cremations. In *Interacting with the Dead: Perspectives on Mortuary Archaeology for the New Millennium* (G.F.M. Rakita, J.E. Buikstra, L.A. Beck, and S.R. Williams, eds.), pp. 150–154. University Press of Florida, Gainesville, FL.

Brose, D.S. and Greber, N. (1979). *Hopewell Archaeology*. Kent State University Press, Kent, OH.

Byers, A.M. (2005). The mortuary "laying-in" crypts of the Hopewell Site: Beyond the funerary paradigm. In *Interacting with the Dead: Perspectives on Mortuary Archaeology for the New Millennium* (G.F.M. Rakita, J.E. Buikstra, L.A. Beck, S.R. Williams, eds.), pp. 124–141. University Press of Florida, Gainesville, FL.

Charles, D.K. and Buikstra, J.E. (1983). Archaic mortuary sites in the Central Mississippi Drainage: distribution, structure, and behavioral implications. In *Archaic Hunters and Gatherers in the American Midwest* (J.L. Phillips and J.A. Brown, eds.), pp. 117–145. Academic Press, New York.

Cochran, D.R., Knight, K.P., Bush, L. (1997). *McCullough's Run (12-B-1036): A Prehistoric Cremation Cemetery, Bartholomew County, Indiana*. Archaeological Resources Management Service, Ball State University, Muncie, IN.

Deller, D.B. and Ellis, C.J. (2001). Evidence for Late Paleoindian ritual from the Caradoc Site (AfHj-104), southwestern Ontario, Canada. *American Antiquity* **66**, 267–284.

Doran, G.H. (ed.) (2002). *Windover: Multidisciplinary Investigations of an Early Archaic Florida Cemetery*. University Press of Florida, Gainesville, FL.

Duncan, W.N. (2005). Understanding veneration and violation in the archaeological record. In *Interacting with the Dead: Perspectives on Mortuary Archaeology for the New Millennium* (G.F.M. Rakita, J.E. Buikstra, L.A. Beck, and S.R. Williams, eds.), pp. 207–227. University Press of Florida, Gainesville, FL.

Greene, T. R. and Schmidt, C.W. (2000). Analysis of cremated human remains from an Early Archaic mortuary site in southern Indiana. *American Journal of Physical Anthropology* **Suppl. 30**, 166 (abstract).

Justice, N.D. (1987). *Stone Age Spear and Arrow Points of the Midcontinental and Eastern United States*. Indiana University Press, Bloomington, IN.

Kimball, L.R. (1998). Early Archaic settlement and technology: lessons from Tellico. In *The Paleoindian and Early Archaic Southeast* (D.G. Anderson and K.E. Sassman, eds.), pp. 149–186. The University of Alabama Press, Tuscaloosa, AL.

Moorrees, C.F.A., Fanning, E.A., and Hunt, E.E. (1963). Age variation of formation stages for ten permanent teeth. *Journal of Dental Research* **42**, 1490–1502.

Muller, J. (1986). *Archaeology of the Lower Ohio River Valley*. Academic Press, Orlando, FL.

Sassaman, K.E. (1998). Early Archaic settlement in the South Carolina Coastal Plain. In *The Paleoindian and Early Archaic Southeast* (D.G. Anderson and K.E. Sassman, eds.), pp. 58–83. The University of Alabama Press, Tuscaloosa, AL.

Schmidt, C.W. and Larsen, C.S. (2002). Demographic and health reconstruction of the Santa Catalina de Guale Ossuary, Amelia Island, Florida. *American Journal of Physical Anthropology* **Suppl. 34**, p. 136 (abstract).

Tomak, C.H. (1970). *Aboriginal Occupations in the Vicinity of Greene County, Indiana*. M.A. thesis, Indiana University, Bloomington, IN.

Tomak, C.H. (1979). Jerger: an Early Archaic mortuary site in southwestern Indiana. *Proceedings of the Indiana Academy of Science* **88**, 62–69.

Tomak, C.H. (1980). Scherschel: a Late Archaic occupation in southern Indiana with appended chert descriptions. *Central States Archaeological Journal* **27(3)**, 104–111.

Tomak, C.H. (1983). A proposed prehistoric cultural sequence for a section of the valley of the West Fork of the White River in southwestern Indiana. *Tennessee Anthropologist* **8(1)**, 67–94.

Tomak, C.H. (1987). *Associations of cherts and point types in Greene County, Indiana*. Paper prepared for the Annual Meeting of the Indiana Academy of Science, Indiana State University, Terre Haute, IN.

Tomak, C.H. (1991). *The Jerger Phase and Early Archaic Mortuary Ceremonialism in Southwestern Indiana*. Paper presented at the Midwest Archaeological Conference, La Crosse, WI.

Wagner Mires, A.M. (1991). Sifting the ashes: reconstruction of a complex Archaic mortuary program in Louisiana. In *What Mean These Bones* (M.L. Powell, P.S. Bridges, and A.M. Wagner Mires, eds.), pp 114–130. University of Alabama Press, Tuscaloosa, AL.

Webb. W.S. (1946). Indian Knoll, Site Oh 2, Ohio County, Kentucky. *The University of Kentucky Reports in Anthropology and Archaeology*, Vol. IV, No. 3, Part I.

15

Towards an Archaeology of Cremation

Howard Williams, Ph.D.

Department of History and Archaeology, University of Chester, Chester, UK

INTRODUCTION

How can we begin to understand and explain the changing significance of cremation in past societies? From many parts of the world and for many periods of human history from as early as the Upper Palaeolithic (e.g., Bowler *et al.*, 1970) to recent centuries, archaeologists have uncovered and investigated material evidence for the use of fire as a means of transforming and disposing of the dead. This paper argues that in contrast to the rich and widespread evidence for cremation in the archaeological record, theoretical approaches in the archaeology of cremation have been relatively thin on the ground. This failure to adequately engage with the complexity and the variability of cremation practices across cultures seems connected to the fact that most of the theoretical debates and developments in mortuary archaeology have been primarily geared to the investigation of unburned human remains. Therefore, alongside increasingly refined methodologies for investigating burnt bones, it is argued that archaeologists need to redress this imbalance by developing explicit theoretical approaches to the phenomenon of cremation. Such theories need to engage with broad cross-cultural themes and also remain sensitive to the considerable variety of mortuary procedures involving fire employed at different times and in different places.

THE PERCEPTION OF CREMATION

Cremation has tended to receive a bad press in both academic studies of past mortuary behaviour. Archaeologists often regard the wealth of data provided by primary or secondary inhumation burials and mummified human remains as 'superior' to the fragmented evidence left behind from cremation ceremonies. This is reflected in the low profile of cremation in textbooks in mortuary archaeology other than reference to its destructive nature. Cremation 'denies' archaeologists a view of both the mortuary rituals that preceded the burial and the society that created the funeral (e.g., Parker Pearson, 1999).

Cremation is equally overshadowed by almost every other method of corpse disposal in popular science and archaeology books. These tend to be dominated by graphic images of tombs, unburned skeletons, bog bodies, mummies and human sacrifice (e.g., Bahn, 1996; 2002). The archaeological remains of cremation are perhaps simply not photogenic enough for such studies! Consequently, the reader of such texts might be forgiven for gaining the false impression that archaeological discoveries of cremation are either rare or, when found, of limited archaeological interest.

The low esteem in which cremated remains seem to be held in both academic and popular perceptions of mortuary archaeology is perpetuated by the assumption that cremation is primarily a taphonomic and technical process requiring specialist analysis (e.g., McKinley and Roberts, 1993). While this is by no means incorrect, this perspective limits the appreciation of cremation as a ritual performance that requires theoretical consideration. Hence cremation is assumed to be 'poor' archaeological data. Certainly, the burning of the dead can make it so much harder to reconstruct both life ways and death ways from mortuary remains, but this perspective denies the very significance of cremation as *transformation*, a ritual sequence of value for study in its own right rather than as merely a 'formation process' denying the archaeologist a direct view of past 'reality'. Some archaeologists have naïvely taken the destructive nature of cremation over-literally as an indication that little care, attention, or investment was afforded to those disposed of by this manner in the past; cremation is often regarded as the 'poor man's disposal method'. In fact the opposite scenario might equally be true in some instances, given the resources necessary to dispose of the dead by fire (McKinley, 2006).

Overcoming such misconceptions involves not only embracing the wealth of evidence that cremated human remains can now reveal to the specialist investigator, but also moving beyond the bones to recognize the rich evidence provided by the range of artefacts, contexts, monuments and landscapes associated with cremated remains in the archaeological record. It equally demands a theoretical focus on the transformation and the ephemeral quality of cremation practices rather than simply upon the material that survives.

In this light, it is argued that an 'archaeology of cremation' needs to be explicitly developed. To do this, what is required detailed methodologies (e.g., McKinley, 1994, 1997; Sigvallius, 2005) as well as a theoretical basis for understanding the character and the variability of cremation in the past mortuary practices. To this end, the paper aims at sketching the elements that will allow a theorized approach to be developed to the archaeology of cremation. This approach is explored in relation to one recent case study: excavations on the Westhampnett Bypass in West Sussex, UK. This study was selected to illustrate the high quality of data, analysis and interpretation possible even within projects limited by the tight constraints of time and resources within commercial archaeology (Fitzpatrick, 1997; 2000). However, let us begin by identifying key themes that constitute the basis for an archaeology of cremation.

THEMES IN THE ARCHAEOLOGY OF CREMATION

While archaeologists talk of 'cremation', in fact this term embraces many things. Cremation can involve a complex *rite of passage* (van Gennep, 1960) from rites of separation, liminal rites and rites incorporating the dead and the living participants into new states of Being. In many societies, this is a transition involving the mourners, the soul and the body itself (Hertz, 1960; Metcalf and Huntingdon, 1991) and has ontological, social and cosmological dimensions (Vitebsky, 1993).

Archaeological evidence can reveal direct and indirect evidence for many stages of the cremation process from the preparation of the body, the construction of the pyre, the burning of the body, the sorting, selection and treatment of the ashes, the burial of the cremated remains and the raising of monuments to the dead (McKinley, 1994). All of these actions have a spatial as well as a temporal dimension: some can be located in close proximity to each other while in other contexts, each stage of the cremation ritual may have dedicated and separate locales of activity in the landscape (see Downes, 1999; Oestigaard, 1999; Pearce, 1997). For example, at the later Romano-British cemetery at Brougham, Cumbria, excavations revealed pyre material indicating the presence of nearby pyres. The evidence from the graves suggested how the pyre was constructed, and how a wealth of vessels and sacrificed animals were placed with the clothed cadaver. The graves also illustrated how the burial rituals were conducted. Meanwhile, the foundations of stone buildings and fragments of tomb stones represent the remains of monuments raised to commemorate the dead (Cool, 2004).

Yet cremation is not only a multi-staged ritual process; the technology of cremation can vary considerably both within and between periods and sites. Moreover, there can be complex relationships with other means of disposing of the dead within the same communities (e.g., Lucas, 1996). Cremation can be the dominant rite of a society or a group or it can be a rite performed in only exceptional circumstances, but it is almost always one among many options for disposing of the dead available at any given time. This is a theme as applicable to prehistory (e.g., Larsson, 2004) as to historical periods (Morris, 1992) and, indeed, to the modern world (Davies, 2002; Jupp, 1993). In this light we can consider the variety of evidence that together constitute the 'archaeology of cremation'.

CREMATED BONE

The burning of the body can be a complete process in which the whole corpse is reduced to ashes, or it can be partial by design or by accident, with only parts of the body involved or the body subject to only cooking, roasting or charring (Binford, 1963). For example, the colour and fragmentation of the cremated bone from early Anglo-Saxon (later fifth and sixth centuries AD) cinerary urns excavated at Spong Hill, Norfolk, suggests a very efficient and complete cremation process (McKinley, 1994). In contrast, burned human remains from certain later Iron Age Scandinavian cremation graves suggest low firing temperatures interpreted as either deliberate or accidental partial cremation. In some instances it is possible to suggest that the intention may have only been to 'cook' or 'roast' the dead (Oestigaard, 2000; see also Noy, 2000). In the ritual process, the conflagration itself is but one element. It can be a 'primary rite' (a rite of separation: van Gennep, 1960) following closely upon biological death. This seems to have been the case at Spong Hill mentioned above (McKinley, 1994). Alternately, cremation can serve as a 'secondary rite' (a rite of incorporation) taking place following other ritual practices such as exposure, temporary inhumation, mummification or excarnation as suggested for Neolithic and Chalcolithic Iberia (Weiss-Krejci, 2005). Consequently, cremation is a process that can be paired with numerous other methods for managing and transforming the body.

Armed with a battery of osteological and scientific techniques, archaeologists can reconstruct aspects of the identity of the living people from the burnt bone, including their age at death, sex and some of the pathological conditions they suffered from. In large samples of several thousand cremation burials, as with the early Anglo-Saxon cremation burial studied by Jackie McKinley at Spong Hill, we can begin to identify patterns in demography (McKinley, 1994). Moreover, increasingly sophisticated osteological methods serve to challenge the stereotype that cremation can yield limited evidence (e.g., Cuijpers and Schutkowksi, 1993). By correlating this osteological data with mortuary variability, differing uses of the rite for different social groups can be revealed (Ravn, 2003; Richards, 1987).

These data often provide a valuable addition to our understanding of the ritual technology of cremation. For example, in considering cremated materials from archaeological contexts from the British Bronze Age, Jackie McKinley (1997) warns against lumping all cremated materials into the label 'cremations'. Cremated bones can derive from many different contexts connected to contrasting stages of the cremation process. The identification of cremated remains does not necessarily mean the recognition of a burial; it may instead reflect a pyre, materials connected to the pyre site, or even sweepings and deposits in adjacent features (Arcini, 2005; McKinley, 1997). The careful excavation of pyre features, deposits of pyre debris and cremation burials can reveal diverse information about the technology and ritual practices of past societies.

To compliment such variability in the cremation process, what constitutes a 'cremation burial' can also vary considerably. In some cases this can be the pyre site itself, as in Roman *bustum* graves (McKinley, 2000). In other instances the majority, if not all, of the bones are collected from the pyre for burial. Yet there are many examples where only a 'token' (a sample) of the ashes is retrieved from the pyre for burial (McKinley, 1997). In such cases we can speculate as to how the remaining remains were treated: were they left on the pyre or retrieved for circulation or disposal elsewhere? This question is appropriate for the archaeology of the Southwest United States where ethno-historic and archaeological evidence indicates a variety of post-cremation treatments including the reburning of the ashes as a secondary rite to end the mourning ceremonies (Beck, 2005).

By definition, cremation provides many options of how to dispose of the ashes, and archaeologists must be wary of assuming 'natural' or 'common sense' explanations for the evidence they find. For example, for cremating societies in the ethnographic present, the interment of the ashes is only one option and makes no more or less 'common sense' than leaving the ashes at the pyre site (see Beck, 2005). If the decision is made to retrieve the ashes, burial is but one option. Ashes may be scattered, immersed in water, displayed above ground, stored in the dwellings of the living, interred in cemeteries and sacred sites or used to compose effigies, statues and portable artefacts. Mourners may even consume them as part of a funerary feast! These options are not mutually exclusive, and different portions of the ashes might have contrasting destinations. While many of these practices are archaeologically invisible, the practice of the partition and the burial of portions of the cremated remains in different locations have been recognized archaeologically (e.g., Bowman, 1991).

PYRE GOODS AND GRAVE GOODS

In many societies in the past and the present, material culture can be deployed at every stage of the cremation process. A crude but useful distinction can be identified between 'pyre goods' (including the costume and offerings placed with the dead on the pyre) and 'grave goods' (objects placed with the cremated remains in the grave; McKinley, 1997). By comparing and contrasting these different objects, it is sometimes possible to recognize the contrasting use of artefacts in successive stages of the cremation process. For the Gallo-Roman site of Septfontaines-Dĕckt, Luxembourg, Michael Polfer observed that vessels used to serve food were found most commonly in ustrina (pyre sites) while vessels used to serve drink were more common as pyre goods suggesting their selective removal from the pyre site for burial with the dead. Moreover, drinking vessels are even more common as grave goods (Polfer, 2000: 35).

Zooarchaeological investigations have revealed the centrality of animals in the cremation rituals of many past societies. For example, Julie Bond has discussed how the placing of animals on pyres in early Anglo-Saxon England comprised a mixture of food offerings and animal sacrifice (Bond, 1996; see also Iregren, 1995; Sigvallius, 1995). Even seemingly 'natural' materials may be deliberately placed on cremation pyres and deposited with cremation burials. These might include seemingly mundane objects such as stones, pebbles and burnt earth that in some instances may be as important as ritual deposits as the placing of artefacts with the dead (Artelius and Lindkvist, 2005). In different societies, these materials can gain their significance through their association with the cadaver and their transformation by fire. However, it is equally important to carefully distinguish between artefacts and materials that may have been deliberately placed with the dead and unintentional products of the cremation process (e.g., Henderson *et al.*, 1987).

In some instances, we can identify selected categories of 'grave goods' with specific roles in the burial ritual following cremation. For example, collections of high-status dining sets were placed over graves during the backfilling of a series of rich early Roman cremation burials at Alton, Hampshire (Millett, 1986). In other instances, the absence of certain artefacts in cremation graves may be regarded as significant. For example, the stark contrast between the adult male weapon burial rite of early Anglo-Saxon England and the complete absence of weapons from contemporary cremation burials of all ages and genders strongly suggests that weapons and other iron implements were deliberately removed from association with the dead in the final stages of the cremation rite (Williams, 2005). These examples illustrate how grave goods associated with the dead are always a *selection* whilst others are disassociated from the dead in various ways to be disposed of elsewhere or recycled among the survivors.

While often small and seemingly mundane, grave goods can have special roles in the symbolism of the cremation burial rite. For example, the prevalence of artefacts connected with the management of hair is evident in early Anglo-Saxon cinerary urns by the frequent occurrence of combs and toilet implements. This might suggest a specific connection between the post-cremation rites and the reconstitution of the social 'body' following cremation symbolized by the deceased's hair (Williams, 2003). A complimentary theme has been pursued by Gunnar Andersson (2005) in his discussion of Thor hammer rings in the Viking-period cremation graves of central Sweden. These iron rings with miniature hammers suspended upon them are interpreted as pre-Christian symbols of regeneration and fertility appropriate for placing with the dead following the dissolution of the cadaver during the cremation process. In other cases, grave goods placed with ashes can have more to do with dispersing death pollution rather than with protecting or regenerating the dead. For example, the spears found thrust into the cremation burial of an adult female of Viking-period date beside Lake Dalstorp in Västergötland, Sweden, have been interpreted by Tore Artelius as indicating a 'revenant's' grave; the spears serve to dedicate the dead to an afterlife existence and disperse the fear and pollution associated with the liminality of the funeral (Artelius, 2005).

VESSELS, CONTAINERS AND GRAVE STRUCTURES

The huge diversity of cinerary urns and containers are integral to the archaeology of cremation. Cinerary urns can be simply domestic vessels given a secondary use. For instance, in some Romano-British cremation cemeteries as from the legionary fortress of Caerleon, Wales, it seems likely that the vessels employed to contain ashes were cheap and available urns, some may even be 'wasters' unfit for any other use (Evans and Maynard, 1997). Yet even modest, broken and badly made vessels might gain a new significance, even holding sacred connotations, through their use to contain the remains of the dead. For Iron Age Norway, the use of cooking vessels as cinerary urns may reveal an important metaphor for how the dead were understood as sacrificial food offerings by the living (Oestigaard, 1999, 2000).

In other cultures and contexts, particular types of organic, ceramic, glass or metal containers were carefully selected or even purpose-built as urns. For example, in early Anglo-Saxon England, the size, shape and the decorative schemes incised, stamped and embossed upon cinerary urns in some instances correlate with the social identity of the deceased in terms of osteological sex and age. This evidence suggests that a careful process of either selection or design was employed to communicate the deceased's social identity for those participating and observing the collection, transportation and burial of the ashes (Richards, 1987). Moreover, Julian D. Richards (1992) interprets the emphasis towards the decoration of the top sides of cinerary urns as related to the display of the pot within the grave, the decoration speaking to the mourners at the final stage of the burial sequence.

In some instances, the form and design of cinerary urns was closely connected to the symbolism of the cremation process and the commemoration of the dead. The 'house urns' of Bronze Age northern Europe have been interpreted as the representations of granaries and hence symbols of fertility and regeneration. This choice of design may have been appropriate for the post-cremation rituals that frequently emphasize the incorporation of the dead into an ancestral state with reference to regenerative symbolism (Bradley, 2002).

Equal attention needs to be paid to grave structures, for while cremation graves may often appear to be mundane pits, the repositories for ashes can be carefully designed to facilitate specific statements about the dead and engagements with the ashes. In some past cultures, cremated remains can be placed within elaborate structures, as with the cists of early Bronze Age Orkney investigated by Jane Downes (1994). Such structures may be intended for single funerary episodes but certain arrangements of stones and slabs facilitate their reopening to receive successive cremation deposits (e.g., McKinley, 1994). In other situations, the context of the ashes might be best seen as a persistent arena for social interaction between the living and the dead rather than as a 'final resting place'. An example of this can be seen in the early Roman 'pipe burials' where a conduit from the surface into the cinerary container allowed repeated libations to be poured by mourners to appease the spirit of the deceased and to commemorate the anniversary of the death (e.g., Wheeler, 1929).

MONUMENTS

Cremation burials can be associated with many different forms of below- and above-ground structures and monuments. Some memorials might be constructed for single interments, yet they often have evolving biographies of successive burial episodes as well as monumental enlargement, elaboration and reuse. In some instances, the archaeologist can identify a complex interplay between cremation and other disposal methods within 'monumental genealogies'. This is evident for early Bronze Age mortuary practices from southern England. In some instances, both the individual mounds themselves and the evolution of barrow cemeteries indicate an evolving performative space for cremation ceremonies and burial rites (Barrett, 1990; see also Brittain, 2006). Alternately, the modest size of cremated remains encourages their integration into communal monuments designed to receive multiple burials. Numerous examples of this phenomenon are known from the monumental tomb architecture of the ancient world, such as Roman columbaria or the well-known roadside house tombs of Isola Sacra from the Roman port of Ostia (Hope, 1997).

Another example of repeated cremation burials associated with communal monuments is discussed by the Swedish osteoarchaeologist Berit Sigvallius. Her osteological analysis focused upon the burnt human remains from within a ship-shaped stone setting of Iron Age date from southern Sweden. The cremated remains scattered within the monument indicate that it was used as a communal mortuary monument for the burial and scattering of cremated remains for many centuries (Sigvallius, 2005).

Cremation can also be closely connected to the 'life histories' of monuments. In early Neolithic Britain and Ireland, cremation is associated with complex sequences of mortuary activity in and around a diversity of monuments of earth, timber and stone. The biographies of these monuments include phases of building, decay and, in some cases, concerted destruction by fire. In these instances, a metaphorical relationship can be suggested between the monument's history and the transformation of the dead by fire, as for the early Neolithic earthen long barrows of East Yorkshire (Lucas, 1996: 106).

In further cases, the symbolism, sensory and experiential qualities of monumental architecture seem to relate to the cremation process. The late Neolithic and early Bronze Age passage graves of the British Isles are characterized by restricted access along passages into elaborate chambers built upon solar alignments. For the famous Irish passage grave of Newgrange, light would pour into the chamber via a roof box upon midwinter sunrise. When excavated, the chamber was found to contain burnt and unburnt human remains, suggesting that, as with other Irish passage graves, the architecture and seasonal solar experience was connected to the final resting places of the dead following complex mortuary procedures, possibly associated with the veneration and commemoration of ancestors (O'Kelly, 1982: 126). The connections between light, heat and seasonality of the sun's interaction with the monument may have been connected to the symbolism of the cremation rituals conducted around and within these monuments, evoking the widespread theme of regeneration through transformation.

A later permutation of the same theme can be recognized for the Early Bronze Age cairns at Balnuaran of Clava near Inverness in northern Scotland. The alignment and the orientation of both passage graves and ring cairns focused on the midwinter sunset. Moreover, the contrasting use of white and red stones in the composition of the monuments may have emphasized the impact of sunlight and fire upon the monuments during ceremonies. Such solar associations may have provided apt symbolism for mortuary and commemorative rituals focusing on the transformation of the dead by fire, and certainly these tombs were used and reused for cremation burials during the Bronze Age (Bradley, 2000).

Therefore, monuments have diverse interactions with cremation practices in different societies, and this can be extended to a consideration of 'ephemeral monuments'. These can be defined as monuments created for temporary display or public destruction and therefore can be particularly apposite for cremation practices, given the destructive nature of burning the dead. Ephemeral monuments deliberately destroyed during funerary rituals can include the houses of the living and mortuary structures built for the preparation or the display of the corpse prior to cremation. However, the funerary pyre can itself be an important ephemeral monument, built to contain and display the body and also serving as part of the spectacle of the funeral with its conspicuous and spectacular fiery demise (see Williams, 2004a).

In some cases, the archaeologists can identify the interplay between multiple ephemeral monuments within the same sequence of funerary obsequies. At Folly Lane, St Albans, Rosalind Niblett identified the remains of a high-status multi-staged cremation ceremony dating to around the time of the Roman conquest. The sequence involved a number of ephemeral monuments. While the timing of events remains unclear, the funeral could have lasted several months or even years. The excavations show that a subterranean chamber was built, possibly to provide a mortuary chamber in which the body was afforded temporary residence. Adjacent to it, a mound was constructed upon which a pyre was raised and subsequently set alight. Some of the pyre remains were thrown into the subterranean chamber during its deliberate destruction. The destroyed chamber was then capped with a mound. Later, a Romano-Celtic temple was built over the site of the pyre mound. In this instance, it appears that the ephemeral monuments not only were important in the mortuary process but also gave rise to more permanent commemorative foci. The exceptional funerary sequence is interpreted by Niblett as providing the focus for an ancestor cult that endured through the Roman period (Niblett, 2000; Williams, 2004a).

In other contexts, the cremation pyre is less ephemeral but forms a monumental focus for repeated mortuary rituals. This is evident in the *ustrina* of Roman cremation practices. The role of pyre sites as monumental foci also applies to the stone pyre settings dated to the early medieval period from Hermisgarth, Sandy, Orkney (Downes, 1997). Ethnographic parallels can be found in the repeated use of cremation platforms in Bali (Downes, 1999). A similar view can be seen of modern crematoria: these are 'transformatory monuments' with architectural features designed to transform the dead rather than primarily or exclusively serving as final resting places for ashes.

The monumentality associated with cremation burials has numerous knock-on effects, enabling different uses of cemetery space and hence different monument forms in contrast to other disposal mechanisms. For example, the large 'urnfields' of later prehistoric northern Europe provide a close proximity between the modest-sized mounds that would not have been possible for communities inhuming their dead. This may have engendered a distinctive perception of a community of ancestors for communities disposing of the dead in this way (see Parker Pearson, 1993). This leads us to consider the distinctive landscape contexts associated with cremation in the past.

LANDSCAPES

Cremation can create a distinctive engagement between mortuary practice, space and place. Funerals can be regarded as sequences of ritual display, and the act of cremation as but one of these events that projected the identity of the dead and mourners across the landscape. Different sites might be chosen for successive stages of the funeral. With small-scale archaeological investigations, the proximity of different stages of the cremation process can often be missed and more extensive excavation strategies should be encouraged to place cremation burials in context. For example, geophysical surveys have allowed the identification of pyre sites associated with cremation burials and monuments (Marshall, 1998) while extensive excavations around cremation graves can lead to the recognition of pyre deposits and pyre-related features (McKinley, 1997; Petrov, 2002).

Studies can also investigate cremation on a landscape scale. For example, for the Iron Age of southern Jutland, Mike Parker Pearson (1993) has discussed the changing spatial distances of cremation cemeteries from contemporary settlement sites. He also considers the spatial association of urnfields with ancient monuments of Neolithic and Bronze Age date as well as the topographical locations of cemeteries. These spatial relationships are regarded by Parker Pearson as evidence of shifting social and power relationships between the living and the dead within Iron Age societies. Meanwhile, Roymans and Kortlang (1999) have addressed the landscape locations of Bronze Age and early Iron Age urnfields in the lower Rhine region. They argue that the urnfields actively ordered the social and cosmological landscape both through their location and internal organization. The importance of cemeteries as places for the ancestors was emphasized through the allusions to domestic architecture in the forms of selected funerary monuments.

The power of place created by both the cremation process and the interment of the ashes can be identified for historic periods as well. In the early Roman world, pyre sites and cremation created a distinctive zone for the dead, encircling towns. The dead physically surrounded the living and were encountered on journeys both into and out of the city. The dead collectively accrued a broader significance, defining urban identities and social memories (Hope, 1997; Esmonde Cleary, 2000). Likewise, this writer has highlighted evidence indicating that the large cremation cemeteries of early Anglo-Saxon eastern England may have served as significant ceremonial and ancestral 'central places' used for gatherings by many different social groups (Williams 2002).

The landscapes of cremation extend beyond the pyre and the grave. The potential afforded by cremation for the retrieval and circulation of the ashes means that many places and spaces from temples to houses can be incorporated into the mortuary process and subsequent commemorative and ancestral rites. For example, cremated remains can sometimes be found circulating 'beyond the grave' in Bronze Age Scandinavia where cremated and unburnt human remains have been found upon settlement sites. Hence in some societies, there were complex and lengthy interactions between the ashes of the dead and the communities of the living rather than fixed landscape locations (Eriksson, 2005).

CREMATION IN CONTEXT

A final theme concerns interpreting cremation in relation to a wider context of other fiery transformations in past societies. Cremation is often associated, both physically and conceptually, with other ritual uses of fire to transform and commemorate the dead. For instance, early Anglo-Saxon cremation burials at Snape in Suffolk were associated with other fire rituals. These included burnt stone features interpreted as cooking pits that may have been associated with funerary feasts. Also, charred wood, flints and cremated bones were placed over contemporary inhumation burials, perhaps as a means of 'smoking' and hence purifying the grave (Filmer-Sankey and Pestell, 2001: 243–244, 259–261).

Other uses of fire may be connected to cremation rituals in spatial, technological and symbolic terms. The parallel ritualized transformations associated with both metal-working and cremation rituals have been considered for Iron Age Scandinavia (Gansum, 2004a,b). Meanwhile, transformation by fire can provide a theme linking the treatment of the dead to the transformation of other forms of material culture. Lars Larsson, when discussing the Swedish Neolithic, has argued that stone axes could be regarded as analogous to humans and afforded cremation rites (Larsson, 2000). Meanwhile, buildings and monuments can be subject to deliberate ritualized fiery destruction. Sometimes this may not be accidental but might indicate a form of cremation, either burning the dead within their homes or treating the house as symbolically equivalent to a dead human being. This perspective can be usefully applied to the early first-century BC structure uncovered at Navan Fort, Armagh, Northern Ireland. The elaborate multi-ring timber structure (possibly a temple) was filled with stones and burned down soon after construction. It was subsequently covered with a turf mound in a planned ritual process (Waterman, 1997: 224–230).

Therefore, the archaeology of cremation encapsulates much more than the cremated bones themselves; it involves contexts, artefacts, spaces, monuments and even the landscape context of, and relationships between, different stages of the cremation process. Having introduced the range of archaeological evidence pertaining to the study of cremation in past societies, the study will now move on to address the diverse theoretical perspectives that archaeologists have applied to the study of cremation in the archaeological record.

THEORETICAL APPROACHES FOR THE ARCHAEOLOGY OF CREMATION

Given the complexity and the diversity of the evidence for cremation rituals in past societies, from the Palaeolithic to recent times, it is a challenge to develop theories to understand and explain the reasons why human societies have employed fire as a means of transforming human cadavers into heat, light, smoke, ashes and burnt bones. One of the earliest detailed descriptions of the discovery of cremation burials was recorded by the famous seventeenth-century scholar Sir Thomas Browne in his *Urn-Buriall* of 1658:

> In a Field of old *Walsingham*, not many moneths past, were digged up between forty and fifty Urnes, deposited in a dry and sandy soile, not a yard deep, nor farre from one another: Not all strictly of one Figure, but most answering these described: Some containing two pounds of bones, and teeth, with fresh impressions of their combustion. Besides the extraneous substances, like peeces of small boxes, or combes handsomely wrought, handles of small brasse instruments, brazen nippers, and in one some kinde of *Opale* (Browne, 2005 [1658]: 9).

Browne thought that the cinerary urns were Roman in date although subsequently they have been attributed to the early Anglo-Saxon period. Browne regarded the cinerary urns as the traces of past people, evidence of their beliefs and practices, and their attempts to commemorate the dead. Since Browne's day, cremated remains have been investigated within very contrasting theoretical paradigms. This subsequent section takes these approaches in broadly chronological sequence while accepting that there are overlaps between each approach, particularly for the most recent studies.

EARLY PERSPECTIVES – EVOLUTION, RACE, CULTURE AND RELIGION

Since the development of archaeology as a defined discipline in the nineteenth century, cremation has frequently been regarded as a 'normative' racial or cultural trait. Burning the dead was sometimes seen as an inherent predisposition of a race or as indicative of stages in evolution from savagery through barbarity to civilisation. Evolutionary, racial and diffusionist perspectives are often combined in nineteenth-century writings on burial customs. For instance, in *Horae Ferales* of 1863, John Kemble regarded cremation as a barbarian trait among different races as far-flung as the Etruscans, the Celts and the Germans. For the Germans, cremation indicated their pagan and primitive origins. Its abandonment displayed their inexorable rise towards Christian Victorian 'Anglo-Saxon' civilisation (Kemble, 1863).

With the emergence of what has been dubbed 'culture-historic' archaeologies in the early twentieth century, cremation was often regarded as a fashion, the evidence of the spread of ideas and religious beliefs, or alternately, the invasion or migration of people. For example, Britain's late Iron Age 'Aylesford-Swarling' culture was regarded as the result of Belgic settlers mentioned by Julius Caesar (Fitzpatrick, 1997). Following Kemble's precedent, the same is true of the interpretation of early Anglo-Saxon cinerary urns as an indication of pagan Germanic settlers in eastern England in the fifth century AD (e.g., Myres, 1969; see Richards, 1987).

SOCIAL PERSPECTIVES AND MORTUARY VARIABILITY

Culture-historic perspectives in mortuary archaeology continue to hold credence in some quarters and have been reinvented in different guises. Yet with the self-dubbed 'New Archaeology' from the 1960s and 1970s, this normative treatment of burial customs was discounted (Binford, 1971). Instead, cremation was among the practices that became increasingly interpreted in *social* and *economic* rather than in *cultural* and *religious* terms. Based on the assumption that the dead were interred according to the scale and the character of their social roles and the social network they enjoyed in life, the emphasis was focused upon the variability in mortuary practices. This approach had the advantage of considering whether the composition and organization of past societies could be discerned in the variable treatment afforded to the dead (Chapman *et al.*, 1981). While these studies favoured inhumation burials, they were also applied to the variability found within the cremation traditions (e.g., Ravn, 2003; Richards, 1987) and societies that, over time, fluctuated between using cremation and inhumation (e.g., Morris, 1992).

The scientific emphasis of the New Archaeology provided the theoretical impetus for more sophisticated methodologies for investigating cremated human remains and the taphonomic processes associated with cremation. However, this encouraged a tendency to regard cremation as simply a *bias* that obscures the straightforward reconstruction of social organization rather than a ritual process worthy of investigation in its own right.

SYMBOLISM, ESCHATOLOGY AND COSMOLOGY

As with culture-historic approaches, social approaches to cremation burials remain popular. However, from the 1980s, the rise of post-processual and contextual archaeologies critiqued the social approach of the New Archaeology. The attempts to directly read social organization were questioned and challenged from multiple perspectives.

The post-processual studies retained a bias towards the study of inhumation graves (e.g., Pader, 1980) and collective secondary burial rites (e.g., Shanks and Tilley, 1982). The studies of cremation burials, while addressing the symbolism of urns and grave goods, retained a social focus on variability (e.g., Richards, 1987). Yet cremation began to be regarded as a symbolic, eschatological and cosmological practice rather than simply an index of either culture, religious belief or social structure (e.g., Pearce, 1997).

Early attempts to address the meaning of cremation in past societies were somewhat restricted to its 'levelling' and 'masking' qualities (Parker Pearson, 1982). More recently, these post-processual studies have developed more sophisticated interpretations of the meaning of cremation. Such approaches seem particularly popular in Scandinavian archaeology where cremation ritual has been linked to concepts of afterlife journeys by the soul (Gräslund, 1994) and studied in terms of eschatology and cosmology in relation to other fire rituals. Historical and ethnographic analogies have been extensively employed to support these interpretations (Kaliff, 1998, 2005; Gansum, 2004a, b; Oestigaard, 1999, 2000, 2005; Parker Pearson, 2004).

In Britain and the United States, cremation has been investigated as a further means of 'ancestralization': transforming the dead into an ancestral state. This was often seen (inspired by Robert Hertz's work) in terms of a transformation of the cadaver and the soul between structural dichotomies of wet to dry, from decomposing to inert (Hertz, 1960; Metcalf and Huntingdon, 1991). To this end, cremation was interpreted as a distinctive means to an end: the placing and commemorating of ancestors and connections between the living and the dead (but see also Rakita and Buikstra, 2005). These approaches also used extensive ethnographic and literary analogies. The focus was on the meanings and metaphors inherent in cremation rituals and attempts to understand how the cremation rituals related to other contemporary social and ritual practices in which fire is employed.

IDEOLOGY, POWER, AND AGENCY

The meaningful nature of cremation will never have existed in a power vacuum. As a public display, cremation rituals were enmeshed in strategies of domination and resistance. Within post-processual approaches, this led to a consideration of how cremation served as a context for the negotiation of power relations and the constitution of ideologies (Parker Pearson, 1982). More developed ideological approaches have emphasized the performance of cremation and the agency of mourners. In this sense, the cremation process can be considered as a 'field of discourse' (Barrett, 1994) in which power relations and social identities are negotiated through ritual performance (e.g., Barrett, 1990, 1994). John Barrett developed this argument through a discussion of the sequence of cremation and post-cremation rites found in Early Bronze Age Wessex, including the pyre site and cremation burial beneath a round barrow from Edmonsham in Dorset (Barrett, 1994). Meanwhile, Jane Downes (1994) has explored related ideas through her excavation of a Bronze Age burial mound at Mousland, Stromness, Orkney. She suggested that different artefacts associated with the cist were connected to successive ritual actions by the mourners in transforming the identities of the dead (see also Downes, 1999).

This perspective adopted by Barrett and Downes upon ritual performance and the agency of the mourners has implications for the significance of cremation in early historic archaeology. For instance, the choice of cremation in the 'princely' barrow burials found at Sutton Hoo date to the late sixth and early seventh century AD. Within these rites that also incorporated boat burial and chamber graves, Carver regarded cremation as an element in a shifting permutation of theatrical performances aimed at asserting the pagan and Scandinavian affiliations of the incipient East Anglian kingdom (Carver, 2005).

A further theme in the agency and power of cremation concerned the agency of the dead body itself. Rather than regarding the cadaver as an inert substance, past communities may have perceived the corpses as an animate agent in the cremation process, demanding proper treatment and prompting the commemoration of the deceased. Drawing upon forensic, experimental, ethnographic and ethno-archaeological research, this author has suggested that cremation concerns not only the performance and agency of living but also the mnemonic agency of the cadaver in early Anglo-Saxon England (Williams, 2004b).

COMMEMORATING PERSONHOOD

Recent studies have addressed the nature of *personhood* and *social memory* in mortuary archaeology. Outside of the modern West, the concepts of the individual are far from universal. In the past, personhood was likely to have been relational and 'dividual' – made of sets of materials and substances shared and exchanged between people (Fowler, 2004). Considerations of personhood in relation to mortuary practices are concerned with the transformation of the person in death. Mortuary practices can be seen as ways of transforming, fragmenting and reconstituting personhood. This can be achieved through a variety of technologies, actions and metaphors (Brittain, 2006; Fowler, 2001; Brück, 2004).

Few studies have directly applied these ideas to cremation rites but one exception is Joanna Brück (2001). She has explored the communities of the middle and later Bronze Ages of southern England (*c*. 1500–700 BC), identifying metaphorical connections between the life histories of human bodies, houses, pots and quernstones. Cremation ceremonies drew upon parallels between the transformation of people and objects. Brück regards that cremation in this period is best understood as a means of fragmenting and regenerating the dead, incorporating them into a social and cosmological order. She argued that the symbolism of fertility associated with cremation was linked to the cycle of households and material culture. This theme is supported not only by the material culture of the cremation burials, but also by their systematic location in the same south-east quadrant of burial mounds, mirroring the spatial organization of contemporary houses.

Other studies have focused on mortuary practices as ways of selectively remembering and forgetting the dead. For this, Andy Jones has usefully coined the term 'technologies of remembrance', emphasizing the practical and sequential nature of cremation procedures rather than regarding them simply as a performed set of symbolic meanings or metaphors (see Jones, 2001a, b, 2003). Here cremation is considered as a means of making funerals distinctive from, but alluding to, earlier interments and hence linking monuments, ritual performance and place (see also Last, 1998; Mizoguchi, 1993). Jones (2001b) has discussed how 'citations' of personhood were inscribed onto place in early Bronze Age Scotland mediated by the choice of cinerary urn and the location chosen for the burial (see also Williams, 2004b). By seeing social memories constituted and reproduced through the cremation process, shifting sequences of burial rites over time and the choices made by the survivors over whether to respect or diverge from existing practices, these approaches address the relationship between agency and structure in past mortuary traditions and the importance of the past in the past.

SOURCES FOR ANALOGY

Alongside these changing approaches has been a long-running debate over the use of ethnographic analogy and cross-cultural generalizations in the interpretation of cremation in the past (see Carr, 1995; Ucko, 1969). Certainly, key anthropological theories used in archaeological research have been heavily inspired by cremating societies in recent times and the present. For example, Robert Hertz's (1960) work on secondary burial incorporated Balinese cremation rituals. His theory was subsequently developed by Metcalf and Huntingdon (1991) who regarded cremation as an extreme form of 'secondary burial' (see Rakita and Buikstra, 2005). Meanwhile, Arnold van Gennep (1960) was inspired by Indian tribal societies' cremation rituals in regarding death as a rite of passage. The regenerative symbolism of funerals and their role in consolidating an ideal social order are themes regularly drawn from Jonathan Parry's influential work on northern Indian cremation rituals (Bloch and Parry, 1982; Parry, 1994). Putting many problems with ethnographic analogy aside, to date archaeologists have yet to fully exploit the rich and varied data available that provide insights into the technology and the material culture of cremation. In this regard can be cited ethno-historical studies of cremation on the American North-West coast (Kan, 1989) and the diversity of cremation rituals in India outside of the high-caste Brahmin rites usually studied (Århem, 1989; ManiBabu, 1994; Vitebsky, 1993). Equally there are informative variations in cremation rites in south-east Asia to compliment the repeated study of Balinese cremation (e.g., Hudson, 1966). To date, these studies have had a restricted use in reconstructing formal procedures (e.g., McKinley, 1994) but there remain many other levels by which analogies might be of relevance to the study of the social significance, the symbolism and the contexts of cremation in past societies.

Even less consideration has been given by archaeologists to the increasing body of literature by sociologists exploring cremation in Western death ways. These studies can provide profitable perspectives on the relationships between cremation, landscape, monuments and architecture (e.g., Davies, 1996, 2002; Grainger, 2005; Teather, 1999). Of particular value for archaeologists are sociological studies into a range of commemorative material cultures associated with cremation. From photographs, urns, flowers, grave stones and cemetery spaces, these studies have direct implications for the archaeology of cremation (e.g., Prendergast, *et al.*, 2006). Sociological studies also allow archaeologists to theorize the social, economic and religious factors influencing the choice to cremate alongside other disposal choices (e.g., Jupp, 1993; Parker Pearson, 1982). Sociological research also considers the perception and the treatment of ashes as simultaneously both human remains and material culture that facilitate diverse responses, treatments and options for disposal (Davies, 2002; Hallam and Hockey, 2001; Prendergast *et al.*, 2006; see also Harvey, 2004).

Historical and literary studies provide evidence for cremation in past societies that can aid and enhance archaeological interpretations. The innumerable descriptions of cremation provided by ancient authors address the cremations of historical personages. For example, Suetonius' account of Caesar's funeral encourages an appreciation of the socio-politics and the symbolism of cremation in the Roman world (Graves (trans.), 1957: 51–53). There are also valuable brief commentaries on neighbouring cultures such as the Roman historian Tacitus' account of cremation among the *Germani* (Mattingly (trans.), 1948: 123–124). Literary sources also record cremations in history and legend, whether remembered or imagined. Examples of these include the funerals portrayed in Virgil's *Aeneid* (e.g., Day Lewis (trans.), 1952: 162), the hero's funeral in the Anglo-Saxon poem *Beowulf* (Heaney (trans.), 2002: 70–78) or the cremation of the Norse god Baldr in Sturluson's *Edda* (Faulkes (trans.), 1987: 49–50). Such accounts are not valuable primarily for their historicity but because they illustrate the central place in mythology and cosmology that cremation can be afforded. Such accounts demonstrate elaborate procedures, vivid spectacles and the rich symbolism that might accompany ancient cremation rituals (e.g., Noy, 2000).

The most famous written source used to interpret cremation burials comes from the unique first-hand account of a Rus funeral on the Volga by the tenth century traveller Ibn Fadlan. He describes the funeral culminating in the cremation of a Rus chief within a boat followed by the raising of a mound topped by a grave marker. The pre-cremation obsequies are described in some detail with particular attention given to both animal sacrifices and the rape and killing of a slave girl to accompany her master into the afterlife (Warmind, 1995). This source has been extensively used to interpret the symbolism and the eschatology of cremation in the Viking world. The comparison and the integration of historical and archaeological sources can provide complimentary evidence about mortuary variability and the nature of cremation rituals in a particular period, as illustrated by a recent study of medieval Mongolia (Crubézy *et al.*, 2006). Similarly, through a consideration of historical and archaeological sources in tandem, it is possible to identify and reappraise connections between socio-political change and religious conversions, for instance for Sung China (Ebrey, 1990), the Roman Empire (Morris, 1992), Carolingian Saxony (Effros, 1997) and Viking Scandinavia (Andersson, 2000, 2005).

Considerable potential for analogy can be realized through the study of the affects of fire on human bodies conducted within the forensic sciences. While these studies are conducted to assist in criminal investigations, they can, in different ways, give us insights into what cremation involves and provide insights into the interpretation of the archaeological record (e.g., Bohnert *et al.*, 1998). While the observation of modern cremations raises a range of ethical questions and the technical process is very different from open-air cremations of the past, the data can be of direct relevance to the study of past ritual practices (McKinley, 1994; Williams, 2004b).

Archaeologists have the ability to pursue direct research into cremation practices in contemporary societies to generate analogies for the study of the past. Ethno-archaeological work provides effective analogies to assist in theory-building in relation to the cremation process and the material culture. Mike Parker Pearson's influential analogy from the use of modern cremation in Cambridge had subsequent implications his studies of mortuary practices, ideology and society in prehistoric southern Scandinavia (Parker Pearson, 1982, 1984, 1993). Similarly, the ethno-archaeological fieldwork of Terje Oestigaard in Nepal has inspired his interpretations of Bronze and Iron Age Scandinavian cremation rites (Kaliff and Oestigaard, 2004; Oestigaard, 1999, 2000, 2005). Equally, Jane Downes (1999) has employed a discussion of ethno-archaeological fieldwork in Bali to show its potential for understanding the ephemeral material traces left by elaborate rituals and to suggest relationships between cremation, cosmology and society in Bronze Age Orkney.

Finally, a further option for analogy and theory-building is experimental archaeology (e.g., Shipman *et al.*, 1984). Outdoor cremations, usually using animal carcasses, have provided a range of insights into both the practical and the technological aspects of cremation (see Leinweber, 2005; Marshall, 1998; McKinley, 1997). Such experiments also provide parameters for appreciating cremation as a ritual display and the range of stages and elements that may have been attributed with meanings and significance in past societies.

The archaeology of cremation is more than a series of theoretical ideas; it also involves a raft of analogies from cognate disciplines and interdisciplinary research. In this regard, a traditional focus on ethnography (see Ucko, 1969) can be augmented by a range of other source materials and disciplines to create a fully developed archaeology of cremation.

CASE STUDY: A LATE IRON-AGE CREMATION CEMETERY NEAR WESTHAMPNETT, SUSSEX

No single study can encapsulate and illustrate all of the types of theoretical approaches and archaeological data in current use. Indeed, the very methods employed by archaeologists from survey to excavation will affect the theoretical approaches being asked and the quality of the data being identified (Nicklasson, 1999). However, it is necessary to illustrate the potential for a fully contextual archaeology of cremation that integrates osteological evidence into a broader social interpretation of burial data. Such approaches require an exploration of mortuary variability, mortuary symbolism as well as an appreciation of the mnemonic significance of the cremation process in building traditions and social identities. To this end, the final section of the paper explores a single case study in the study of cremation in past societies.

The developer-funded excavation ahead of the construction of the Westhampnett Bypass, east of Chichester in West Sussex in southern England demonstrates the potential for integrating theory, method and data in the analysis of cremation in the past (Fitzpatrick, 1997). The cemetery is one of the broader phenomena of late Iron Age cremation cemeteries found across south-east Britain in the first century BC. This distinctive and novel mortuary tradition (the Aylesford-Swarling culture, as mentioned above) used to be regarded as the evidence of intrusive people and customs brought from the Continent. Strong connections still appear to have existed between these rites and northern French burial rites of the period, but interpretations have shifted to address the social and ideological significance of the rite for indigenous communities with long-distance contacts (Fitzpatrick, 1997: 208–209; Fitzpatrick, 2000; Hill, 1999: 264–268; Pearce, 1997).

An area of around 100 m (west to east) × 75 m (north–south) was uncovered to examine a site revealed by a machine-cut evaluation trench (Fitzpatrick, 1997: 1–9). The excavations found a multi-phased site consisting of around 200 graves of late Iron Age, Romano-British and Anglo-Saxon date, although here the discussion will focus upon the largest, late Iron Age phase. In this period, cremation was the only mode of disposal of the dead. The study combines an investigation of the cremated bone by Jackie McKinley with a full consideration of the artefacts, contexts and the spatial parameters of the site by Andrew Fitzpatrick.

THE CEMETERY AS A RITUAL SPACE

A striking feature of the Westhampnett Bypass site is that different stages of mortuary process were identified during the excavations (Figure 15.1). These included 11 features interpreted as pyre sites recognized from their distinctive

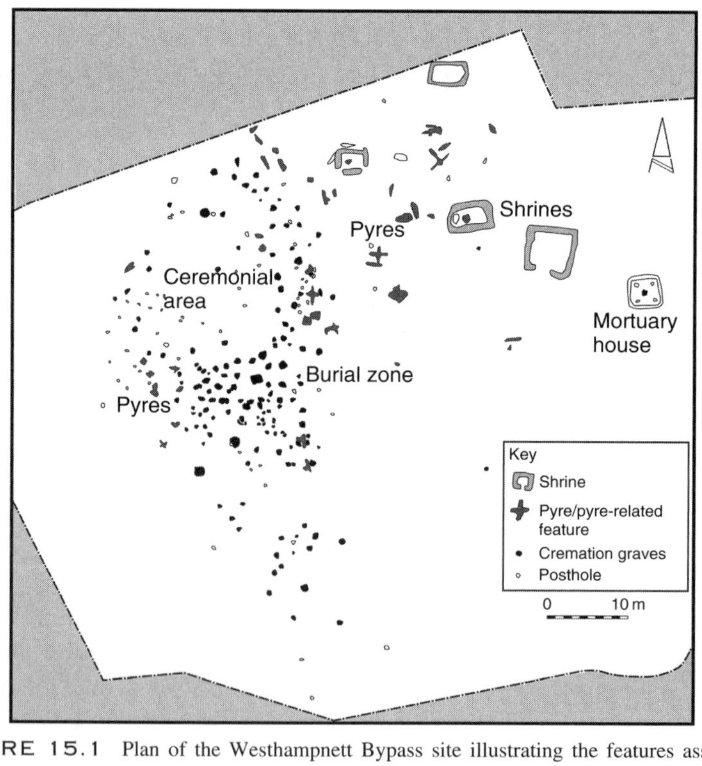

FIGURE 15.1 Plan of the Westhampnett Bypass site illustrating the features associated with mortuary practices during the early first century BC (redrawn after Fitpatrick 2000: 24).

X-, Y-, and T-shaped ditches. These features were interpreted as flues cut to assist the flow of air and assist in the cremation process. They contained charcoal, burnt soil, flint and gravel, burnt human and animal bone as well as the remains of burnt objects. In some instances there was evidence of *in situ* burning of the ground surface associated with these features. There were a further 35 pyre-related features identified in association with these pyres, some of which are likely to be pyre sites, others the deposition of pyre material in natural hollows or deliberate burials of pyre debris (Fitzpatrick, 1997: 18–32; Figure 15.2).

To the north-east of the cemetery were five small and ditched enclosures in a line from west–north–west to east–south–east. The most easterly of these was square and surrounded a single cremation burial located separately from the rest of the cemetery (Fitzpatrick, 1997: 13–18; Figure 15.1). These structures might be interpreted as mortuaries for the temporary preparation and display of the cadaver prior to cremation. An alternative explanation is that they were shrines to ancestors or deities. Equally, they could have served as above-ground cinerary repositories – an alternative to the burial of the ashes for selective individuals or groups. Similarly, there were numerous post-holes from the site, numbering almost 50 in total. These cannot be attributed to a clear function although some appear in line that may suggest their roles, boundaries, screen or façades during mortuary procedures

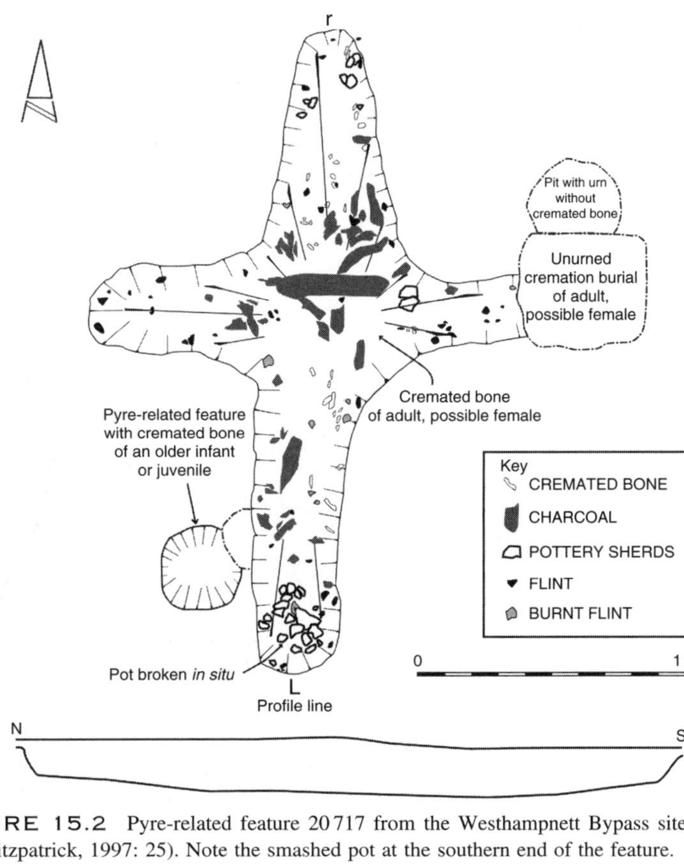

FIGURE 15.2 Pyre-related feature 20 717 from the Westhampnett Bypass site (redrawn after Fitzpatrick, 1997: 25). Note the smashed pot at the southern end of the feature.

(Fitzpatrick, 1997: 234). An alternative is that they had a role in the mortuary commemoration, perhaps serving to suspend bags containing cremated remains.

Finally, there were 161 burials of cremated materials and objects, 140 of which were concentrated within a defined area encircled by the pyre sites and respecting a circular area. This area, strikingly free of graves, has two alternate interpretations. One is that here there was a mortuary mound and the cremations inserted into it had been dispersed when the mound was subsequently levelled during agricultural activities. Alternately, the space may have served as a stage for ritual performances connected with the surrounding burials (Figure 15.1). Many of the burials were truncated by the later activity, but what is notable is that only 10 of the grave cuts overlapped. This suggests that graves were marked in some way – perhaps by small mounds – and were usually respected in subsequent burial rituals. Where the disturbance of earlier burials took place it was in the most popular area for burial on the south side of the respected circular area, suggesting a particular desire for interment in this area (Fitzpatrick, 1997: 13–15, 234). The graves varied in size and shape (round, oval and square) and contained unurned human cremation deposits intermingled with cremated animal remains and burnt objects. Given the discrete locations of the burnt remains within graves, it is likely that they were originally contained in leather or that were not preserved.

CREMATING THE DEAD

The pyre goods found from the graves at the Westhampnett Bypass site included brooches, bracelets, rings, knives, a key, a coin, wooden vessels revealed by their metal fittings and bone artefacts. Some of these were recovered from the pyre sites also (Figures 15.3 and 15.4). Pots were placed on

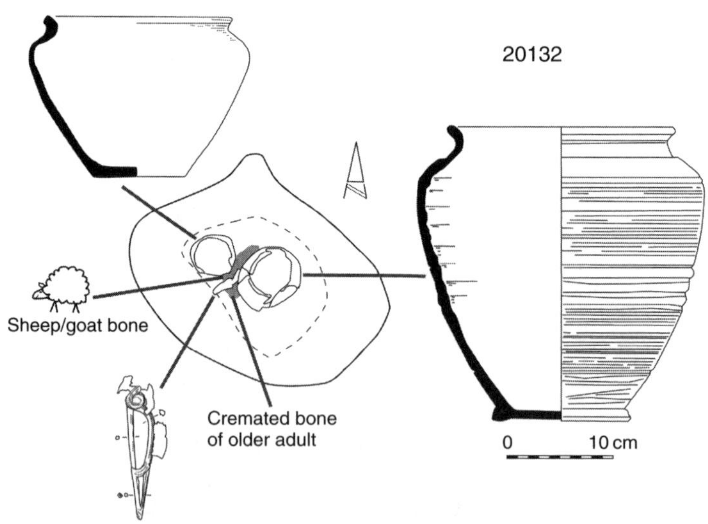

FIGURE 15.3 Cremation burial 20 132 of an unsexed older adult from the Westhampnett Bypass site (redrawn after Fitzpatrick, 1997: 155). Pyre goods included an iron brooch and sheep/goat bone. Grave goods consisted of two high-shouldered, necked jars (Fitzpatrick, 1997: 152).

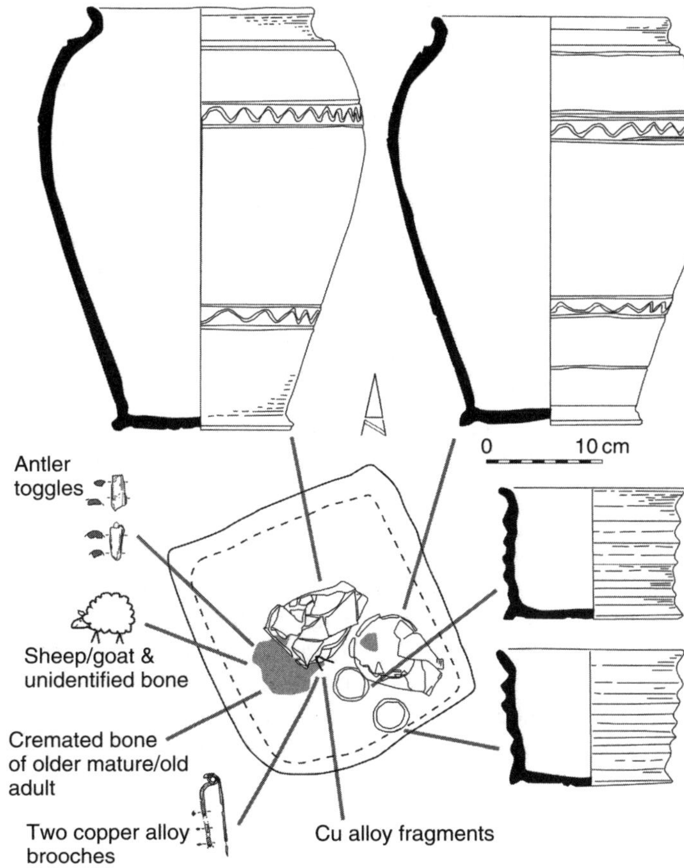

FIGURE 15.4 Cremation burial 20 484 of an older mature/older adult from the Westhampnett Bypass site (redrawn after Fitzpatrick, 1997: 184). Pyre goods included two copper alloy brooches, two fragments of antler toggles and sheep/goat bones. Grave goods consisted of two corrugated bowls and two high-shouldered, necked jars (Fitzpatrick, 1997: 182).

the pyres, and further pots seem to have been deliberately smashed after the fire had died down and incorporated into pyre-related features (Mepham in Fitzpatrick, 1997: 137). The species of sacrificed animals whose remains were placed with the dead were mainly pig and sheep/goat with some cattle. Animals were sacrificed at their prime meat age and joints of meat were then placed with the dead; however, there remains the possibility that some whole animals joined the pyre (McKinley in Fitzpatrick, 1997: 73–77). Rowena Gale examined the rich charcoal remains from the pyre debris and burials. Her results allowed aspects of pyre structure and technology to be reconstructed. It was noticed that oak and ash were selected as pyre timber while willow, hazel, cherry, maple and poplar were among the species that could have formed brushwood, biers or other structures (Gale in Fitzpatrick, 1997: 77–82). The evidence of nails from the pyre sites suggests the reuse of timbers from domestic structures or (alternately) that biers, coffins or catafalques adorned the pyres (Gale in Fitzpatrick, 1997: 78). It is possible that plant material and cereals formed part of the pyre or were deliberate offerings placed with the dead (Hinton in Fitzpatrick, 1997: 85–86). Textile elements and indications of fur pelts recovered from mineralized traces upon iron objects suggest that wrappings and clothing adorned the cadavers for cremation (Walton and Rogers in Fitzpatrick, 1997: 111).

Drawing upon experimental and ethnographic analogies as well as the osteological evidence from both pyre-related features and the cremation burials, the osteoarchaeologist Jackie McKinley was able to assess the ritual process (McKinley in Fitzpatrick, 1997: 55–73). McKinley assigned many of the cremated remains to broad-age categories and in a minority of instances could suggest a tentative sexing, both of which facilitated the mortuary analysis (see below). She also identified two definite and one possible instance of dual cremations – when two individuals were cremated together (McKinley in Fitzpatrick, 1997: 70).

From the bone colour, weight and degree of fragmentation, McKinley was able to suggest that the cremation technology was efficient and complete, but that even accounting for bone-loss due to taphonomic factors, most burials represent the retrieval and the burial of only small portions of the cremated body (Figure 15.5). The rest may have been left on the pyre sites, as cremated bones were found within these. Yet equally it is clear that the pyre debris was cleaned from the pyre site for its repeated reuse. This would have provided the opportunity for the close examination of pyre debris and the removal by mourners of any elements required for subsequent rituals (McKinley in Fitzpatrick, 1997: 65–66). Therefore, it may not be enough to suggest that the cremation was sufficient and that post-cremation rituals were merely of

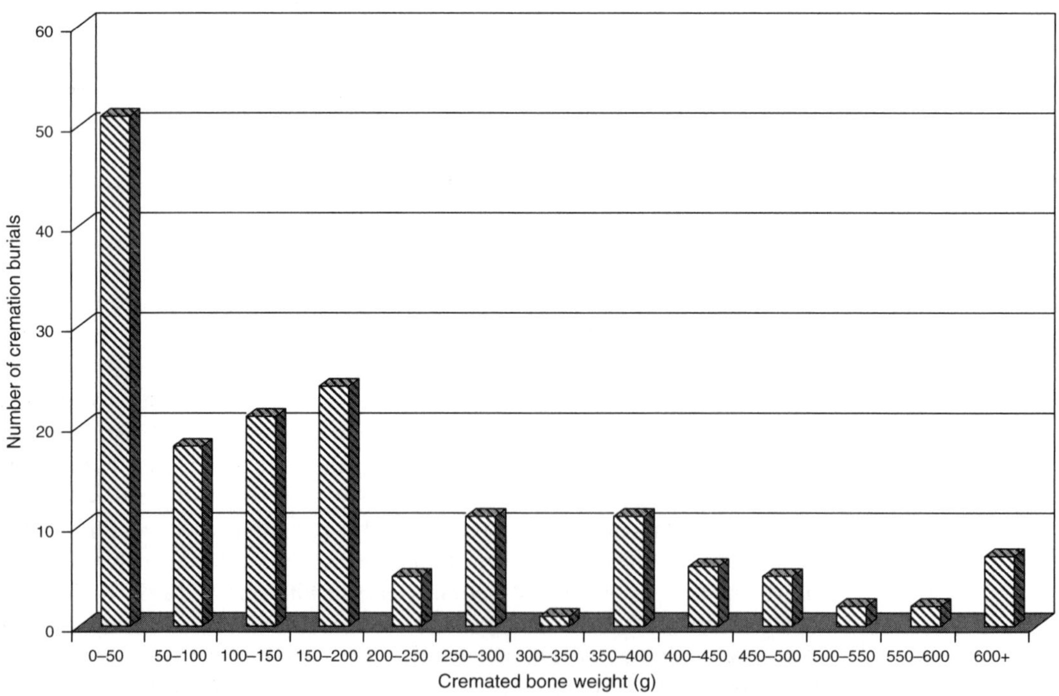

FIGURE 15.5 Weight of cremated bone in identifiable Iron Age graves (redrawn after Fitzpatrick, 1997: 213).

secondary importance in the ritual sequence. Moreover, there are five instances of pyre debris from within graves, suggesting that the collection and the burial of the ashes were closely related in time and therefore integral aspects of the same ritual process (McKinley in Fitzpatrick, 1997: 71). Hence, the token interments were evidently an important ritual act following the cremation.

Grave goods had an important role in the Westhampnett Bypass burial rites. Whereas pots were smashed on the pyre itself, unburned pots were added to the graves as the principal grave goods. These consisted of a variety of ceramic jars and bowls, either placed singly (56% of cases) or in a range of combinations of up to five vessels (Mepham in Fitzpatrick, 1997: 133–134). They were rarely used as cinerary urns and presumably contained offerings of food and drink for the dead. The fact that the grave pits were not always full may hint that further organic grave goods were inserted that had not survived. However, it was equally notable that unburned portable artefacts were absent from the graves.

MORTUARY VARIABILITY

Following an assessment of the chronological span of the cemetery's use from around 90 BC to 50 BC, Powell and Fitzpatrick situated the burial rites in relation to the late Iron Age 'Aylesford-Swarling' culture that represents new ideology in southern Britain in the first century BC. They analyzed the mortuary practices at Westhampnett Bypass, drawing upon the specialist reports, and suggested that the cemetery was used communally by a number of farmsteads for several generations (Powell and Fitzpatrick in Fitzpatrick, 1997: 213). The authors aimed to address mortuary variability, but challenged the possibility that the personae of the deceased might be read directly and simplistically from the surviving mortuary remains. Moreover, the destructive nature of cremation impeded the reconstruction of how the deceased was portrayed on the pyre, and hence the status and the identity of the person as portrayed by the mourners. Even so, the mortuary practices were likely to indicate the decisions of family, friends and allies accumulating wealth for the deceased rather than the specific identity of the dead person. Furthermore, the motivation for the treatment of the dead need not have been the display of a static social identity but the idealized portrayal of the dead in relation to shared ancestral ideals and cosmological principles.

With reference to the low frequency of pyre debris from the burials, it is suggested that selecting out bones for burial is an efficient task. About one quarter of the intact cremation burials had no, or very little (under 25 g) of, cremated bones surviving, which, as argued above, might indicate that they represent memorials rather than true burials. Looking at a chart of weight against the number of graves (Figure 15.5), it is clear that there are two peaks, one in which the graves contained small amounts of bones and those with a larger amount of bones. These represent one and two handfuls of cremated materials. This could be seen as laziness on the part of mourners, or evidence that these handfuls were symbolic of the whole, but it may equally suggest that cremated remains were divided up: some being buried, others kept in storage or circulated among mourners.

The burial space incorporated some important patterns as well. Looking at the spatial arrangement of age categories (the limited number of sexed individuals made patterns difficult to be ascertained), Powell and Fitzpatrick noted that individuals aged over 45 years concentrated around the inner circle closest to the 'empty' area, a pattern that does not appear to be chronological (Powell and Fitzpatrick in Fitzpatrick, 1997: 214). This is thought to indicate differential access for older individuals in proximity to the 'ritual space' absent from graves. There are also a series of 'focal graves' that are larger, square in shape, contain large numbers of pots, and in one case the remains of multiple individuals (e.g., Figure 15.4).

Comparing the burial assemblages with nearby settlement sites suggested the absence of household goods, tools and items of personal adornment such as rings as well as weaponry. The fact that some of the metalwork was found among the pyre debris, such as brooches, indicated that the dead were dressed on the pyre. However, it seemed that these items were retrieved from the pyre prior to the burial rites. Pyre goods appear to be age-related, with the largest numbers placed with adults. The same applies for grave goods; the number of artefacts present is related to age, with the highest number found in 'elders' over 45 years and the least with children. Age-related grave goods were indicated by vessel forms; bowls without accompanying jars were found in the majority of child graves and increasingly fewer in subsequent age categories, while the combination of a jar and a bowl showed a reverse pattern, being most common with elders (Powell and Fitzpatrick in Fizpatrick, 1997: 221). Jars in isolation also appear to have been biased towards adults.

INTERPRETING THE MORTUARY PRACTICES

How does this evidence lead us towards an interpretation of mortuary practices in the late Iron Age phase of the Westhampnett Bypass site? Cremation appears to have been the dominant rite for the Iron Age community using the site. Moreover, burning the dead clearly embodied shared values and principles through its repeated performance and forged links between the community and the place. Fitzpatrick (1997, 2000) considered the possible alignment of the cemetery in relation to cosmological principles governing Iron Age societies. He linked this to the ritual process as a rite of passage following the ideas of van Gennep (1960) and Hertz (1960) in which cremation initiated a transition process that ended with the closing of the grave and the erection of a grave marker (Figure 15.6). Drawing on the concept of liminality, Fitzpatrick suggested that the cremation rituals served to create a ritualized sense of *communitas* through repeated rituals that transformed the dead. These in turn encouraged the conflation of individual identities through transformation by fire and burial in a communal burial site, perhaps creating a community of ancestors embodied by the cemetery.

It is possible to develop Fitzpatrick's ideas further. Concerning the construction of personhood, it appears that the cremation rituals are a means of fragmenting artefacts, animals and the body. A new identity for the dead was constructed through further offerings of food and drink contained within vessels. It is possible that rather than simply acts of incorporation, these votive

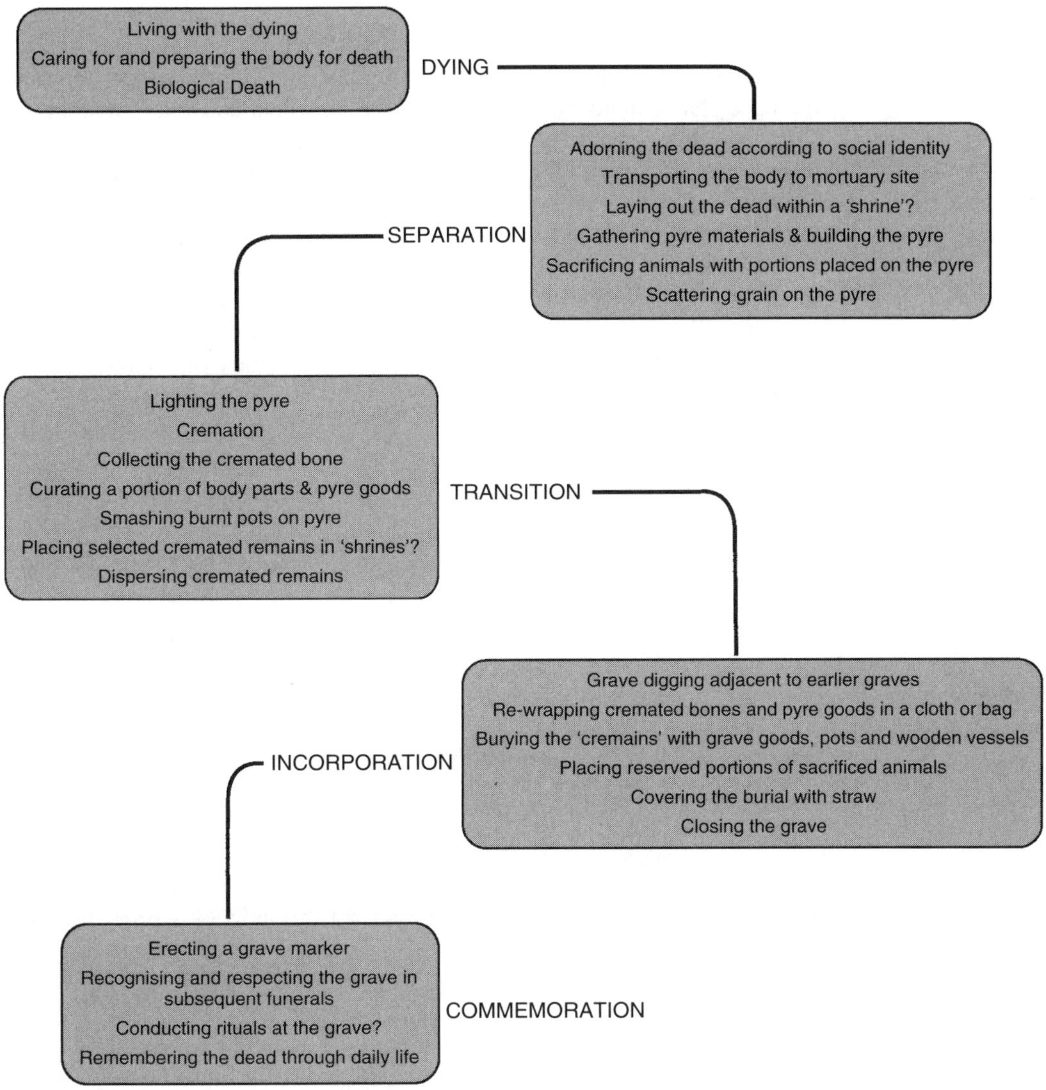

FIGURE 15.6 The ritual process at Westhampnett (redrawn and developed after Fitzpatrick, 1997: 241).

acts serve to secure further ongoing transformations for the dead. Moreover, the vast majority of the remains were distributed elsewhere or left at the pyre site. This argument seems to be supported by an exception that proves the rule. In burial 20252, the cremated bone was found in four separate deposits at the corners of the graves. Could this represent a rare instance where the decision was made to inter all of the separate retrieved elements from the person, elements that in most other cases were circulated elsewhere? Certainly when cremated remains were interred, they were incorporated into a ritual space where a communal identity may have been more important than the celebration of the individual death. In this light, cremation in the late Iron Age was a process of remembering and forgetting of the social person through the transformation, circulation and burial of objects and bodies.

CONCLUSION

Through this brief discussion of the Westhampnett Bypass excavation report, the paper has concluded by exploring the rich data and interpretations possible from the contextual study of cremation in the past. It has been argued that cremation can be regarded as more than second-rate mortuary evidence. Equally, the archaeology of cremation is much broader than the analysis of burned bones. The landscapes, spaces, monuments, contexts, artefacts and bones associated with cremation practices need to be integrated into theorized interpretations of social structure, symbolism, power, personhood and social memory. The challenge for the future is for archaeologists to develop the integration of theories, methods and data in the study of cremation in past societies.

ACKNOWLEDGMENTS

The author would like to thank the editors and Julia Wiecken for discussions related to this article.

BIBLIOGRAPHY

Andersson, G. (2000). West and east: two sides of the same coin? On variations and their significance in inhumation burials in Eastern Central Sweden c. 950–1150 AD. In *Early Medieval Religion: Archaeological Review from Cambridge*, Vol. **17**(2) (A. Pluskowki, Ed.), pp. 133–147. University of Cambridge, Cambridge.

Andersson, G. (2005). With Thor on our side: the symbolism of the Thor Hammer-Ring in Viking Age burial ritual. In *Dealing with the Dead: Archaeological Perspectives on Prehistoric Scandinavian Burial Ritual* (T. Artelius and F. Svanberg, eds.), pp. 45–62. National Antiquities Board, Stockholm.

Arcini, C. (2005). Pyre sites before our eyes. In *Dealing with the Dead: Archaeological Perspectives on Prehistoric Scandinavian Burial Ritual* (T. Artelius and F. Svanberg, eds.), pp. 63–72. National Antiquities Board, Stockholm.

Århem, K. (1989). Into the realm of the sacred: an interpretation of Khasi funerary ritual. In *On the Meaning of Death: Essays on Mortuary Rituals and Eschatological Beliefs* (S. Cederroth, C. Corlin and J. Linström, eds.), pp. 257–299. Almqvist and Wiksell, Stockholm.

Artelius, T. (2005). The revenant by the lake: spear symbolism in Scandinavian Late Viking Age burial ritual. In *Dealing with the Dead: Archaeological Perspectives on Prehistoric Scandinavian Burial Ritual* (T. Artelius and F. Svanberg, eds.). National Antiquities Board, Stockholm.

Artelius, T. and Lindkvist, M. (2005). Bones of the earth – imitation as meaning in Viking Age burial ritual. *Current Swedish Archaeology* **13**, 25–39.

Bahn, P. (ed.) (1996). *Tombs, Graves and Mummies*. Weidenfeld and Nicolson, London.

Bahn, P. (ed.) (2002). *Written in Bones*. David and Charles, Newton Abbot.

Barrett, J. (1990). The monumentality of death: the character of Early Bronze Age mortuary mounds in southern Britain. *World Archaeology* **22(2)**, 179–189.

Barrett, J. (1994). *Fragments from Antiquity*. Blackwell, Oxford.

Beck, L.A. (2005). Secondary burial practices in Hohokam cremations. In *Interacting with the Dead: Perspectives on Mortuary Archaeology for the New Millennium* (G.F.M. Rakita, J.E. Buiktra, L.A. Beck and S.R.Williams, eds.), pp. 150–154. University of Florida Press, Gainesville, FL.

Binford, L.R. (1963). An analysis of cremations from three Michigan sites. *Wisconsin Archaeologist* 44, 98–110.

Binford, L.R. (1971). Mortuary practices: their study and their potential. In *Approaches to the Social Dimensions of Mortuary Practices* (J. Brown, ed.), pp. 6–29. Memoir of the Society of American Archaeology 25, Washington, DC.

Bloch, M. and Parry, J. (1982). Introduction: death and the regeneration of life. In *Death and the Regeneration of Life* (M. Bloch and J. Parry, eds.), pp. 1–44. Cambridge University Press, Cambridge.

Bohnert, M., Rost, T. and Pollak, S. (1998). The degree of destruction of human bodies in relation to the duration of the fire. *Forensic Science International* **95**, 11–21.

Bond, J. (1996). Burnt offerings: animal bone in Anglo-Saxon cremations. *World Archaeology* **28**(1), 76–88.

Bowler, J.M., Jones, R., Allen, H. and Thorne, A.G. (1980). Pleistocene human remains from Australia: a living site and human cremation from Lake Mungo, Western New South Wales. *World Archaeology* **2**, 39–60.

Bowman, B. (1991). The Foster Site cremation: a single individual, partition cremation in Milam County, Texas. *Plains Anthropologist* **36**, 31–42.

Bradley, R. (2000). *The Good Stones: A New Investigation of the Clava Cairns.* Society of Antiquaries of Scotland Monograph Series No. 17, Edinburgh.

Bradley, R. (2002). Death and the regeneration of life: a new interpretation of house urns in Northern Europe. *Antiquity* **76**, 372–377.

Brittain, M. (2006). Technologies of disclosure: posthuman practices and cremation in Neolithic and Bronze Age Wales. *Archaeological Review from Cambridge* **2**, 76–97.

Browne, T. (2005) [1658]. *Urne Burial.* Penguin, London.

Brück, J. (2001). Body metaphors and technologies of transformation in the English Middle and Late Bronze Age. In *Bronze Age Landscapes: Tradition and Transformation* (J. Brück, ed.), pp. 149–160. Oxbow, Oxford.

Brück, J. (2004). Material metaphors: the relational construction of identity in early Bronze Age burials in Ireland and Britain. *Journal of Social Archaeology* **4**, 307–333.

Carr, C. (1995). Mortuary practices: their social, philosophical-religious, circumstantial, and physical determinants. *Journal of Archaeological Method and Theory* **2**, 105–200.

Carver, M. (2005). *Sutton Hoo: A Seventh-Century Princely Burial Ground and its Context.* British Museum, London.

Cool, H. (2004). *The Roman Cemetery at Brougham, Cumbria, Excavations 1966–7.* Society for the Promotion of Roman Studies, Britannia Monograph Series No. 21, London.

Crubézy, E, Ricaut, F.X., Martin, H.,*et al.* (2006). Inhumation and cremation in medieval Mongolia: analysis and analogy. *Antiquity* **80**, 894–905.

Cuijpers, A.G.F.M. and Schutkowski, H. (1993). Histological age determination of the cremation human bones from the urnfields of Deventer 'T Bramelt and Markelo Friezenberg'. *Helinium* **33**, 99–107.

David, N. and Kramer, C. (2001). *Ethnoarchaeology in Action.* Cambridge University Press, Cambridge.

Davies, D.J. (1996). The sacred crematorium. *Mortality* **1**, 83–94.

Davies, D.J. (2002). *Death, Ritual and Belief,* 2nd ed. Continuum, London.

Day Lewis, C. (1952). *Virgil: The Aeneid.* Oxford University Press, Oxford.

Downes, J. (1994). Excavation of a Bronze Age burial at Mousland, Stromness, Orkney. *Proceedings of the Society of Antiquaries of Scotland* **124**, 141–154.

Downes, J. (1997). Hermisgarth, Sanday: the investigation of pyre settings and Pictish cist burials in Orkney. *Proceedings of the Society of Antiquaries of Scotland* **127**, 609–626.

Downes, J. (1999). Cremation: a spectacle and a journey. In *The Loved Body's Corruption: Archaeological Contributions to the Study of Human Mortality* (J. Downes and T. Pollard, eds.), pp. 19–29. Cruithne, Glasgow.

Ebrey, P. (1990). Cremation in Sung China. *American Journal of History* **95**, 406–428.

Effros, B. (1997). *De partibus Saxoniae* and the regulation of mortuary custom: a Carolingian campaign of Christianization or the suppression of Saxon identity? *Revue Belge de Philologie de D'Histoire* **75**, 267–86.

Eriksson, T. (2005). Human bones in the Bronze Age of Uppland. In *Dealing with the Dead: Archaeological Perspectives on Prehistoric Scandinavian Burial Ritual* (T. Artelius and F. Svanberg, eds.), pp. 237–260. National Antiquities Board, Stockholm.

Esmonde Cleary, S. (2000). Putting the dead in the place: burial location in Roman Britain. In *Burial, Society and Context in the Roman World* (J. Pearce, M. Millett and M. Struck, eds.), pp. 127–142. Oxbow, Oxford.

Evans, E. and Maynard, D.J. (1997). Caerleon Lodge Hill Cemetery: the Abbeyfield site 1992. *Britannia* **28**, 169–244.

Faulkes, A. (trans.) (1987). *Snorri Sturluson: Edda.* Dent, London.

Filmer-Sankey, W. and Pestell, T. (2001). Snape Anglo-Saxon Cemetery: Excavations and Surveys 1824–1992, Ipswich, East Anglican Archaeology Report No. 95.

Fitzpatrick, A. (1997). *Archaeological Excavations on the Route of the A27 Westhampnett Bypass, West Sussex, 1992.* Wessex Archaeology Report No. 12, Salisbury.

Fitzpatrick, A. (2000). Ritual, sequence, and structure in Late Iron Age mortuary practices in north-west Europe. In *Burial, Society and Context in the Roman World* (J. Pearce, M. Millett and M. Struck, eds.), pp. 15–29. Oxbow, Oxford.

Fowler, C. (2001). Personhood and social relations in the British Neolithic with a study from the Isle of Man. *Journal of Material Culture* **6**, 138–163.

Fowler, C. (2004). *The Archaeology of Personhood.* Routledge, London.

Gansum, T. (2004a). The Archaeology of earth. *Current Swedish Archaeology* **12**, 7–21.

Gansum, T. (2004b). Role the bones – from iron to steel. *Norwegian Archaeological Review* **37**, 41–57.

van Gennep, A. (1960). *Rites of Passage.* Routledge, London.

Grainger, H. (2005). *Death Redesigned: British Crematoria: History, Architecture and Landscape.* Spire, Reading, MA.

Gräslund, B. (1994). Prehistoric soul beliefs in Northern Europe. *Proceedings of the Prehistoric Society* **60**, 15–26.

Graves, R. (trans.) (1957). *Suetonius: The Twelve Caesars.* Penguin, London.

Hallam, E. and Hockey, J. (2001). *Death, Memory and Material Culture.* Berg, Oxford.

Harvey, G. (2004). Endo-cannabilism in the making of a recent British ancestor. *Mortality* **9**, 255–267.

Heaney, S. (trans.) (2002). *Beowulf: A Verse Translation.* Norton, London.

Henderson, J., Janaway, R.C. and Richards, J.D. (1987). Cremation slag: a substance found in funerary urns. In *Death, Decay and Reconstruction* (A. Boddington, A.N. Garland and R.C. Janaway, eds.), pp. 81–100. Manchester University Press, Manchester.

Hertz, R. (1960). A contribution to the study of the collective representation of death. In *Death and the Right Hand* (R. Needham and C. Needham [trans.]). Free Press, New York.

Hill, J.D. (1999). The Hinxton Rings – A Late Iron Age cemetery at Hinxton, Cambridgeshire, with a reconsideration of Northern Aylesford-Swarling distributions. *Proceedings of the Prehistoric Society* **65**, 243–273.

Hope, V. (1997). A roof over the dead: communal tombs and family structure. In *Domestic Space in the Roman World: Pompeii and Beyond* (R. Laurence and A. Wallace-Hadrill, eds.), pp. 69–88. Journal of Roman Archaeological Supplementary Series Number 2, Portsmouth.

Hudson, A. B. (1966). Death ceremonies of the Padju Epat Ma'anyan Dayaks'. *Sarawak Museum Journal* **13**, 341–416.

Iregren, E. (1995). Why animal bones in human graves – an attempt to interpret animals in Iron Age cremations in Sweden. In *Cremation Studies in Archaeology* (E. Smits, E. Iregren and A.G. Drusini, eds.), pp. 9–32. Saonara, Lagos.

Jones, A. (2001a). Drawn from memory: the archaeology of aesthetics and the aesthetics of archaeology in Earlier Bronze Age Britain and the present. *World Archaeology* **33**, 334–356.

Jones, A. (2001b). Enduring images? Image production and memory in Earlier Bronze Age Scotland. In *Bronze Age Landscapes* (J. Brück, ed.), pp. 217–228. Oxbow, Oxford.

Jones, A. (2003). Technologies of remembrance. In *Archaeologies of Remembrance: Death and Memory in Past Societies* (H. Williams, ed.), pp. 65–88. Kluwer/Plenum, New York.

Jupp, P. (1993). Cremation or burial? Contemporary choice in city and village. In *The Sociology of Death: Theory, Culture, Practice* (D. Clark, ed.), pp. 169–97. Blackwell, Oxford.

Jupp, P. (2006). *From Dust to Ashes: Cremation and the British Way of Death.* Palgrave, London.

Kaliff, A. (1998). Grave structures and altars: archaeological traces of Bronze Age eschatological conceptions. *Journal of European Archaeology* **1(2)**, 177–198.

Kaliff, A. (2005). The Vedic Agni and Scandinavian fire rituals: a possible connection. *Current Swedish Archaeology* **13**, 77–97.

Kaliff, A. and Oestigaard, T. (2004). Cultivating corpses: a comparative approach to disembodied mortuary remains. *Current Swedish Archaeology* **12**, 83–104.

Kan, S. (1989). *Symbolic Immortality: The Tlingit Potlatch of the Nineteenth Century*. Smithsonian Institute Press, Washington, DC.

Kemble, J.M. (1863). *Horae Ferales*. Lovell Reeve, London.

Larsson, L. (2000). The passage of axes, fire transformation of flint objects in the Neolithic of southern Sweden. *Antiquity* **74**(285), 602–610.

Larsson, L. (2004). The Mesolithic Period in Southern Scandinavia: with special reference to burials and cemeteries. In *Mesolithic Scotland and its Neighbours* (A. Saville, ed.), pp. 371–392. Society of Antiquaries of Scotland, Edinburgh.

Last, J. (1998). Books of life: biography and memory in a Bronze Age barrow. *Oxford Journal of Archaeology* **17**(1), 43–53.

Leineweber, R. (2005). Bradneu: Verbrennung auf dem Scheiterhaufen – oder – Studie über bradderformierte Beigräben aus römischen Kaiserzeit Innergermaniens, In *Von Der Altsteinzeit Über „Ötzi" Bis Zum Mettelalter* (Both, F. and Vorlauf, D., eds.), pp. 319–331. Isenzee, Oldenburg.

Lucas, G. (1996). Of death and debt: a history of the body in Neolithic and Early Bronze Age Yorkshire. *Journal of European Archaeology* **4**, 99–118.

ManiBabu, M. (1994). Post-cremation-urn-burial of the Phayengs (Manipur): a study of mortuary behaviour. *The Eastern Anthropologist* **47**, 157–171.

Marshall, A. (1998). Visualising burnt areas: patterns of magnetic susceptibility at Guiting Power 1 Round Barrow (Glos. UK). *Archaeological Prospection* **5**, 159–177.

Mattingly, H. (trans.) (1948). *Tacitus: Agricola and Germania*. Penguin, London.

McKinley, J. (1994). *The Anglo-Saxon Cemetery at Spong Hill, North Elmham. Part VII: The Cremations*. East Anglian Archaeology 69, Dereham.

McKinley, J. (1997). Bronze Age barrows and funerary rites and rituals of cremation. *Proceedings of the Prehistoric Society* **63**, 129–145.

McKinley, J. (2000). Phoenix rising: aspects of cremation in Roman Britain. In *Burial, Society and Context in the Roman World* (J. Pearce, M. Millett and M. Struck, eds.), pp. 38–44. Oxbow, Oxford.

McKinley, J. (2006). Cremation... the cheap option? In *Social Archaeology of Funerary Remains* (R. Gowland and C. Knüsel, eds.), pp. 81–88. Oxbow, Oxford.

McKinley, J. and Roberts, C. (1993). *Excavation and Post-excavation Treatment of Cremated and Inhumed Human Remains*. Institute of Field Archaeologists Technical Paper No. 13, London.

Metcalf, D. and Huntingdom, R. (1991). *Celebrations of Death – The Anthropology of Mortuary Ritual*, 2nd ed. Cambridge University Press, Cambridge.

Millett, M. (1986). An Early Roman cemetery at Alton, Hampshire. *Proceedings of the Hampshire Field Club and Archaeological Society* **42**, 43–87.

Mizoguchi, K. (1993). Time in the reproduction of mortuary practices. *World Archaeology* **25**(2), 223–235.

Morris, I. (1992). *Death-Ritual and Social Structure in Classical Antiquity*. Cambridge University Press, Cambridge.

Myres, J.N.L. (1969). *Anglo-Saxon Pottery and the Settlement of England*. Clarendon, Oxford.

Niblett, R. (2000). Funerary rites in Verulamium during the early Roman period. In *Burial, Society and Context in the Roman World* (J. Pearce, M. Millett and M. Struck, eds.), pp. 97–104. Oxbow, Oxford.

Nicklasson, P. (1999). Excavating postprocessually – a theoretical utopia or a practical reality? *Lund Archaeological Review* **5**, 59–72.

Noy, D. (2000). Half-burnt on an emergency pyre: Roman cremations which went wrong, *Greece and Rome* **47(2)**, 186–196.

Oestigaard, T. (1999). Cremations as transformations: when the dual cultural hypothesis was cremated and carried away in urns. *European Journal of Archaeology* **2(3)**, 345–364.

Oestigaard, T. (2000). Sacrifices of raw, cooked and burnt humans. *Norwegian Archaeological Review* **33(1)**, 41–58.

Oestigaard, T. (2005). *Death and Life-Giving Waters: Cremation, Caste, and Cosmogony in Karmic Traditions*. BAR International Series 1353, Oxford.

O'Kelly, M. (1982). *Newgrange: Archaeology, Art and Legend*. Thames and Hudson, London.

Pader, E.J. (1980). Material symbolism and social relations in mortuary studies. In *Anglo-Saxon Cemeteries 1979* (P. Rahtz, T. Dickinson and L. Watts, eds.), pp. 143–160. BAR British Series 82, Oxford.

Parker Pearson, M. (1982). Mortuary practices, society and ideology: an ethnoarchaeological study. In *Symbolic and Structural Archaeology* (I. Hodder, ed.), pp. 99–113. Cambridge University Press, Cambridge.

Parker Pearson, M. (1993). The powerful dead: archaeological relationships between the living and the dead. *Cambridge Archaeological Journal* **3**(2), 203–229.

Parker Pearson, M. (1999). *The Archaeology of Death and Burial.* Stroud, Sutton.

Parker Pearson, M. (2004). Earth, wood and fire: materiality and stonehenge. In *Soils, Stones and Symbols* (N. Boivin and M.A. Owoc, eds.), pp. 71–89. UCL, London.

Parry, J.P. (1994). *Death in Banaras.* Cambridge University Press, Cambridge.

Pearce, J. (1997). Death and time: the structure of late Iron Age mortuary ritual. In *Reconstructing Iron Age Societies* (A. Gwilt and C. Haselgrove, eds.), pp. 174-80. Oxbow, Oxford.

Petrov, N. (2002). Cremation outside the grave: searching for the traces of a funeral pile. In *Fire in Archaeology* (D. Gheorgiu, ed.), pp. 143–146. BAR International Series 1089, Oxford.

Polfer, M. (2000). Reconstructing funerary rituals: the evidence of *ustrina* and related archaeological structures. In *Burial, Society and Context in the Roman World* (J. Pearce, M. Millett and M. Struck, eds.), pp. 30–37. Oxbow, Oxford.

Prendergast, D., Hockey, J. and Kellaher, L. (2006). Blowing in the wind? Identity, materiality, and the destinations of human ashes. *Journal of the Royal Anthropological Institute* **12**, 881–898.

Rakita, G.F.M. and Buikstra, J.E. (2005). Corrupting flesh: reexamining Hertz's perspective on mummification and cremation. In *Interacting with the Dead: Perspectives on Mortuary Archaeology for the New Millennium* (G.F.M. Rakita, J.E. Buiktra, L.A. Beck and S.R.Williams, eds.), pp. 97–106. University of Florida Press, Gainesville, FL.

Ravn, M. (2003). *Death Ritual and Germanic Social Structure.* BAR International Series 1164, Oxford.

Richards, J.D. 1987. *The Significance of Form and Decoration of Anglo-Saxon Cremation Urns.* BAR British Series 166, Oxford.

Richards, J.D. (1992). Anglo-Saxon symbolism. In *The Age of Sutton Hoo* (M. Carver, ed.), pp. 131–149. Boydell, Woodbridge.

Roymans, N. and Kortlang, F. (1999). Urnfield symbolism, ancestors and the land in the Lower Rhine Region. In *Land and Ancestors: Cultural Dynamics in the Urnfield Period and the Middle Ages in the Southern Netherlands* (F. Theuws and N. Roymans, eds.), pp. 33–62. Amsterdam University Press, Amsterdam.

Shanks, M. and Tilley, C. (1982). Ideology, symbolic power and ritual communication: a reinterpretation of Neolithic mortuary practices. In *Symbolic and Structural Archaeology* (I. Hodder, ed.), pp. 129–154. Cambridge University Press, Cambridge.

Shipman, P., Foster, G., Schoeninger, M. (1984). Burnt bones and teeth: an experimental study of color, morphology, crystal structure and shrinkage. *Journal of Archaeological Science* **11**, 307–325.

Sigvallius, B. (1995). Animals in Iron Age cremations in Central Sweden. In *Cremation Studies in Archaeology* (E. Smits, E. Iregren and A.G. Drusini, eds.), pp. 39–50. Saonara, Lagos.

Sigvallius, B. (2005). Sailing towards the afterlife: analysis of a ship-formed burial monument at Hellerö by the Baltic Sea. In *Dealing with the Dead: Archaeological Perspectives on Prehistoric Scandinavian Burial Ritual* (T. Artelius and F. Svanberg, eds.), pp. 159–172. National Antiquities Board, Stockholm.

Teather, E.K. (1999). High-rise homes for the ancestors: cremation in Hong Kong. *Geographical Review* **89(3)**, 409–430.

Ucko, P. (1969). Ethnography and archaeological interpretation of funerary remains. *World Archaeology* **1(2)**, 262–280.

Vitebsky, P. (1993). *Dialogues with the Dead – The Discussion of Mortality among the Sora of Eastern India.* Cambridge University Press, Cambridge.

Warmind, M.L. (1995). Ibn Fadlan in the context of his age. In *Thye, The Ship as Symbol* (O. Crumlin-Pedersen and B. Munch, eds.), pp. 131–137. National Museum, Copenhagen.

Waterman, D. (1997). *Excavations at Navan Fort 1961–71.* The Stationary Office, Belfast.

Weiss-Krejci, E. (2005). Formation processes of deposits with burned human remains in Neolithic and Chalcolithic Portugal. *Journal of Iberian Archaeology* **7**, 37–73.

Wheeler, R.M. (1929). A Roman pipe-burial from Caerleon, Monmouthshire. *Antiquaries Journal* **9**, 1–7.

Williams, H. (2001). An ideology of transformation: cremation rites and animal sacrifice in early Anglo-Saxon England. In *The Archaeology of Shamanism* (N. Price, ed.), pp. 193–212. Routledge, London.

Williams, H. (2002). Cemeteries as central places: landscape and identity in early Anglo-Saxon England. In *Central Places in the Migration and Merovingian Periods. Papers from the 52nd Sachsensymposium* (B. Hardh and L. Larsson, eds.), pp. 341–362. Almqvist, Lund.

Williams, H. (2003). Material culture as memory: combs and cremation in early medieval Britain. *Early Medieval Europe* **12(2)**, 89–128.

Williams, H. (2004a). Ephemeral monuments and social memory in early Roman Britain. In *TRAC 2003: Proceedings of the Thirteenth Annual Theoretical Roman Archaeology Conference* (B. Croxford, H. Eckardt, J. Meade and J. Weekes, eds.), pp. 51–61. Oxbow, Oxford.

Williams, H. (2004b). Death warmed up: the agency of bodies and bones in early Anglo-Saxon cremation rites. *Journal of Material Culture* **9(3)**, 263–291.

Williams, H. (2005). Keeping the dead at arms length: memory, weaponry and early medieval mortuary technologies. *Journal of Social Archaeology* **5(2)**, 253–275.

INDEX

PLATE 1 This figure illustrates a burned distal femur. The epiphyseal end has fractured off and has been pulled up the shaft by shrinking muscles, while the distal shaft is left exposed. This photograph is an example of a bone shaft that exhibits obvious fracturing as a result of fire (see Figure 2.1, p. 23).

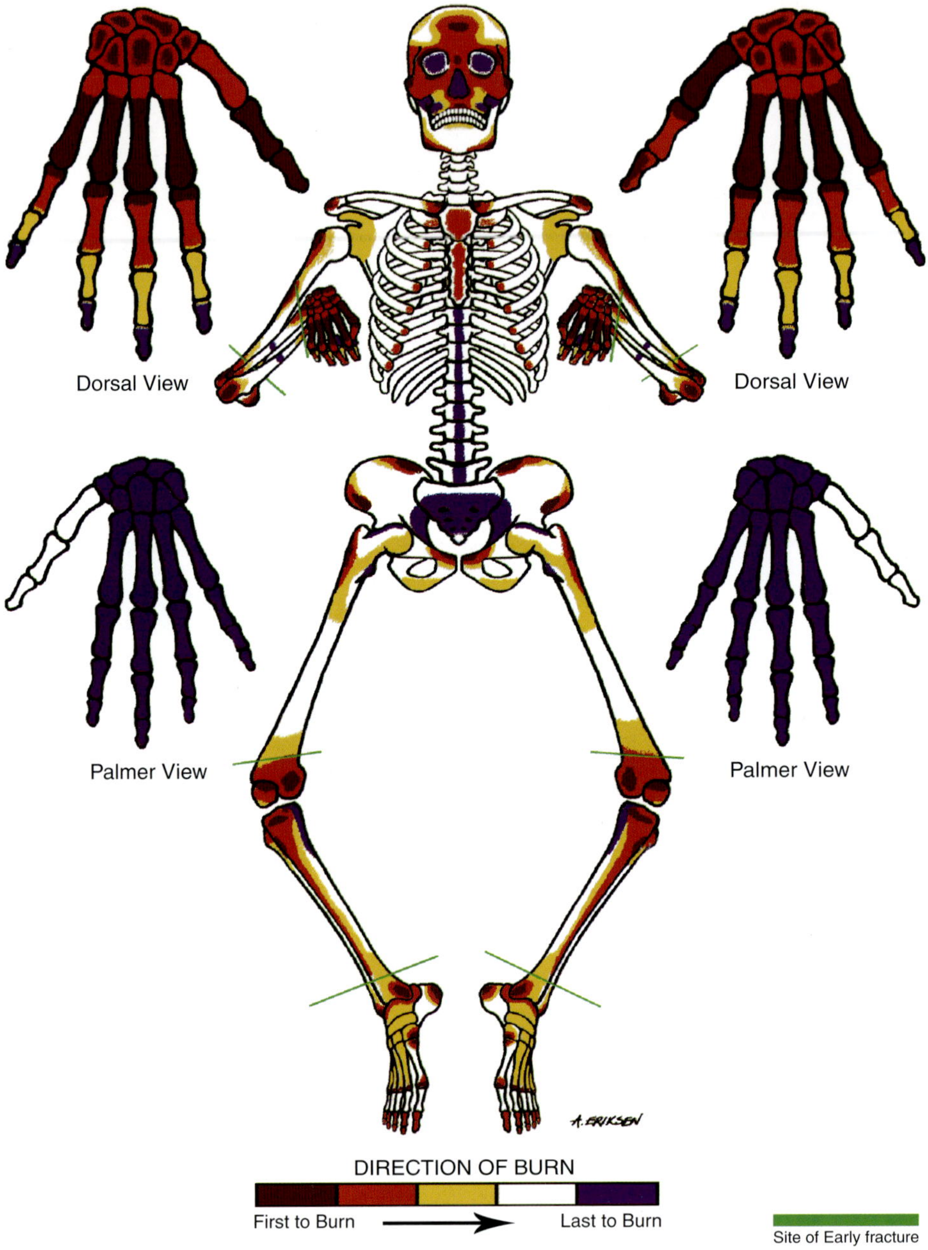

PLATE 2 This is a diagram of an anterior skeleton in pugilistic posture highlighting the initial, secondary, and final areas to express burning on bone. The figure also includes dorsal and palmar views of the pattern of burning on the hand. The green lines indicate common areas of fracture (see Figure 2.7, p. 32).

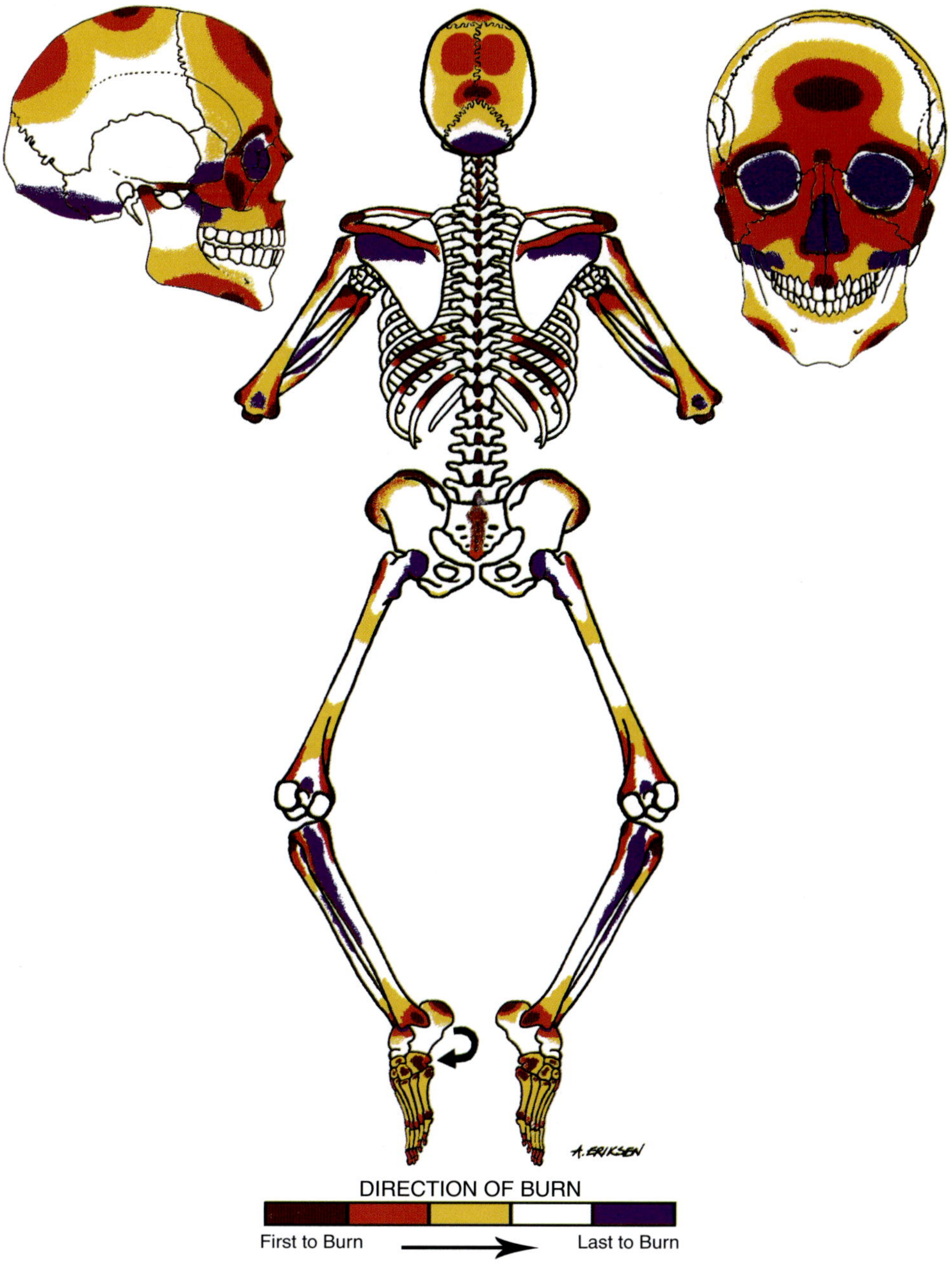

PLATE 3 This is a posterior view of the skeleton diagram in pugilistic posture highlighting the initial, secondary, and final areas to burn on bone. This figure also includes a magnified view of the burn patterns on the frontal and lateral skull (see Figure 2.8, p. 33).

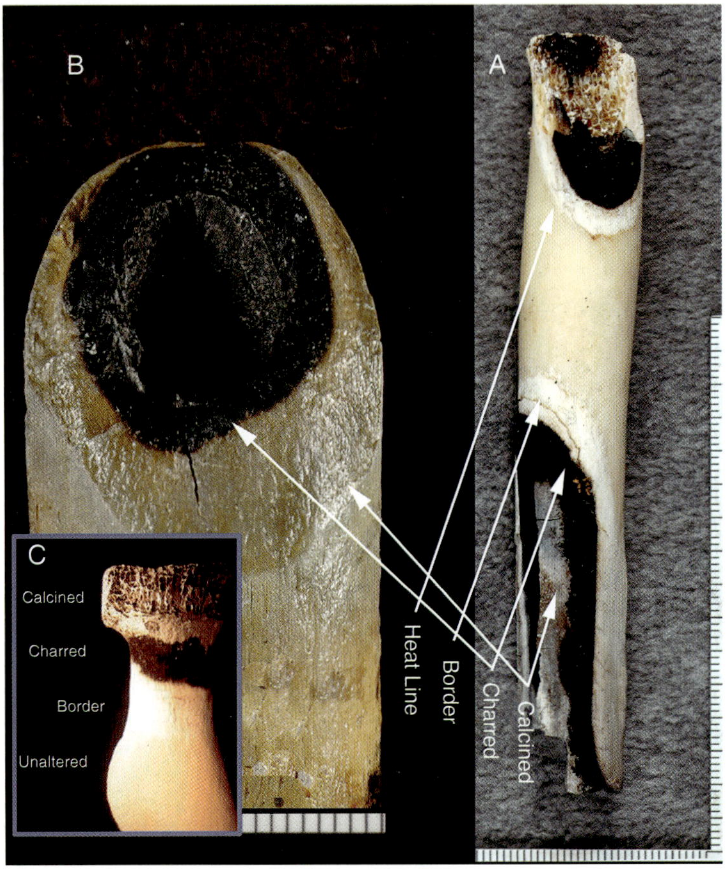

PLATE 4 This figure illustrates color changes in a three partially burned bone shafts with fractures due to fire. Rather than suggesting that color change in burned bone is arbitrary and indecipherable, this varied coloration demonstrates how bones burn predictably. Inset A demonstrates color changes in a proximal radius. Inset B demonstrates color differences from the outside to the inside of this long bone (before the shaft is compromised); the lighter calcined bone is external and the black charred (protected) bone internal. Inset C is another example of color change in a proximal radius (see Figure 2.10, p. 36).

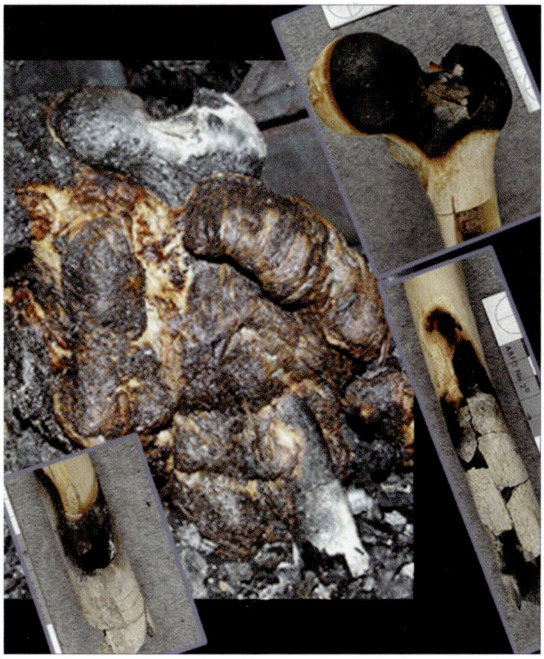

PLATE 5 As tissue burns, muscles contract somewhat predictably to form patterned destruction of the bone. This figure illustrates an upper leg before and after (insets) processing (see Figure 2.21, p. 45).

PLATE 6 This illustration demonstrates tissue shrinkage of the nuchal muscle lines of a burned occipital bone on the back of a skull. The photograph exhibits how soft tissues are pulled back on the head due to strong neck muscles. The creeping of the soft tissues registers in repetitive tissue line fractures (see Figure 2.22, p. 46).

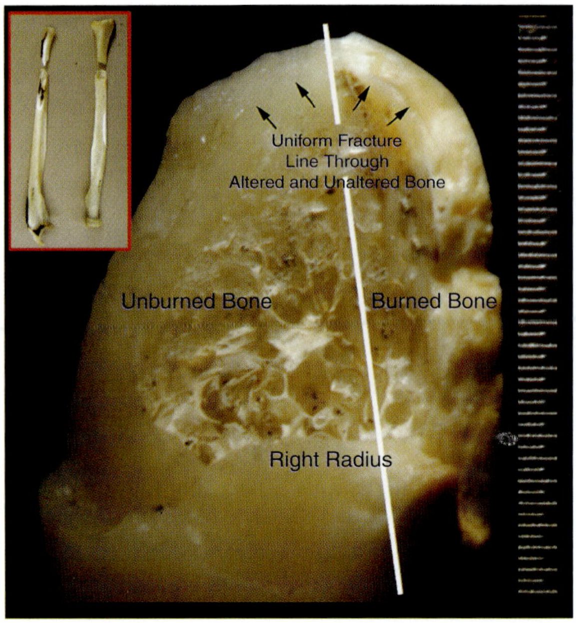

PLATE 7 The distal radius demonstrates a consistent fracture pattern traversing burned and unburned bone uniformly (line separates the two). All evidence appears consistent with a wet bone fracture (see Figure 2.24, p. 48).

PLATE 8 This figure illustrates a bone fracture of a partially burned ulna. The figure shows how the fracture contour changes from wet to dry bone. This pattern indicates that the fracture occurred after the arm burned (see Figure 2.26, p. 49).

PLATE 9 Calcined tooth with large patch of dark gray (see Figure 3.1, p. 58).

PLATE 10 Transverse fracturing on root fragment (see Figure 3.2, p. 59).

PLATE 11 Darkly colored crown of an unerupted premolar (see Figure 3.3, p. 61).

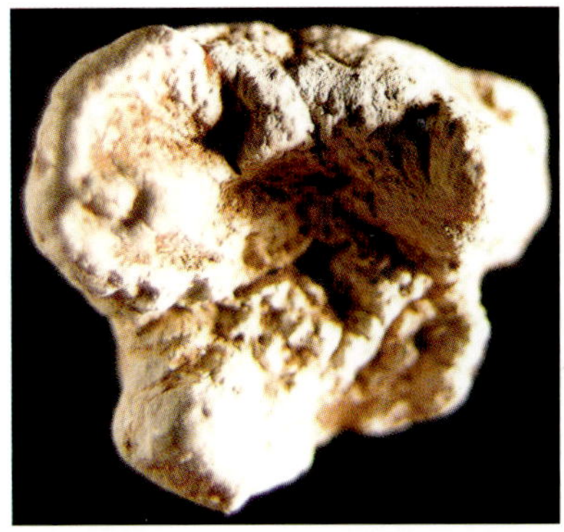

PLATE 12 Chalky appearance of a calcined, immature crown (see Figure 3.4, p. 62).

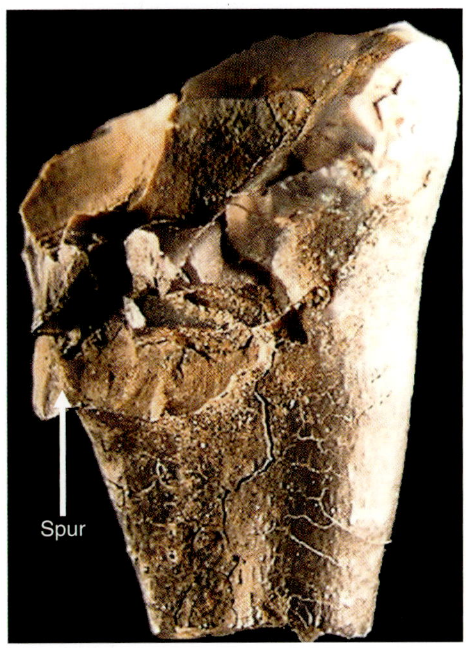

PLATE 13 Spur projecting from single molar radical, not created in burned single-rooted teeth (see Figure 3.6, p. 64).

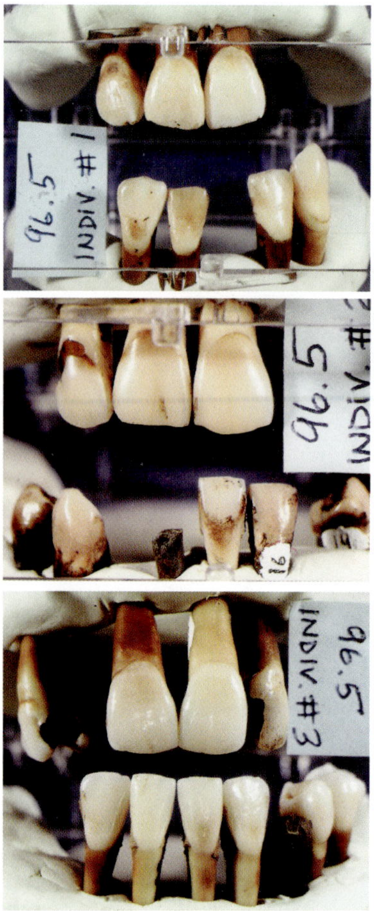

PLATE 14 Three dentitions reconstructed from the Baumeister serial homicide case (photos by S.P. Nawrocki) (see Figure 3.7, p. 70).

PLATE 15 View inside the retort showing the location of the cremation burner (front) in the cremation chamber and the afterburner (back) located in a section of the after-chamber (see Figure 4.2, p. 78).

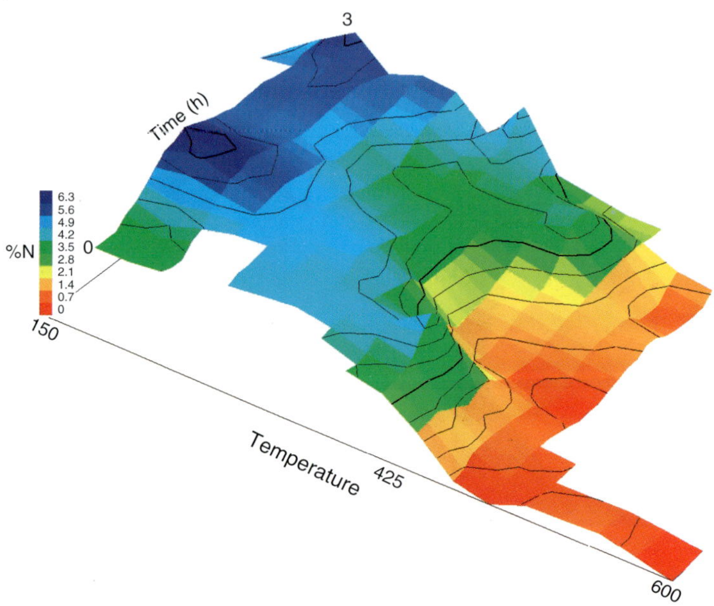

PLATE 16 Changes in the percentage of nitrogen with heating time and temperature (see Figure 5.1, p. 98).

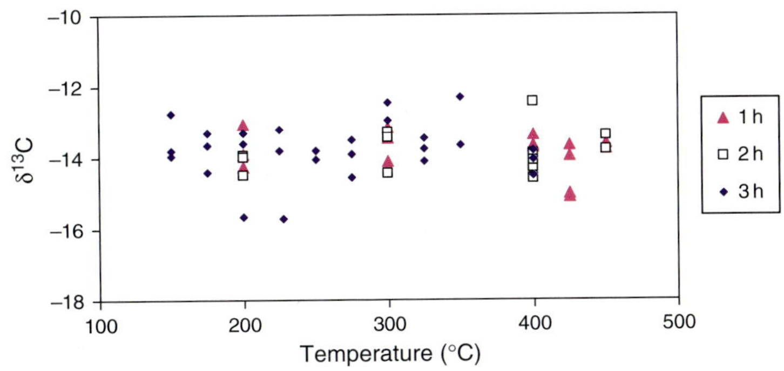

PLATE 17 Changes in stable carbon isotope ratios with temperature (see Figure 5.2, p. 98).

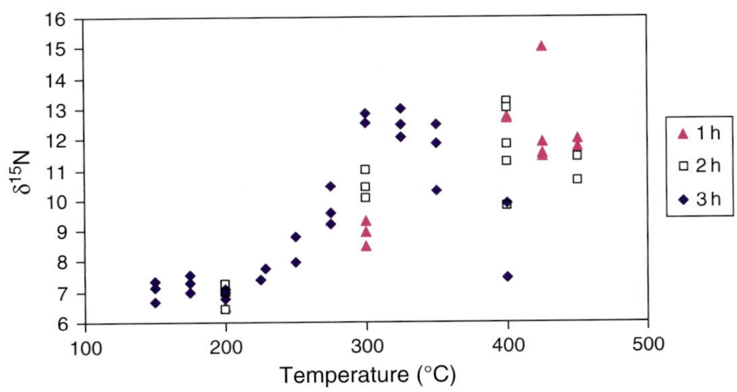

PLATE 18 Changes in stable nitrogen isotope ratios with temperature (see Figure 5.3, p. 99).

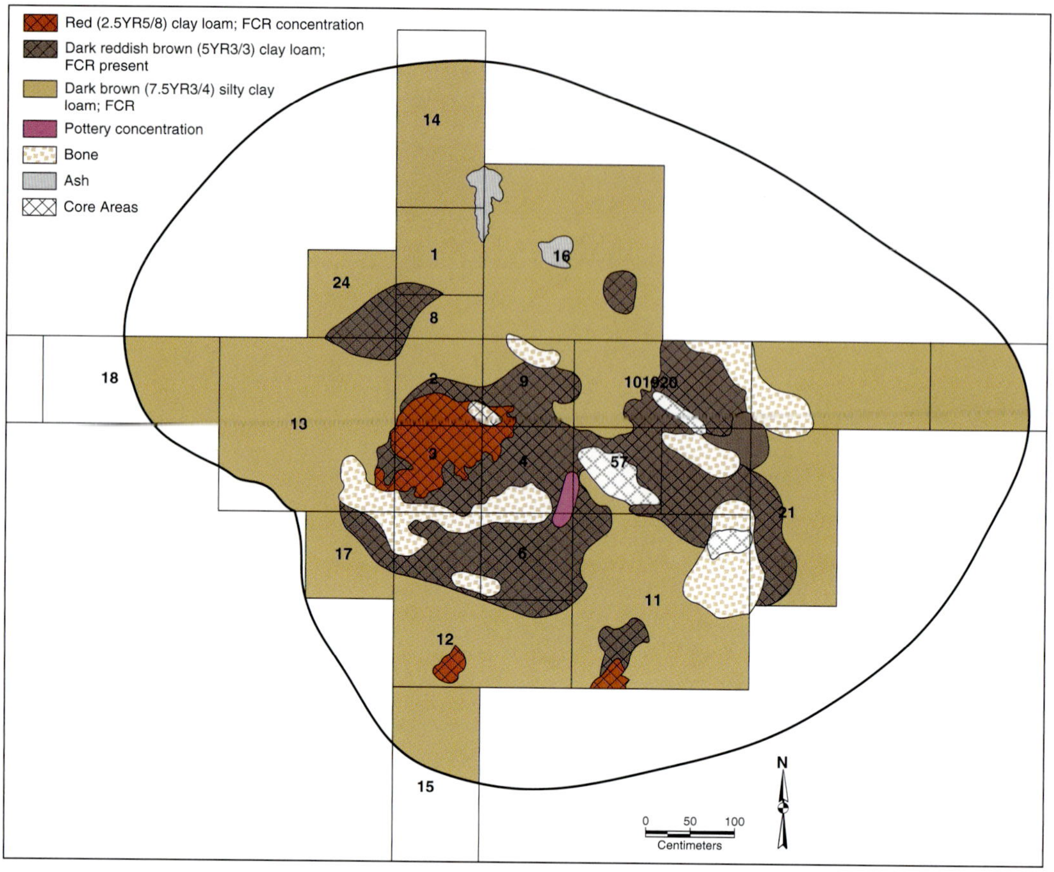

PLATE 19 Walker-Noe site plan map (modified from Pollack *et al.* 2005) (see Figure 6.2, p. 114).

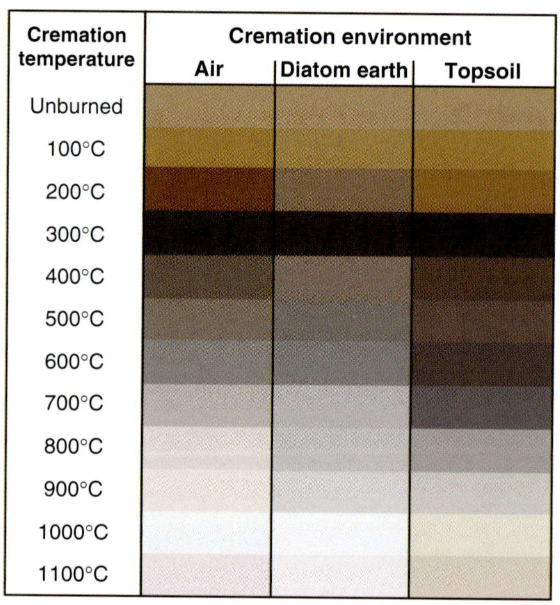

PLATE 20 Colors generated using the average RGB color values of experimental bones samples cremated from 1–3 h in a furnace while surrounded by air, diatomaceous earth, and topsoil (see Figure 7.1, p. 132).

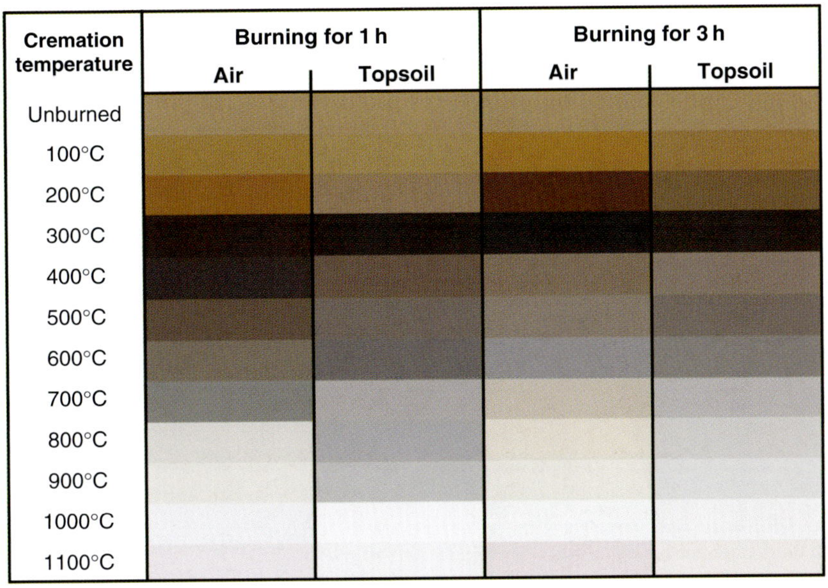

PLATE 21 Colors generated using the average RGB color values of experimental bones samples burned for 1 or 3 h in a furnace while surrounded by air or topsoil with a high organic content (see Figure 7.2, p. 132).

PLATE 22 Crown patina and flaking/separation of crown at cementoenamel junction (CEJ) (see Figure 8.1, p. 140).

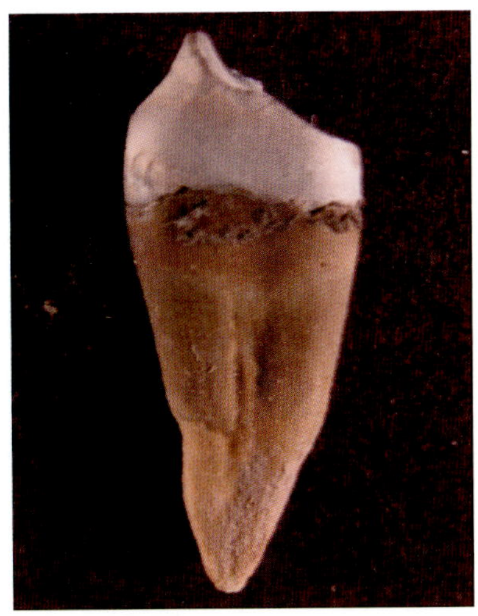

PLATE 23 Differences in dentin color above and below the cementoenamel junction (CEJ) (see Figure 8.2, p. 141).

PLATE 24 Long bone shaft fragment from a Romano-British burial showing the 'sandwich' effect of differential levels of oxidation through the bone thickness (see Figure 10.2, p. 167).

PLATE 25 Experimental pyre cremation undertaken in Shetland, northern Scotland, showing pyre structure and effects of veering strong winds. (NB: The pyre structure is more open than normal due to half the fuel in this case comprised a central peak stack.) (see Figure 10.3, p. 168).

PLATE 26 Indentified elements of lower limb from the Romano-British burial from Purton, Wiltshire, showing varying degrees of oxidation to the bone (see Figure 10.6, p. 177).

PLATE 27 Section of cremated sheep spine from an experimental pyre cremation; cremated vertebral held articulated by charred longitudinal ligament (see Figure 10.8, p. 179).

PLATE 28 Burial No. 66 showing charred skeletal elements. Photo by A. Weber (see Figure 11.3, p. 196).

PLATE 29 Burial No. 54 showing more extreme evidence of fire. Photo by A. Weber (see Figure 11.4, p. 197).

PLATE 30 Bones recovered from E.U. 13, level 4, illustrating the density of deposits (see Figure 12.3, p. 203).

PLATE 31 (a) Lightly burned skull, reconstructed from 64 pieces (see Figure 12.4a, p. 205).

PLATE 31 (b) Calcined long bone fragment, reconstructed from six pieces (see Figure 12.4b, p. 206).